CAMBRIDGE LIBRARY COLLECTION

Books of enduring scholarly value

Botany and Horticulture

Until the nineteenth century, the investigation of natural phenomena, plants and animals was considered either the preserve of elite scholars or a pastime for the leisured upper classes. As increasing academic rigour and systematisation was brought to the study of 'natural history', its subdisciplines were adopted into university curricula, and learned societies (such as the Royal Horticultural Society, founded in 1804) were established to support research in these areas. A related development was strong enthusiasm for exotic garden plants, which resulted in plant collecting expeditions to every corner of the globe, sometimes with tragic consequences. This series includes accounts of some of those expeditions, detailed reference works on the flora of different regions, and practical advice for amateur and professional gardeners.

The Trees of Great Britain and Ireland

Although without formal scientific training, Henry John Elwes (1846–1922) devoted his life to natural history. He had studied birds, butterflies and moths, but later turned his attention to collecting and growing plants. Embarking on his most ambitious project in 1903, he recruited the Irish dendrologist Augustine Henry (1857–1930) to collaborate with him on this well-illustrated work. Privately printed in seven volumes between 1906 and 1913, it covers the varieties, distribution, history and cultivation of tree species in the British Isles. The strictly botanical parts were written by Henry, while Elwes drew on his extensive knowledge of native and non-native species to give details of where remarkable examples could be found. Each volume contains photographic plates as well as drawings of leaves and buds to aid identification. The species covered in Volume 6 (1912) include spruce, juniper, laburnum, cherry, mulberry and gum trees.

Cambridge University Press has long been a pioneer in the reissuing of out-of-print titles from its own backlist, producing digital reprints of books that are still sought after by scholars and students but could not be reprinted economically using traditional technology. The Cambridge Library Collection extends this activity to a wider range of books which are still of importance to researchers and professionals, either for the source material they contain, or as landmarks in the history of their academic discipline.

Drawing from the world-renowned collections in the Cambridge University Library and other partner libraries, and guided by the advice of experts in each subject area, Cambridge University Press is using state-of-the-art scanning machines in its own Printing House to capture the content of each book selected for inclusion. The files are processed to give a consistently clear, crisp image, and the books finished to the high quality standard for which the Press is recognised around the world. The latest print-on-demand technology ensures that the books will remain available indefinitely, and that orders for single or multiple copies can quickly be supplied.

The Cambridge Library Collection brings back to life books of enduring scholarly value (including out-of-copyright works originally issued by other publishers) across a wide range of disciplines in the humanities and social sciences and in science and technology.

The Trees
of Great Britain
and Ireland

VOLUME 6

HENRY JOHN ELWES
AUGUSTINE HENRY

CAMBRIDGE
UNIVERSITY PRESS

CAMBRIDGE
UNIVERSITY PRESS

University Printing House, Cambridge, CB2 8BS, United Kingdom

Published in the United States of America by Cambridge University Press, New York

Cambridge University Press is part of the University of Cambridge.
It furthers the University's mission by disseminating knowledge in the pursuit of
education, learning and research at the highest international levels of excellence.

www.cambridge.org
Information on this title: www.cambridge.org/9781108069373

© in this compilation Cambridge University Press 2014

This edition first published 1912
This digitally printed version 2014

ISBN 978-1-108-06937-3 Paperback

THE TREES OF GREAT BRITAIN AND IRELAND

The Trees

of

Great Britain

& Ireland

BY

Henry John Elwes, F.R.S.

AND

Augustine Henry, M.A.

VOLUME VI

Edinburgh: Privately Printed

MCMXII

CONTENTS

Contents

ILLUSTRATIONS

NOTICE TO SUBSCRIBERS

THE favourable manner in which the Notice to Subscribers issued with the Fifth Volume of this work (*The Trees of Great Britain and Ireland*) was received, leads us to hope for their approval of the course we are now about to adopt, after consultation with some of our best supporters.

In consequence of the great additions to our knowledge, and the extreme difficulty of some of the genera described, the last part of the work has increased to over 500 pages, which, if published in one volume, would far exceed in size and cost those hitherto published. We have therefore decided to complete the work in two volumes. Vol. VII. will be issued, together with a general Index to the whole work, before the end of the year. In accordance with what was indicated in our last Notice to Subscribers the cost of these two volumes and the Index will be Five Guineas.

H. J. ELWES.
A. HENRY.

PICEA

THE characters of the Genus Picea and of the two sections into which it is divided have been given in Vol. I. pp. 75-76, with a description of the species belonging to the section *Omorica*. At that time the Sikkim spruce (*P. spinulosa*), one of this section, was imperfectly known, and a full account of it is now given at the end of this article. See p. 1392.

In the section *Eu-picea*, the leaves are quadrangular or rhombic in section, and bear stomatic lines on all their four sides. About fifteen species of quadrangular-leaved spruces are known,[1] which may be readily distinguished by the following key, based on the characters of the branchlets, buds, and leaves.

KEY TO SECTION EU-PICEA

I. *Branchlets quite glabrous.*
   * *Leaves on lateral branches radially arranged, spreading uniformly on all sides.*
   1. *Picea Smithiana*, Boissier. Western Himalayas. See p. 1366.
      Branchlets pendulous, grey. Buds large, resinous, pointed. Leaves slender, about 1½ in. long.
   2. *Picea Maximowiczii*, Regel. Japan. See p. 1374.
      Branchlets not pendulous, reddish brown. Buds small, resinous. Leaves, ⅜ to ½ in. long.
   ** *Leaves on lateral branches in an imperfect radial arrangement, not pectinate in two sets on the lower side of the branchlets, which are not pendulous.*
   3. *Picea Schrenkiana*, Fischer and Meyer. Central Asia, in the Alatau and Thianshan ranges. See p. 1364.
      Branchlets ashy grey. Terminal buds subglobose, girt with a ring of keeled pointed pubescent ciliate scales. Leaves rigid, sharp-pointed, ¾ to 1¼ in. long.
   4. *Picea pungens*, Engelmann. Wyoming, Colorado, Utah, New Mexico. See p. 1389.
      Branchlets at first glaucous, becoming reddish brown. Buds with the tips of their upper scales usually loose and reflexed. Leaves stout, rigid, with a hard sharp-pointed apex, ¾ to 1¼ in. long.

---

[1] Not including the spruces of China, of which two or three species introduced by Wilson are in cultivation at Coombe Wood, but are too young to describe.

5. *Picea polita*, Carrière.   Japan.   See p. 1370.
Branchlets yellow.   Buds shining reddish brown, with closely imbricated scales.   Leaves rigid, stout, curved, ending in a spine-like point.

*** *Leaves on lateral branches, imbricated on the upper side of the branchlet; those below, pectinate and spreading laterally in two sets.*
6. *Picea alba*, Link.   North America.   See p. 1380.
Branchlets greyish or pale brown, usually glaucous.   Buds with glabrous nonciliate bifid scales.   Leaves disagreeable in odour when bruised, about $\frac{1}{2}$ in. long.
7. *Picea bicolor*, Mayr.   Japan.   See p. 1372.
Branchlets yellow, glabrous on lateral branches, pubescent in the furrows on leading shoots.   Buds with scarious scales.   Leaves, with two conspicuous white stomatic bands, each of five to six lines, on the two dorsal sides, and two bands of two lines on the two ventral sides.

II. *Branchlets*[1] *variable, quite glabrous or with slight scattered pubescence.*
8. *Picea excelsa*, Link.   Europe.   See p. 1337.
Branchlets reddish, usually quite glabrous, or with slight pubescence often confined to the grooves between the pulvini.   Terminal buds conical, acute, without resin, girt with a ring of keeled pubescent ciliate pointed scales. Leaves, usually $\frac{3}{4}$ to 1 in. long, with two to three stomatic lines on each of the four sides.
9. *Picea albertiana*, Stewardson Brown.   Alberta, British Columbia, Montana, Wyoming.   See p. 1385.
Branchlets greyish yellow, either glabrous or with minute pubescence usually confined to the pegs from which the leaves arise.   Buds slightly resinous with rounded entire scales.   Leaves, in an imperfect radial arrangement on the lateral branches, $\frac{1}{2}$ to 1 in. long.

III. *Branchlets always plainly pubescent.   Leaves arranged on lateral branches, as in* P. excelsa.

* *Terminal buds with a ring of conspicuous long subulate scales.*
10. *Picea nigra*, Link.   North America.   See p. 1375.
Branchlets covered with dense short glandular pubescence.   Leaves bluish or glaucous green, about $\frac{1}{2}$ in. long.
11. *Picea rubra*, Link.   North America.   See p. 1377.
Branchlets, as in *P. nigra*.   Leaves yellowish green or dark green, not glaucous, $\frac{1}{2}$ to $\frac{5}{8}$ in. long.
12. *Picea Glehnii*, Masters.   Saghalien, Yezo.   See p. 1369.
Branchlets reddish, with short non-glandular pubescence, confined to the furrows between the pulvini.   Leaves slender, $\frac{1}{3}$ to $\frac{1}{2}$ in. long.

[1] Cf. *P. bicolor*, No. 7, which has pubescent leading shoots and glabrous lateral branches.

** *Terminal buds without long subulate scales.*

13. *Picea orientalis*, Carrière.   Asia Minor, Caucasus.   See p. 1362.

Branchlets slender, pale brown, covered with dense short non-glandular pubescence.   Leaves, ¼ to ⅔ in. long, shining dark green, blunt and bevelled at the tip.

14. *Picea Engelmanni*, Engelmann.   Western North America.   See p. 1387.

Branchlets greyish yellow, with a sparse minute glandular pubescence. Leaves disagreeable in odour when bruised, bluish green, ⅞ to 1 in. long.

15. *Picea obovata*, Ledebour.   Northern Scandinavia, Russia, Siberia ; sporadic at high altitudes in the mountains of Central Europe.   See p. 1359.

Branchlets reddish brown, covered with a dense minute non-glandular pubescence.   Leaves, ⅔ to ⅜ in. long, short-pointed, with three to four stomatic lines on each side.

PICEA EXCELSA, COMMON SPRUCE

Picea excelsa, Link, in *Linnæa*, xv. 517 (1841); Willkomm, *Forstl. Flora*, 67 (1887); Mathieu. *Flore Forestière*, 540 (1897); Ascherson and Graebner, *Syn. Mitteleurop. Flora*, i. 196 (1897); Schröter, in *Vierteljahrs. Naturf. Ges. Zürich*, xliii. 125-252 (1898); Kent, Veitch's *Man. Conif.* 432 (1900); Kirchner, Loew and Schröter, *Lebengesch. Blütenpfl. Mitteleuropas*, 99 (1904); Clinton-Baker, *Illust. Conif.* ii. 38 (1909).

Picea rubra, Dietrich,[1] *Fl. Berol.* 795 (1824).

Picea vulgaris, Link, in *Abhand. Akad. Berlin*, 1827, p. 180 (1830).

Picea Abies, Karsten, *Pharm. Med. Bot.* 324 (1881).

Pinus Abies, Linnæus, *Sp. Pl.* 1002 (1753).

Pinus Picea, Du Roi, *Obs. Bot.* 37 (1771) (not Linnæus).

Pinus excelsa, Lamarck, *Fl. Franc.* ii. 202 (1778).

Abies Picea, Miller, *Dict.*, 8th ed., No. 3 (1768).

Abies excelsa, De Candolle, in Lamarck, *Fl. Franc.* iii. 275 (1805); Loudon, *Arb. et Frut. Brit.* iv. 2293 (1838).

Abies carpatica, Lawson, *Pinet. Brit.* ii. 137, t. 20 (1867).

A tree, often attaining in Britain 120 to 140 ft. in height and 10 to 12 ft. in girth, in central Europe attaining 200 ft. high and 15 to 20 ft. in girth.   Bark on young stems brownish, thin, smooth ; on older trees thick, and scaling off on the surface in thin small scales.   Young branchlets, reddish or yellowish brown, glabrous or with a minute scattered non-glandular pubescence, often confined to the furrows between the pulvini.   Buds conical, acute, reddish brown, without resin, with rounded scarious scales ; terminal bud girt with a few acuminate keeled pubescent ciliate scales.

Leaves on erect shoots radially spreading, more or less appressed to the twigs with their tips directed upwards : on lateral branches, pectinate below, the lower side of the twig being laid bare, most of the leaves being directed forwards and outwards ; while on the upper side of the twig, the leaves in the middle line are more or less

---

[1] Dietrich's name and description apply to the common European spruce, and not to the American red spruce, as is often erroneously supposed.

appressed, with their tips directed forwards and slightly upwards. Leaves, variable in size, usually $\frac{1}{2}$ to $\frac{3}{4}$ in., occasionally 1 in. long, rigid, straight or curved, ending in a short callous point, rhombic in section, with two or three stomatic lines on each of the four sides; resin-canals variable, occasionally absent or only one present, usually two, one at each end of the transverse axis of the rhomb, close to the epidermis.

Staminate flowers solitary in the axils of the leaves of the branchlets of the preceding year, rarely terminal on lateral branchlets, ovoid, about an inch long; stamens numerous, spirally arranged, reddish, each with two pollen-sacs directed downwards and dehiscing longitudinally, and a prominent denticulate connective; pollen grains, each with two air-vesicles.

Pistillate flowers, appearing in summer as brown buds at the tips of the branchlets of the current year, developing in the following spring, about 2 in. long; sessile, erect, cylindrical, purplish red; scales carmine red, oval, with a truncate erose apex; bracts about half the length of the scales, not increasing in size after the time of flowering, ovate-lanceolate, denticulate, with a long acuminate apex. After fertilisation the young cones leave the erect position, and gradually become pendulous, their scales becoming closely imbricated, and in the usual form of the species green in colour.

Cones ripe in October, when they turn brownish; cylindrical, pendulous, variable in size, about 4 to 6 in. in length; usually opening in spring and letting the seeds escape when a dry east wind is blowing; falling from the tree in the subsequent summer or autumn; scales thin and flexible, rhombic, with a truncate emarginate or dentate apex, variable in size, $\frac{5}{8}$ to $\frac{3}{4}$ in. wide, 1 to $1\frac{1}{4}$ in. long, pale brown and glabrous on the exposed part, dark reddish brown and minutely pubescent on the concealed part; bract about $\frac{1}{5}$ in. long, lanceolate, denticulate at the acute or acuminate apex. Seed about $\frac{1}{6}$ in. long, dark dull brown; seed with wing about $\frac{3}{5}$ in. long; wing broadest near the obliquely rounded denticulate apex.

Seedling.—Seeds sown in spring germinate in four or five weeks, the radicle first making its way out of the seed coats, and the caulicle carrying up the cotyledons, which are at first enveloped as with a cap by the albumen of the seed. The cap is soon cast off, and the cotyledons spread in a whorl. The cotyledons are six to ten in number, united at their base by a sheath, about $\frac{1}{2}$ in. long, triangular in section, with the upper edge faintly serrate, without resin-canals, stomatic on the two inner surfaces, deciduous at the end of the second year. The plant at the end of the first year is about 2 to 3 in. high, the young stem bearing, in addition to the whorl of cotyledons, spirally arranged primary needles, which are rhomboidal in section, serrulate on the four angles, with two resin-canals, and inserted on raised pulvini. Branching occurs in the third or fourth year, when the leaves assume their adult form, being entire and not serrulate. No tap root is formed, the root dividing into numerous branches spreading in all directions. Throughout the life of the tree the absence of the tap root, seen in the seedling, persists; and the roots of the spruce are usually spreading and do not penetrate the soil to any great depth.

The spruce is normally monœcious, but instances have been known of individuals which always bear staminate flowers; and hermaphrodite flowers have been observed. The flowers are pollinated by the wind, the pollen being carried to an immense distance; as far as' eight miles in a case which was noticed near Munich. In the vicinity of spruce forests the pollen often descends in enormous quantity, covering the ground and the surface of lakes and rivers with yellow patches.

I. The following variations occur in the form of the scales of the cone :—

1. Var. *europæa*, Schröter, *op. cit.* 142 (1898).

> Var. *montana*, Ascherson and Graebner, *op. cit.* 198 (1897).
> *Picea vulgaris*, Link, var. *europæa*, Teplouchoff, in *Bull. Soc. Nat. Mosc.*, xli. pt. ii. 249 (1869).

Cone-scales rhombic, gradually narrowing in the upper third to a truncate, slightly inflexed, emarginate or denticulate apex. This is the common form of *P. excelsa*, widely distributed throughout central Europe, and also occurring in southern Sweden. In the Alps it is rarely found over 5000 feet elevation.

2. Var. *acuminata*, Beck, in *Ann. Nat. Hofm. Wien*, ii. 39 (1887).

Cone-scales, contracted suddenly into a long bifid recurved undulate apex. This variety is of rarer occurrence in central Europe than the preceding; but is found in the Jura[1] and the Alps, and is said to be common in eastern Prussia and in southern Sweden.

3. Var. *triloba*, Ascherson and Graebner, *op. cit.* 199 (1897).

Scales of the cone trilobed at the apex. This is a much less common variation, which has been noticed in a few trees growing at Blankenburg[2] in the Harz Mountains, at Soglio[3] to the north of Lake Como, and in Moravia.[4]

II. There appear to be two races of the common spruce in the continental forests, which are mainly distinguishable by the colour assumed by the unripe cones in August.

4. Var. *chlorocarpa*, Purkyne, in *Allg. Forst. u. Jagdzeit*, liii. 1 (1877). Cones remaining green in August.

5. Var. *erythrocarpa*, Purkyne, *loc. cit.* Cones becoming dark violet in August.

Purkyne considered that important differences in the growth of the tree, in the character of the wood, in the staminate and pistillate flowers, and in the soil occupied by each form, were correlated with the differences in the colour of the cones; but Schröter considers that these are not established, and suggests further investigation.

III. The spruce varies much in habit in the wild state, and several remarkable sports have been described.

6. Var. *viminalis*, Caspary, in *Schr. Phys. Oekon. Ges. Königsberg*, xiv. 126 (1873).

> *Pinus viminalis*, Sparrman, *ex* Alstroemer, in *Vet. Ac. Handl. Stockh.* xxxviii. 310 (1777).
> *Pinus hybrida*, Liljeblad, in *Svensk Fl.* (1792).

---

[1] Cf. Aubert, *Flore de la Vallée de Joux*, 345 (1900).
[2] A. Braun, in *Verh. Bot. Verein Prov. Brandenburg*, xviii. Sitzb. 13 (1876).
[3] Ascherson and Graebner, *ex* Schröter, *op. cit.* 204, fig. 31 (1898).
[4] Wilhelm, in *Oesterr. Forstzeit.* 1888, p. 169.

Branches in remote almost horizontal whorls, with very long and slender branchlets (often 10 ft., occasionally 20 ft. long) without or with very few lateral branchlets. Leaves radially spreading.

This remarkable form of the weeping spruce was considered by Linnæus[1] to be a hybrid between the spruce and *Pinus sylvestris*. It has been observed in about twenty places in Sweden, where it is vulgarly called *Tysk gran* or German spruce, in about the same number of localities in Norway, and in isolated cases in Livland, East Prussia, Poland, Thuringia, Tyrol, Styria, Carinthia, Carniola, and Switzerland.[2] When sown, the peculiar habit is occasionally reproduced.[3]

7. Var. *pendula*, Jacques and Hérincq, *Man. Gén. Plantes*, iv. 340, 341 (1857).

A remarkable form of the weeping spruce, narrow and columnar in habit, with pendulous branches almost appressed to the stem. Conwentz[4] has described this form, known to him as a single tree[5] in the Stellin forest near Elbing in West Prussia, another[6] at Jegothen, near Heilsberg in East Prussia, and two others[7] near Schierke in the Harz Mountains. Kraemer[8] found another in a forest near Kreut in Bavaria. Solitary examples have also been found in Switzerland,[9] in northern Hungary,[10] and in the Bukowina.[11] The seed of the weeping spruce near Jegothen, when sown by Conwentz,[4] gave twelve trees, only one of which showed a tendency to the weeping habit.

A similar tree with longer leaves, lighter in colour than the typical form, was discovered[12] about the year 1860 by Mr. R. Smith Carrington in a plantation near Kinlet Hall, Shropshire, which was propagated by R. Smith and Company, Worcester, who sold it under the name *Abies excelsa inverta*,[13] Gordon, *Pinet. Suppl.* 4 (1862), a name scarcely worth keeping distinct from var. *pendula*, Jacques and Hérincq, which antedated it a few years. A fine example, about 30 ft. high, was growing[14] in 1897 at Ide Hill, Sevenoaks, Kent; and a good specimen exists at Murthly Castle. There is also a good example[15] at Barbier's nursery, Orleans.

Other kinds of weeping spruce, probably including *Abies excelsa pendula*, Loudon, a form introduced by Booth, are irregular in habit and much more spreading. A very fine example occurs at Durris.

---

[1] Linnæus refers to it as *Abies procera viminalis* in *Fl. Suec.* 288 (1745).

[2] Cf. Schröter, *op. cit.* 151, who draws attention to the fact that *P. Breweriana*, of the Siskiyou Mountains, has this habit as a constant specific character.

[3] Cf. Wilhelm, in *Verh. K. K. Zool. Bot. Ges. Wien*, xxxvii. (1887).

[4] *Beob. Seltene Waldbäume W. Preussen*, 135 (1895).

[5] Figured in *Gartenflora*, 1899, p. 618, fig. 86 ; and by Conwentz, *op. cit.* 141, figs. 12, 13.

[6] Figured by Conwentz, *op. cit.* 147, fig. 14.

[7] Figured in *Gartenflora*, 1901, p. 315, figs. 48, 49 ; and by Conwentz, *op. cit.* 150, 152, figs. 15, 16.

[8] In *Flora*, 1841, p. 700.

[9] Schröter, *op. cit.* 156 (1898).

[10] Schilberszky, in *Kertészeti Lapok*, vii. (1892), describes a weeping spruce near Leutschau.

[11] Cf. *Oesterr. Forst. u. Jagdzeit.* 1897, p. 356.

[12] Nicholson, in *Woods and Forests*, 1884, p. 691 ; and *The Garden*, xxv. 229 (1884).

[13] *Picea excelsa inversa*, Beissner, *Nadelholzkunde*, 361 (1891).

[14] *Gard. Chron.* xxii. 368, fig. 109 (1897). Cf. also *Gard. Chron.* xxix. 263, fig. 98 (1901).

[15] Figured in *Gartenflora*, 1899, p. 617, fig. 87 ; and by Conwentz, *op. cit.* 163, fig. 17.

The "Cornish fir" which was mentioned by Hayes[1] as growing in 1794 at Avondale in Co. Wicklow, was pendulous in habit and bore large cones, sometimes nearly a foot in length. The remarkable pendulous spruce[2] at Shelsley Walsh, in the Teme Valley in Worcestershire, bears cones 9 in. in length, and appears to be identical with Hayes' variety.

8. Var. *columnaris*, Carrière, *Conif.* 248 (1855). Narrowly columnar in habit, with short horizontal branches, clothed with dense short branchlets and foliage.

This form, which has been known a long time in cultivation, exists in the wild state in Switzerland, where six trees are known by Schröter in the five localities of Stanserhorn, Stockhorn, la Brévine, Chavannes, and la Berboleuse, all at high altitudes between 4000 and 5800 ft.

The columnar spruce[3] is to be carefully distinguished from the narrow spruce, known as the *spitzfichte*,[4] in which the habit does not result as a sport, but is due to a severe climate, which checks the growth of the branches. The *spitzfichte* is similar to the columnar spruce in form, being narrowly cylindrical, but the stem is sparingly clad with short branches, wide apart, and forming a thin crown of foliage. The *spitzfichte* is never seen at low levels in the Alps and Jura, but occurs near the timber line, often forming small groves in exposed situations. This climatic form is much more common in *P. obovata* in Lapland, Finland, and northern Scandinavia.

9. Var. *pyramidata*, Carrière, 247 (1855).

Var. *stricta*, Schröter, *op. cit.* 158 (1898).

Branches ascending at a narrow angle, forming a nearly fastigiate tree. Trees of this kind are occasionally seen in the forests of central Europe, and are rarely found in the seed bed in nurseries.

10. Var. *strigosa*, Christ, in *Garden and Forest*, ix. 252 (1896).

A form with numerous slender horizontal branchlets, spreading from all sides of the branches, giving the tree the habit of the common larch. This variety occurs in one locality in the canton of St. Gall in Switzerland.

11. Var. *eremita*, Carrière, in Jacques and Hérincq, *Man. Gén. Plantes*, iv. 341 (1857).

A tree of slender pyramidal habit with numerous branches, directed upwards at a small angle with the stem, short stout branchlets, large buds, and distant short thick sharp-pointed needles.

Var. *Remonti*, said by Kent[5] to be a dwarf modification of this, is described by Masters[6] as of dense compact pyramidal habit, recalling that of *Cupressus Lawsoniana*, var. *erecta viridis*.

[1] *Planting*, 165 (1794). It is first mentioned apparently in *London Catalogue of Trees* (1730), as the long-coned Cornish fir, said to have been "brought from America some years previously and planted in Devon and Cornwall."
[2] Erroneously referred to *P. Smithiana* (as *P. Morinda*) in *Gard. Chron.* 1869, p. 713, and xix. 132 (1896).
[3] Dr. Christ, in *Garden and Forest*, ix. 252 (1896), uses the term columnar spruce for the *spitzfichte*, which is not strictly accurate.
[4] First named and described by Berg, in *Jahrbuch K. Sächs. Akad. Forst. Tharand.* xiii. 83 (1859).
[5] Veitch's *Man. Conif.* 433 (1900). [6] In *Gard. Chron.* vii. 578 (1890).

12. Var. *virgata*, Caspary, *op. cit.* xiv. 125 (1873).

Abies excelsa, De Candolle, var. *virgata*, Jacques, in *Ann. Soc. Hort. Paris*, xliv. 653 (1853).
Abies excelsa Cranstonii, Knight and Perry, *Syn. Conif.* 36 (1850).
Picea excelsa, Link, var. *denudata*, Carrière, in *Rev. Hort.* iii. 102, fig. 7 (1854).

Branches very few and usually not in whorls, elongated, straight or curved, with very few or without branchlets. Leaves radially arranged, either longer or shorter than in the common spruce, persistent ten or twelve years.

This variety,[1] which is known as the snake spruce, owes its peculiarities to the arrest of nearly all the buds, which do not develop. Most of the examples recorded are young trees, but one[2] forty years old at Buttes, near Neuveville in the Swiss Jura, was 40 ft. high in 1898. The snake spruce is not uncommon in Norway, where Schübeler found it in seventeen localities between lat. $59\frac{1}{2}°$ and $61\frac{1}{2}°$; and also occurs here and there in Sweden between lat. 58° and 63°. Isolated examples are reported from Finland, Livland, and Courland which are probably *P. obovata*; and others occur in different parts of Germany. It is common in Bohemia; and one example is known in Moravia. Schröter mentions seventeen trees growing in ten localities in Switzerland. Carrière knew only one example, growing in Cochet's nursery at Suynes, near Brie-Comte-Robert, in Seine-et-Marne.

Varieties intermediate between the snake spruce and vars. *pendula*, *monstrosa*, and *viminalis* also occur, but are very rare.

13. Var. *monstrosa*, Schröter, *op. cit.* 170 (1898). (Not Carrière.[3])

Abies excelsa, De Candolle, var. *monstrosa*, Loudon, *Arb. et Frut. Brit.* iv. 2295 (1838).
Abies aclada, Salvi, in *Flora*, 1844, p. 519.
Picea excelsa, Link, var. *monocaulis*, Nördlinger, *ex* Willkomm, *Forstl. Flora*, 76 (1887).

This variety, which never develops any lateral branches, has a single thickened stem, bearing leaves near the apex, persistent for many years, and about $1\frac{1}{3}$ in. in length.

This variety was first described by Loudon, who mentions a single specimen growing in the Chiswick garden, twelve years planted, and about 3 ft. in height. A specimen at High Canons, Hertford, produced cones of the ordinary form in[4] 1907. Salvi found in 1842 four specimens, growing wild in the Euganean Hills, west of Padua. One of these which was transplanted to Isola Bella in Lake Maggiore, where I saw it in 1909, is attached to a bamboo, and trained up the wall of the château; it measures about 30 ft. in height and is nearly as thick (1-$1\frac{1}{2}$ inch) at the top as at the bottom, bearing leaves with very sharp points only on the upper two feet of the stem. Schröter records another specimen at Stockach in Baden, another in Bohemia, and another at Ansbach in Bavaria. A form of this variety is recorded

[1] An analogous form of the common silver fir, *Abies pectinata*, var. *virgata*, Caspary, in *Bot. Zeit.* 778, t. ix. (1882), occurs; but only four examples are known—two in Alsace, one in the Bohemian forest, and another in the Swiss Jura near Neuveville. The latter is described and figured by Schröter, *op. cit.* 168, fig. 15 (1898). Cf. vol. iv. p. 722.

[2] The oldest known to Schröter was one near Dorpat, in Livland, said by Berg, in *Schrf. Naturf. Ges. Univ. Dorpat*, ii. t. 2 (1887), to be sixty years old.

[3] Carrière, *Conif.* 248 (1855), wrongly applied the name *monstrosa* to var. *virgata*, Caspary.

[4] According to *Gard. Chron.* xxv. 146 (1886), var. *monstrosa* at Lucombe, Pince and Co.'s Nursery, Exeter, produced cones in 1886 which were similar to those of the ordinary spruce.

from Silesia and Thuringia, which bears a few undivided branches at the base, the upper part being without branches.

14. Var. *globosa*, Berg, in *Schrift. Naturf. Ges. Univ. Dorpat*, ii. 19, 20 (1887).

In this variety, normal growth is replaced by numerous close branches, irregularly dividing into a great number of branchlets, similar to a witches' broom, and forming either a globose bush without any leader, or a conical bush with a leader arising out of a globose base. I saw a remarkable example of the globose spruce in 1909 at the Forestry Experimental Station, Zurich. Seedlings had been raised, one quarter of which had reverted to the habit of the ordinary spruce, the others being very various in appearance and intermediate between the parent form and the normal habit of the species.[1]

In the true dwarf forms[2] of the spruce, the branching is regular, but the growth of the shoots is very small, and the needles are very short. The most important are :—

15. Var. *Clanbrassiliana*, Carrière, in Jacques and Hérincq, *Man. Gén. Plantes*, iv. 341 (1857).

Abies excelsa, De Candolle, var. *Clanbrassiliana*, Loudon, *Arb. et. Frut. Brit.* iv. 2294 (1838).

A compact low dense globose bush, seldom higher than 5 or 6 ft. ; branches and branchlets, much shortened ; leaves about ¼ to ½ in. long ; buds very red in colour. This is supposed to have been found on the Moira estate, near Belfast, about the end of the eighteenth century, when it was introduced into England by Lord Clanbrassil. This dwarf form has been found growing wild in Thuringia, and near Stockholm, and in Jemtland in Sweden. It is always sterile, and is propagated by cuttings.

Elwes found at Tullymore Park, Co. Down, a large bush of this form measuring 10 ft. high and 28 ft. in circumference, which he was informed was either the original or a part of it, and was supposed to be about one hundred and fifty years old. A specimen at Aldenham has reverted to the normal type, and is now growing rapidly into an erect tree.

16. Var. *tabulæformis*, Carrière, *Product. et Fixat. Variétés*, 52 (1865), *Conif.* 333 (1867).

A prostrate form, with slender branchlets spreading horizontally over the ground. This is said by Carrière to have been taken, probably as a cutting, from a witches' broom, growing on an ordinary spruce in the Trianon. Torssander[3] found a similar plant in Södermanland in Sweden, thirty years old, and only 20 in. high.

17. Other dwarf forms have been named, as vars. *pumila*,[4] *pygmæa*,[5] *Gregoryana*,[6] *Maxwelli*,[7] etc.

---

[1] Cf. Engler, in *Mitt. Schweiz. Forst. Versuch.* viii. pt. ii. 117, figs. 8, 9 (1904).
[2] See under *Witches' Brooms*, p. 1345.   [3] In *Södermanland Botan. Notiser*, 1897, p. 169.
[4] Beissner, *Nadelholzkunde*, 365 (1891).   [5] Loudon, *op. cit.* 2295 (1838).
[6] Said by Gordon, *Pinetum*, 9 (1875) to have been raised in the Cirencester Nursery.
[7] Originated as a seedling in Messrs. Maxwell's nursery, Geneva, New York. Cf. *Woods and Forests*, 1884, p. 502, and Rehder, in Bailey, *Cycl. Amer. Hort.* iii. 1333, fig. 1798 (1901).

IV. Several varieties are known in which the leaves are coloured :—

18. Var. *aurea*, Carrière, *Conif.* 246 (1855).

Leaves yellowish white, shining. The golden spruce has been found wild in Carinthia.

19. Var. *finedonensis*, Gordon, *Pinet. Supp.* 4 (1862).

Leaves pale yellow at first, changing to a bronze colour, and ultimately becoming green.[1] This originated at Finedon Hall, Northamptonshire, where it came up accidentally in a bed of common spruce. It often loses its colour in cultivation, and at Colesborne has entirely reverted to the normal green. Var. *mutabilis*[2] has the young shoots creamy yellow in colour, changing to green by the end of the season. Mr. Bean[3] saw in Hesse's nursery, Weener, Hanover, a very beautiful variety, with creamy white young shoots, which is called var. *argenteo-spica*.

20. Var. *variegata*, Carrière, *Conif.* 246 (1855).

Leaves variegated with pale yellow. A variegated form is mentioned by Loudon ; and Wittrock[4] found a tree with leaves variegated white at Helsingfors.

V. The colour of the bark of the common spruce varies from whitish grey to brown, probably due to influence of soil and climate. The following sports have been observed.

21. Var. *corticata*, Schröter, *op. cit.* 184 (1898).

Bark thick, up to $3\frac{1}{2}$ in., longitudinally fissured, and resembling that of a pine in external appearance, though in microscopical structure like the ordinary spruce. Schröter knew in 1898 only six spruces with thick bark, occurring in Austria, Bohemia, Hesse, Bavaria, and Switzerland ; but more than twenty are now known[5] in the latter country alone.

22. Var. *tuberculata*, Schröter, *op. cit.* 190 (1898).

Lower part of the stem covered with corky excrescences, projecting about an inch above the surface of the bark, where side branches are given off.[6] Four examples only were known to Schröter in 1898, two in Austria, one in Bavaria, and one in Switzerland ; but Badoux[5] states that many more have since been found in Switzerland.

VI. In addition to the varieties and sports just described, which are of unknown origin, there are many peculiar forms of the spruce which are due to external influences, and which cannot, properly speaking, be named varieties or sports.

1. The candelabra spruce is often produced, when the leading shoot is broken off by the force of the wind or by other causes. A whorl of secondary branches becomes erect below the broken part of the stem, and forming a series of leaders, grows up, giving the tree a candelabra-like appearance.

2. Dwarf spruces,[7] which are mere bushes, with irregular branches, dense

---

[1] Fowler, in *Gard. Chron.* 1872, p. 76, speaks of the inconstancy of the colour in different parts of the tree.

[2] Cf. Masters, in *Gard. Chron.* vii. 578 (1890).　　　　[3] *Kew. Bull.* 1908, p. 391.

[4] In Hartman, *Skand. Flora*, 35 (1889).

[5] Badoux, in *Journ. Forest. Suisse*, 1907, quoted by Beissner, in *Mitt. Deut. Dend. Ges.* 1910, p. 122.

[6] Cf. Cieslar, in *Centralblatt Gesamte Forstwesen*, xx. Heft 4, pp. 145-149 (1894). Schröter compares these corky excrescences with those developed on the stems of Zanthoxylum, studied by Barber, in *Ann. Bot.* vi. 155 (1892).

[7] *Picea ellipsoconis*, Borbas, *Magyar Bot. Lapok*, i. 26 (1902), a shrub-like spruce growing as scrub near tree-limit in the western Carpathians, with short broad cones, is considered by Pax, in *Pflanzenverb. Karpathen*, ii. 177 (1908), to owe its peculiarities to the high altitude, similar shrubs being recorded for the eastern Alps by Beck.

foliage, and numerous leaders occur in alpine regions, and are due either to the severe climate or to constant cropping by goats and sheep.

3. Witches' brooms on the spruce have hitherto been supposed to be due to the irritation of fungi, bacteria, or mites. Tubeuf,[1] in January 1907, sowed seed which he obtained from a witches' broom that had borne cones. The greater part of the seedlings are normal, but a certain number are dwarf and bushy. Tubeuf supposes that some of the former will in time develop witches' brooms on some of their branches, and that the latter will probably remain dwarf, resembling the varieties[2] *Clanbrassiliana*, *pumila*, etc., already referred to as of unknown origin.

4. Masters[3] gives a figure of a remarkable branch of a spruce, in which the leading shoot had split into two portions for some distance, re-uniting above to form again one stem.

DISTRIBUTION

P. excelsa is a native of Europe, extending from the Pyrenees, Alps, and Balkans northward through south Germany and east Prussia to Scandinavia, and eastward through the Carpathians and Poland to western Russia.

In France, the spruce occurs in the mountains, mixed with the silver fir, and in the zone above it, the lower limit in the Vosges and Jura being about 2000 ft. It is not at all abundant in the Vosges, where it ascends to 4300 ft.; but in the Jura covers large areas, and reaches 5000 ft. altitude. It attains its greatest importance in Savoy, which is the only region in France where the spruce is the dominant tree, forming one half of the whole area occupied by forests. The forest of Thônes,[4] near Annecy, which I visited in 1904, is one of the best examples of a spruce forest in France; and is treated on the selection system. It contains about 320 acres, lying between 2500 and 4300 ft. elevation on a steep slope, and is a mixture of two-thirds spruce and one-third silver fir. The standing timber is estimated at 7000 cubic ft. per acre, the annual felling averaging 53 cubic ft., with a revolution fixed at 144 years.

The spruce is absent in the Cevennes, and is extremely rare on the north side of the Pyrenees, where it is replaced by fine forests of *Pinus montana*. Willkomm records it for the Pyrenees of Catalonia and Aragon, where it is not at all common. Its most southerly point in western Europe is the forest of La Cinca, south of Mt. Maladetta in lat. 42° 30′.

In Germany, the northern limit of distribution, beginning in the Vosges, passes through the Pfalz, and after crossing the Rhine at lat. 50°, makes a bend to the westward through Westphalia, and reaches the Weser Mountains, where, near Minden, the spruce attains its most northerly point as a wild tree in western Germany, lat. 52° 20′. From here the limit passes through Hildesheim, Wolfenbüttel, Walbeck near Magdeburg, and Halberstadt to Altenburg; whence, taking a north-easterly direction, it is continued through Spremburg and Soran to Ostrowo, reaching the Russian frontier at lat. 52°. It then passes northward, parallel with the frontier, to

[1] Cf. Prof. Somerville, in *Quart. Journ. Forestry*, iv. 309 (1910).   [2] Cf. var. *tabulæformis*, Carrière, *ante* p. 1343.
[3] In *Gard. Chron.* xxiii. 274, fig. 52 (1885).   [4] Cf. *Bull. Soc. Forest. Franche-Comté*, vii. 630 (1904).

Gilgenberg in the province of East Prussia, and reaches the Gulf of Danzig at Elbing.   In southern Germany it is scattered, or in small woods in the plains and valleys, being probably planted;[1] and as a wild tree is nearly confined to the mountains, where it occupies a distinct zone, with clearly defined upper and lower limits.   The largest pure forests in Germany occur in the Harz mountains, which are almost entirely covered with spruce, ascending to 3300 ft., and in the Iser and Riesen mountains, up to 3900 ft.   In the other great forest regions of Germany, as in the Thuringian, Bavarian, and Bohemian forests, the Fichtel and Erz mountains, central Saxony, etc., the spruce, in mixture with the silver fir, covers immense areas. In the province of East Prussia, there are very large forests on the plain, in which the spruce grows in company with the common pine, birch, alder, and willows; but it is absent on pure sandy soils, where the common pine reigns supreme.

The spruce is met with throughout the Alps in Switzerland and Italy, ranging in Tessin between 2700 and 6000 feet, and occupying a small outlying area in the Euganean hills in Lombardy.   It is quite unknown in the Apennines.

In Austro-Hungary, extensive forests of spruce, often almost pure, occur in the Carpathians from Silesia to Bukowina, and in the Transylvanian mountains.   The largest spruce recorded[2] is one which grew in the Carpathians, measuring 226 feet in height and $11\frac{1}{2}$ feet in girth at breast height.

In the Balkan peninsula[3] the spruce reaches its most southerly limit, a line extending from the mountains of northern Albania to the Kopaonik mountain in Servia, whence it is prolonged eastward to the Rhodope mountains in Rumelia about lat. 42°.   In Bosnia, Servia, Montenegro, Croatia, and Herzegovina, the spruce usually grows in mixture with the beech and silver fir, occupying a zone on the mountains between 3000 and 6000 feet; but in northern Albania the lower limit rises to 4000 ft.   Huffel[4] states that in Roumania, the spruce attains enormous dimensions, a tree, which was cut down in 1888 in the forest of Tarcau, measuring 195 ft. in height by 3 ft. 3 in. in diameter at breast height; it was 392 years old.   The spruce is much less common in Roumania than the silver fir, ascending to 6000 ft. in Wallachia and to 5000 feet in Moldavia; while in Bukovina the spruce is more abundant than the silver fir, and occupies a zone between 2600 and 5200 ft.

In Russia the southern limit of the spruce (including *P. obovata* and *P. excelsa*) extends from the frontier of Galicia, at lat. 50°, eastwards through northern Volhynia and Starodul in the government of Chernikof, crossing the river Oka at lat. 53° or 54°, to the southern boundary of the government of Kazan.   From this northwards to the Arctic circle, the spruce is prevalent; but the exact boundary of the two species is unknown.   So far as I have seen specimens, the spruce in Finland and near St. Petersburg is *P. obovata*, which all authors agree is the only spruce found in north-eastern Russia, as nothing like *P. excelsa* is seen to the eastward of the rivers Dwina

[1] It is supposed never to be native in situations below 1300 ft., though it thrives when planted.   Left to nature the beech speedily supplants it on all soils at low elevations in southern Germany.

[2] Wessely, quoted by Mathieu, *Fl. Forestière*, 541 (1897).   I have not been able to verify this record; but Schröter and Kirchner, *op. cit.* 115 (1906) state that Enderlin measured in the Grisons two trees as follows:—one, 143 ft. high, 6 ft. 3 in. in diameter, with a volume of 1300 cubic feet; the other, 152 ft. high, 4 ft. 11 in. in diameter, with a volume of 1150 cubic feet.

[3] Beck von Mannagetta, *Veg. Illyrisch. Länd.* 287 (1901).

[4] *Extrait Bull. Minist. Agric. Paris*, 1890, p. 6.

and Viatka. Korskinsky states[1] that the typical form of *P. excelsa* occurs only in western Russia in the region adjoining the German plain; and the varieties which he describes as occurring in central and northern Russia, and linking *P. obovata* with *P. excelsa*, seem to me to be simply *P. obovata* with cones slightly larger than those occurring in the Ural range.

Von Sievers[2] says that the spruce is the only shade-bearing tree in the Baltic provinces, and attains in favourable situations a height of 160 feet. It occurs naturally on better soil than the pine (which occupies poor sandy soil) and competes with the birch and alder on clay, thriving well on deep peat, if this is rich in mineral salts.

The original conifer of Norway was *Pinus sylvestris*, the remains of which are found everywhere in peat mosses. The spruce is a late emigrant from Sweden and Lapland. It occupies in Norway three distinct regions:[3]—

A. The spruce is found in the far north in isolated stations, as on the Varanger-fjord, lat. 69° 30′, at Karasjok, lat. 68° 30′, and at Saltdalen, on the west coast, lat. 67° 10′.[4] The spruce here is *P. obovata*, these stations being outposts of the north Russian spruce, which extends eastwards through Enara Lapland to the Kola peninsula, and through Swedish Lapland from Sulitjelma to Palojuensun on the Muonio river.

B. In the Trondhjem district the spruce reaches the coast, and is connected with the northern Swedish spruce, through four passes in the range separating Norway from the Swedish province of Jemtland. This spruce is *P. obovata*.[5]

C. In southern Norway the spruce, which appears to be *P. excelsa*, occupies a distinct area, separated from the last by the Dovre-fjeld, and continuous with the spruce forests of south Sweden, there being no high mountains intervening between the two countries for a considerable distance north of Svinesund. Throughout this region, no remains of the spruce have been found in peat mosses, though those of the common pine are plentiful; and the spruce is evidently a late emigrant, not having yet reached the west coast. Through Romsdal, Bergenhus, and Stavanger provinces, and the district of Lister, the area covered by forest is not extensive, the principal trees being pine and birch, while the spruce is rarely if ever found wild, except in the inland district of Voss, situated about 40 miles east of Bergen. According to Schübeler, the spruce ascends on the Jotunfjeld to 3250 feet, and in Hallingdal to 3400 feet. South and east of the mountains, the greater part of the very extensive forest area consists of spruce, mixed to some extent with pine and birch. The Norwegian spruce is said to contain a relatively small amount of resin, and is therefore largely used in the production of mechanical and chemical wood-pulp, an industry, which in some places has begun to threaten the continued existence of the spruce forests. The spruce bark is also used for tanning.

---

[1] *Tent. Fl. Ross. Orient.* 494 (1898).   [2] *Forst. Verhält. Balt. Prov.* 18 (1903).

[3] Cf. Sernander, in Engler, *Bot. Jahrbüch.* xv. 3 (1893).

[4] Elwes found it here in 1903 only as a rare isolated tree, and was told that the Ranenfjord, 50 miles south was its real northern limit.

[5] Specimens collected at Trondhjem and at Bräcke in Sweden are identical, and are indistinguishable from specimens gathered in Perm in north-eastern Russia.

In Sweden the exact limits[1] between *P. obovata* and *P. excelsa* are unknown to me ; but there is no doubt that in the northern part of the country, from Jemtland northwards, *P. obovata* is the sole species. In southern Sweden, the tree appears to be *P. excelsa*,[2] and its distribution[8] is peculiar, as it does not extend to the extreme south, not occurring in Skåne north of lat. 56° 10′, and not extending to nearer the west coast, from Strömstad to Halmsted, than fifteen miles at any point. Its remains have not been found in the peat mosses south of Jönköping on Lake Wetter ; and it is supposed to have spread southward of this point in quite recent times ; and this is confirmed by ancient maps of North Skåne, which show that there were no spruce forests in this district at the beginning of the 17th century. (A. H.)

In Scandanavia the spruce is called *gran*, or *rödgran*. From what I have seen in the forests of north and south Trondhjem it is usually found on the better class of land, and even there does not grow so large as farther south, ascending to about 2000 ft. in Tydal and Stordal, and attaining about 100 ft. in height by 8 to 10 ft. in girth. I have not noticed, even near the coast, that the trees are browned by the west wind, as they are sometimes even, far inland, in England, but the tree is rarely seen on the exposed parts of the coast, or on the islands, where the Scots pine grows alone. Schübeler, *Viridarium Norvegicum*, figs. 66, 68, 69, figures three trees remarkable for their habit, one having the branches very short and crowded on the upper part of the stem, and another a good example of the snake spruce, var. *virgata*, Caspary. Figures 73, 74, and 75 show instances of natural layering ; and figures 76 and 77, trees grown from a fallen stem. Figure 78 shows a candelabra-shaped tree growing near Horten in the Christiana fjord.

It is stated that the varieties known in cultivation as vars. *nana*, *inverta*, and *Clanbrassiliana* have all been found wild on the coast of Norway.

The tallest spruce mentioned in Norway by Schübeler was in Hurdalen (lat. 60° 24′), and measured 130 ft. high by 3 ft. 5 in. in diameter ; and I am informed of one recently cut in South Rendalen, which was 125 ft. high, and 15 in. in diameter at 80 ft. from ground, and 25 in. at 20 ft. Five logs over 20 ft. long were cut from this one tree.

The largest spruce I have heard of in Sweden is mentioned by Schübeler (p. 409). It grew in Oster Gotland (lat. 58°) and measured 150 Swedish feet (44.54 m.), with a diameter of 6 ft. (1.78 m.).

In Professor Göppert's memoir[4] on the *Primæval Forests of Silesia and Bohemia* there are many illustrations of the remarkable forms which the spruce assumes when left absolutely in a state of nature, in regions where the snow lies long and deep. These forests are not described in detail, but are above the region of deciduous trees, and consist mainly of spruce and silver fir, with *Sorbus Aucuparia*, *Salix silesiaca*, and *Lonicera nigra* as underwood. Many of the fallen and rotting

---

[1] Wittrock, in *Act. Hort. Berg.* iv. No. 7, p. 69 (1907), agrees with me in laying stress on the character of the twigs in the discrimination of the two species, *P. obovata* and *P. excelsa*, which he considers to exist in Sweden. Cf. his article in Krok, Hartman's *Skand. Flora*, 1889, p. 34.   [2] Cf. Sylven, in *Skog. Tidsk.* 1909, Fack. pp. 201-261.

[3] Cf. Hesselman and Schotte, in *Medd. Stat. Skogsförsöksanstalt*, Heft 3, pp. 1-52, with maps (1906).

[4] Göppert, *Skizzen zur Kenntniss der Urwälder Schlesiens und Böhmens*, in *Nova Acta Acad. Leop. Carol. Nat. Cur.* xxxiv. (1868).

stems are covered with trees which have sprung from seeds germinating in the moss on these trunks. Göppert mentions one about 50 ft. long on which he counted thirty-six living trees of various ages from 4 ft. to 80 ft. high. On another 70 ft. long, there were thirty-two trees from eighty to one hundred years old, all of which had their roots resting on the fallen tree which had given them birth. Such examples are figured in his plates vii., viii., and ix.

Another form is shown in plate ii., figs. 7, 8, and 9, which illustrate trees which have grown from seeds falling on stumps of broken or dead trees at a considerable height from the ground, and which have forced their roots down through the decaying wood, in one case from a height of 16 feet, to the ground. When the stump decayed the roots were strong enough to support the young tree, which eventually was left standing like a Pandanus on a pyramid of its own roots.

In some cases, as shown in plate iv., fig. 11, two trees which had originated separately on the same stump became perfectly inarched at the root. Plate iv., fig. 12, shows a remarkable instance of a stump no less than 6 ft. in diameter, which had become covered with a thick layer of moss, and assumed the appearance of a gigantic mushroom, on the top of which no less than seven young spruces from 2 to 40 ft. high were growing without their roots having reached the ground at all.

Plate ix., fig. 22, proves, according to Göppert, the immense period which may elapse in these forests before the fallen trees are absolutely decayed and resolved into humus. It shows, A, a fallen tree, of which the wood was nearly all dissolved into long brown pieces, only held together by the overgrowing thick moss, into something like the original shape of the trunk; B is a tree which had fallen on the top of it at a later period, and was decayed about half through; C is a living tree estimated at 300 years old, which had germinated on B, and buried its roots partly in and partly on one side of it. Göppert believes that from 1000 to 1200 years may have elapsed since the germination of the lowest tree, A; but it seems to me that even if it was 400 years old when it fell, the second, B, may have fallen soon afterwards, and owe the slower decay of its wood to the comparative dryness of its position above A. Still it proves that the decay of such a comparatively soft wood as spruce or silver fir (the species is not in this case specified) is extraordinarily slow under the conditions prevalent in these forests.

As a proof of its slowness of growth in some instances Forstmeister John remarked that the spruce in the densest parts of the forest attained an age of 120 to 160 years without exceeding 5 to 7 in. in diameter. I have myself cut in Norway a spruce which showed over forty annual rings, and was still thin enough to serve as a walking-stick, which I used through three seasons of elk-hunting before it broke under my weight.

In the Böhmerwald the spruce comes to perfection at a higher level than the beech and silver fir, from 3000 to 3400 feet; and in the Kubany forest there are thousands of trees from 120 to 150 ft. high, and 12 to 16 ft. in girth. It attains a greater age than the silver fir, some trees showing no less than 700 annual rings, though still quite sound. From 3600 ft. up to the highest peaks, which in this range of mountains attain little over 4500 ft., the spruce changes its habit, the stems becoming shorter and the branches more spreading and drooping on account of the heavier snowfall, so that when adult they assume a regular pyramidal or conical shape. At 3500 ft., a tree 3 ft. in diameter showed 420 rings; and on the top of Kubany at 4100 ft., another with a diameter of 2 ft. had 235 rings. On the Arber mountain, the highest peak of all, at 4200 ft., Göppert saw a tree 3 ft. thick, but only 40 ft. in height; but even at this elevation the majority of the trees are neither crippled nor diseased, as is often the case near the limit of trees in the Alps and Riesengebirge, where they are covered with lichens.

Above these altitudes the lower branches often spread on the ground and form natural layers, which grow upright and make a colony of small trees around their parent. Such an instance is shown in plate i. fig. 2, and another even more curious on fig. 3, where the main trunk of a tree about 5 ft. in girth curved to one side and threw up a secondary straight stem from the nearly horizontal part of its bole. Figure 4 shows a fallen stem 32 ft. long, which remained living and bore no less than five erect trees from 10 to 37 ft. high, which apparently drew the whole of their nourishment from the original roots of the parent tree.

Another peculiarity which occasionally appears in these forests are trees with immense swellings on their trunk, in the form of irregular burrs equally developed all round the trunk. Plate i. fig. 5, shows a spruce 18 in. in diameter at the ground, which has a regular swelling, shaped like a flattened orange, no less than 12 ft. in diameter, and from the centre of which the straight trunk again emerges with a diameter of 16 in. Göppert saw in Silesia an even more extraordinary tree (plate i. fig. 6), which was 45 ft. high and 2 ft. in diameter near the ground. At 7 ft. a regular swelling suddenly began (which is described as covered with many branches, but in the drawing shows none on one side) with a diameter of 10 to 12 ft. and 23 ft. high, above which it tapers off into a normal stem. No disease could be found in the bark or wood, which appeared completely sound, and the upper part of the tree is shown crowded with healthy branches.

A remarkable group of four spruces growing at timber line (about 1850 metres altitude), on the north side of the Great Scheidegg, is figured by Dr. Klein,[1] which from their position appear to have all sprung from seeds which have grown on a rotting trunk. Another remarkable illustration of the effect of wind on the growth of the spruce at high elevations is shown in plate 54 of the same work. A group of trees growing at about 4600 ft. on the Feldberg in the Black Forest, from 10 to 16 ft. high, have the branches cut off clean on the west side, which is attributed by Dr. Klein not alone to the drying effect of the wind in winter and spring, but also to the heating of the branches on one side only by the sun. In the same work are several illustrations of the dense spruce bushes, called "feisstannli" by the Swiss, which are common in alpine regions, and are caused by the constant cropping of goats and sheep.

In the virgin forests of the Capella Mountains in Croatia I saw, in 1910, some spruce of immense height;[2] and measured one of about 170 ft. by 12½ ft. I was informed that, in this forest, spruce had been felled 190 ft. high and about 12 ft. in

[1] Karsten and Schenck, *Vegetationsbilder*, II, t. 38 (1905).
[2] Cf. *Quart. Journ. Forestry*, v. 31 (1911).

girth, and that about 300 years is the maximum age at which this tree remains sound.

The spruce is not a native of Britain at the present epoch; but remains of it have been found in pre-glacial beds at Cromer, Mundesley, Bacton, and Happisburgh in Norfolk.[1]

CULTIVATION

It appears to have been introduced early in the sixteenth century, as Turner includes it in his *Names of Herbes* published in 1548; and both Gerard and Parkinson state that it was found in different parts of Britain.

The spruce is easy to raise from seed, but the seedlings grow very slowly for the first three or four years, and are rarely large enough to plant out until they are four to six years old. I have noticed a great deal of variation in the time at which their new growth appears, and it is well to separate the earliest, which are very liable to be injured by spring frost, whilst those which do not start into growth till June remain uninjured. Few conifers are easier to transplant either in spring or autumn, provided the roots are not allowed to become dry; but if exposed to the air in dry or cold weather a good many will die, or languish for two or three years after planting.

The tree grows on almost any soil, but requires a sheltered situation to attain a great height and only comes to perfection where the soil is moist and not liable to dry up in summer. Grown in dense woods, the spruce is liable to be blown down by the wind; but isolated trees make much stronger roots and are moderately storm-firm. In places near the sea the foliage is often injured by the salt contained in the air, and even as far as forty miles from the Bristol Channel I have seen the spruce completely browned on the side exposed to the wind in March.

Though the seed ripens freely in most seasons and germinates readily, the spruce rarely reproduces itself from seed in England owing to its slow growth at first and the weak hold of its young roots on the soil, which cause the seedlings to wither up in summer or to be thrown out of the ground in winter. I only know a few places on my estate where self-sown spruce can be found; and the seedlings have grown so slowly that I am convinced it is not an economic practice to reproduce spruce by seed, except in places where the ground is under snow for a long period. In the Highlands among heather self-sown seedlings are much commoner; and on the shores of Loch Rannoch and in some of the old pine woods at Castle Grant there are considerable numbers of self-sown seedlings, but nothing like the number seen in Scandinavia or in the German forests, where they are protected by deep snow for a long period in winter.

The spruce is a tree which has been planted more largely in England than it deserves to be; for though it will, when established, grow on poor ill-drained soil faster than most conifers, yet the value of its timber when felled is less than that of almost any other tree; and it is, on account of its shallow rooting habit, very likely to be blown down if the wind once gets into the plantation.

[1] C. Reid, *Origin of British Flora*, 151 (1899).

As a nurse it is, on land unsuitable for Scots pine, one of the best we have, if not allowed to overcrowd the hardwoods planted with it; because its branches protect the ground from frost and drought, and its rapid growth acts as a wind-break and draws up the other trees. Its roots are more superficial than those of larch or Scots pine; and it is much cheaper to plant and less liable to injury from frost than silver fir. Loudon quotes (*Arb. et Frut. Brit.* p. 2305) the experience of W. Adam, Esq., of Blair, Kinross-shire, who was a great advocate of planting spruce as a nurse to oak and elm; but it must be cut out or its lower branches lopped before it becomes large.

In the east and north-east of Scotland the tree seems more promising as a forest tree; and in the opinion of Mr. Crozier may produce a more valuable crop than either larch or Scots pine at considerable elevations. He gives me particulars of a plantation of 400 acres on the Durris estate, at an elevation of over 800 ft., which was marketed under his own supervision. The age of the trees was sixty years, the number per acre 560, averaging 10 cubic feet each. Sold standing at 5d. per foot they realised £116 per acre. Some parts of this plantation planted with Scots pine only made £15 per acre, and the best of it under larch was estimated at £70 per acre. In this plantation the spruce was planted in patches, none over three or four acres, usually on sites unsuitable on account of excessive moisture for larch or Scots pine. The greater parts of the area, however, might well have been planted with spruce, as the locality is favourable to its growth, and similar results to the above are the rule rather than the exception on the whole estate. I have lately received from Mr. D. Munro of Banchory a photograph (Plate 371) of this plantation, most of the trees in which were blown down after it was sold. Mr. Crozier states that there is a large demand for home-grown spruce boards for box-making in Scotland, but that the wood must be fairly free from knots and discoloration; and he considers it one of the most useful timbers for house-building. He adds that when planters realise that the limit of altitude for spruce planting lies above the pine belt, and not below it as seems to have been the general idea, and that it must be grown thicker than larch or Scots pine, its economic merits will become more evident than they are at present.

In confirmation of Mr. Crozier's opinion that spruce is a valuable tree for planting for profit at a high elevation, I may refer to Messrs. Robinson and Watt's very full report[1] on the Coombe Plantation, which lies between 900 and 1500 ft. altitude near Keswick in Cumberland. This plantation, which was sixty-one years old in 1910, consists mainly of larch, with a mixture of spruce, amounting to only one or two per cent at the lower levels, but to ten per cent at over 1250 ft. These authors state that here "spruce grows well at all elevations, and everywhere attains a greater volume than larch under the same conditions." At the higher altitudes it much exceeds the larch in volume. The influence of altitude and exposure on the spruce itself is shown in the following table :—

[1] *Journ. Board of Agric.* xvii. 273, 360 (1910).

Elevation : feet.					Height of tree in feet.	Quarter-girth measurement, in cubic feet.
900	.	.	.	.	80	44
1150	.	.	.	.	not stated	32
1250	.	.	.	.	"	26
1450	.	.	.	.	48	8
1520	.	.	.	.	35	4

I have failed to obtain any other exact and reliable figures as to the value of a crop of spruce grown in England, except on such small and isolated patches of land, that they would give no fair criterion.

As a shelter tree it makes a good edging to the roads in a plantation, and can be headed down or clipped when it has grown tall enough to keep the wind out. It bears clipping well and makes a good dense hedge on soils not liable to drought.

Sargent[1] states that as an ornamental tree in America, it loses vigour at twenty-five to thirty years old, except in the most favourable situations; and he only recommends it as a nurse for other trees, as it is very hardy and grows rapidly at first. As a proof, however, of the extremely vigorous growth of the spruce in America, I may say that the tallest tree in Mr. Hunnewell's Pinetum at Wellesley, Mass., which I had the pleasure of visiting with Professor Sargent in 1904, was a Norway spruce which was planted about 1852, and was in 1894 80 ft. high with branches spreading over a circle 60 ft. in diameter. When I saw it, it had increased little in height, but its lower branches had spread to 75 ft. diameter and some of them had rooted; flowers were just showing on May 9, and cones were produced on branches close to the ground, which is rarely the case in Europe. This tree is figured on Plate 340.

According to Pinchot,[2] it thrives throughout the entire north-east of the United States and southward at the higher elevations; but in the west, favourable results have been attained only as far as the eastern part of the prairie region, and then only in the more protected localities. He considers that it should be planted on a large scale in the cut-over land in the north, where the tree will provide a future supply of wood pulp, as it is in every way superior to the native spruces.

REMARKABLE TREES

If I could trust the measurements which have been given me I should say that the tallest spruce in this country is a tree at Rooksbury Park, Hants, the seat of J. C. Garnier, Esq. It is in a densely crowded thicket of rhododendron, surrounded by beech, and was said by Mr. A. Arnold to measure no less than 178 ft.; but after seeing it twice I could not believe that it was over 150 ft., and, owing to its position, could not measure it myself.[3]

In Oates Wood at the top of Cowdray Park, near the superb silver fir figured

---

[1] *Silva N. Amer.* xii. 24, note (1898).   [2] *U.S. Forest Service, Planting Leaflet*, No. 20 (1908).

[3] At my request Mr. Arnold has recently re-measured this tree with a theodolite, and informs me that though he could not get a clear view of its top, he now estimates it at 149 feet.

on Plate 208, there are two remarkably tall trees, which in 1903 I estimated at 140 to 150 ft., but owing to the steep slope on which they grow, and to the adjoining trees, I could not measure a base line. They were 11 ft. 6 in. and 10½ ft. in girth in 1906. Mr. Harold Pearson has recently had these measured by Mr. T. Roberts, forester at Cowdray, who informs me that he levelled a base line, and found the height in January 1911 to be 153 ft.; but this tree which has three leaders is not so handsome as the other, which he thought was about 148 ft. high.

The tallest which I have myself measured are two trees growing on the edge of the lake near Fountains Abbey, Studley Royal, Yorkshire, which Loudon describes as the tallest spruces known to him, and says were 132 ft. high. When I saw them in 1905 I found one to be 140 ft. by 12 ft. 10 in., and the other 137 ft. by 11 ft. They are free from branches for 30 to 40 ft., and seemed in excellent health, though probably over 150 years old (Plate 341).

In Earl Bathurst's woods at Cirencester there are two narrow avenues of spruce known as the Cathedral firs, because they resemble the cross aisles of a cathedral. Of these Plate 342, from a negative taken by Mr. T. E. Gerald Strickland, gives an excellent picture as they were four years ago, but since then several have been blown down, one of which was over 100 ft. high, and showed on the stump 134 annual rings. Those standing average from 110 to 120 ft. high by 8 to 10 ft. in girth.

On my own land at Lyde near Colesborne, in a deep sheltered valley, there is a tree about 125 ft. by 8 ft., but this is beginning to decay at the base, though not much over 100 years old.

At Bowood, Wilts, I saw a very fine tree which measured, in 1908, 125 ft. by 10 ft. 8 in.; but there may be better ones here.

On the Earl of Powis's estate at Walcot, Shropshire, there is a wood of spruce about 100 years old on a steep hill-side next to the Plassey plantation, a photograph of which has been reproduced in the *Quarterly Journal of Forestry*, iii. p. 358. I have seen no spruce plantation in England which equals this, and am indebted to Mr. R. H. Newill, agent for the estate, for the following account :—

"When I came to measure up an area in the Spruce Plantation, near Plassey, I found it difficult to find a piece without any gaps in it, as the wind has been busy of late years, and has blown down many trees. Eventually I chose a piece near the top and squared 1½ chain along the bank and 1 chain down it, an area of 0.15 acre. On this were twenty-two trees standing, of which I enclose measurements.[1]

	Length, Feet.	Quarter-girth, Inches.	Cubic Feet.		Length, Feet.	Quarter-girth, Inches.	Cubic Feet.		Length, Feet.	Quarter-girth, Inches.	Cubic Feet.
No. 1	80	9½	50.1	No. 9	80	12	80	No. 16	78	12	78
2	76	9½	47.7	10	82	13½	103.9	17	80	11½	73.5
3	79	12	79.	11	75	11	63	18	81	8½	40.7
4	80	8	35.6	12	80	14	108.10	19	76	13½	96.2
5	71	9	39.11	13	80	12	80		39	13½	49.4
6	84	10	58.4	14	35	9	19.8	20 {	21	7	7.1
7	81	9½	50.9	15 {	70	13½	88.7	(	15	5	2.7
8	83	9½	52.		12	6½	3.6	21	75	9	42.2
								22	74	10	51.4

Total . . . . 22 trees, 1401 cubic feet.

These vary from 19 to 103 cubic feet, and average 63.7 cubic feet. I found stumps of seven more trees on the area, and taking them at the same average there would be about 1849 feet; or about 193 trees measuring 12,326 cubic feet per acre.

The age of the trees is about 100 years; the rings are well marked to eighty-five years, afterwards so very close together that it is difficult to count them. All the trees are going back very fast, I believe every one is decayed at the butt; and in the lower part of the plantation many are blown down or broken off each year. We could only obtain 3d. to 4d. per foot for this class of timber, and it was in order to turn it to a more profitable use, that I put down the creosoting plant."

Assuming that this plantation had been clean felled at eighty years of age and that it had then contained 10,000 cubic feet per acre, the annual increment would have been 125 feet per acre; and taking the price at 4d. per foot standing the value of the crop would have been £166 per acre.

The trees are facing north-east at an elevation of 600 to 900 ft., and the old red sandstone here seems to suit all kinds of trees, both hardwoods and conifers, as well as any soil in England.

At Kyre Park there is a remarkable old spruce of the candelabra type which has an immense rugged bole broken off at about 30 ft., and 15 ft. 9 in. in girth. One of its upright branches is no less than 10 ft. 4 in. in girth, and twelve others have naturally layered themselves in a circle 64 ft. in diameter, and grown up into trees, two of which are 90 ft. high by 8 ft. in girth.

Another remarkable instance of layering in the spruce is at Langley Park, Slough, where a tree on the lawn has been broken off at about 20 ft. and whose lower branches have formed a complete bower, resembling on a smaller scale that formed by the Whittingehame yew. Some of the small branches, only one to three inches thick, have formed a woody mass and thickened enormously at the point where they have taken root.

In Wales the finest spruce we have seen are in a wood above Gwydyr Castle, where in 1905 I measured two trees in a grove round the bowling green, which were about 125 ft. high by 9 ft. 8 in. and 6 ft. 10 in. in girth. In this grove, which is shown in Plate 343, the spruce seems to clean itself better than in England, and I estimated that there might be 8000 to 10,000 cubic ft. per acre. Mr. Richards, forester to Lord Penrhyn, informed me that at Tyn-y-Coed in the same district of North Wales, a spruce plantation was felled in 1902 and sold to Mr. J. Jones of Liverpool, a tree in which is said to have been 149 ft. high, and that 158 trees in this wood contained 11,937 cubic ft., an average of over 75 cubic ft.; two of them measuring respectively 80 ft. by 23 in. quarter-girth = 294 cubic ft., and 67 ft. by 27 in. = 338 cubic ft. It is evident from these figures that even if the value of the timber is low as compared with imported spruce, yet that it may pay well in this particular district, provided the trees are grown thickly enough.

In Scotland the largest spruce of which we have any record grew at Blair Atholl, and was visited by the Scottish Arboricultural Society in 1879. It was then said to measure 142 ft. high, and to contain over 420 cubic ft. of timber.[1] I

[1] Hunter, *Woods of Perthshire*, 60 (1883).

was informed by the late Mr. Pitcaithley, forester to the late Earl of Mansfield, that this tree was blown down about 1893, when the height above given was verified.

The tallest that we now know of, are probably some trees on the banks of a deep glen at Dupplin Castle, which I saw in 1907, and on account of their leaning to one side could not measure accurately, but thought must be from 130 ft. to 140 ft. high. A fine tree in the same place, dividing into two stems at thirty feet from the ground, measured about 100 ft. by 12 ft. 10 in.

At Methven, Henry measured a tree in 1904 as 125 ft. by 8 ft. 3 inches.

At Inveraray I saw trees over 120 ft. by 8 to 10 ft., and the forester, Mr. Campbell, told me that he had measured one blown down on Ben-y-Cuach 130 ft. long. In the woods of Glenaray the spruce seems to grow very well, being sheltered from the westerly gales ; but I do not remember to have seen such large or thriving trees elsewhere on the west coast of Scotland.

In the east of Sutherlandshire, I am told by Mr. Gillanders that spruce grows well and cleans itself better than in the south.

In Ireland we have not seen any trees of extraordinary size, and as a rule the climate is not adapted to the production of high-class spruce timber. But Mr. A. E. Forbes has sent me an account of a remarkable plantation near Fermoy, which I reproduce verbatim, and am indebted to Lieut. and Quartermaster T. Smith, R.E., for a negative which gives a good idea of this plantation (Plate 344).

"A very fine clump of common spruce is growing in Glenshiskin Wood, which forms part of the property purchased by the War Department a few years ago near Kilworth. This wood occupies a valley running into the Kilworth mountains, a low range of hills formed from the Old Red Sandstone formation. A small mountain stream flows down the centre of this valley, and at one point, about 300 feet above sea-level, an alluvial deposit has been formed along its course of about an acre in extent. In this deposit a clump of almost pure spruce was planted about eighty years ago, and judging from appearances was never thinned or attended to in any way. From time to time poles were doubtless removed from it as required ; but no systematic thinning could have been carried out, as many of the trees still stand within four or five feet of each other.

"This clump probably presents as fine an example of spruce growth in Britain as can be found anywhere. The trees vary in height from 90 to 110 ft., and form long clean poles with little taper, and ranging from $8\frac{1}{2}$ to 18 in. quarter-girth at $4\frac{1}{2}$ ft. from ground. The trees in two-thirds of an acre were carefully measured by Mr. M'Rae, forester at Dundrum, Co. Tipperary, in the spring of 1910, and the summary of the results obtained is given below :—

	No.	Age.	Average Height.	Average Quarter-girth at 4½ ft.	Total Cubic Contents.
Spruce . . .	161	78 years	98 feet	12½ in.	8050
Larch . . .	11	,,	99 ,,	12¼ in.	550
Stumps of Felled Trees	110	...	...	...	...
					8600

" By dividing the trees into three stem classes, a volume of 9600 cubic feet was obtained. Assuming the estimate of 8600 ft. to be correct, the average contents of the trees is about 50 ft., and the total yield per acre would be over 12,000 cubic ft., which for the period of eighty years is higher than anything I have heard of in Great Britain.

" Within recent years, trees similar in size to those still standing have been removed, and the stumps still exist. It is quite possible, therefore, that the existing crop is smaller in volume than that which stood on the ground a few years ago. The high yield is, of course, largely due to the exceptionally favourable soil and situation of the site on which the trees were grown. A fine, rich, and well-drained soil, well provided with soil and atmospheric moisture, and a situation sheltered from all winds, provide ideal conditions for the growth of spruce or any other tree able to thrive with a moderate amount of summer heat. Oak and beech growing in the immediate vicinity of these trees are very poorly developed and covered with lichen and moss, indicating the cool and humid conditions which prevail."

TIMBER

Next to that of the Scots pine, the wood of the spruce is the largest import from the Baltic; and from Norway the proportion of spruce timber is probably greater. On account of climatic and economic causes, it seems probable that this will always be the case, though in the west coast ports American spruce takes its place. For scaffold and ladder poles, small spars and masts, and oars, we cannot hope to compete with the north of Europe; whilst for flooring, joists, and almost all purposes except those for which knotty boards are not objected to, it seems equally hopeless for British growers to attempt to compete with the well-known white deal of commerce.

The reasons why the value of home-grown spruce timber is so low are, first, its very knotty character, caused by the persistence of the branches, which die more slowly than those of other conifers; and, secondly, its want of strength and durability as compared with larch. Continental foresters tell us that the first defect may be obviated by close planting, and cite the large profit which is derived from this tree in Germany and Scandinavia. I have inquired of many of our best practical foresters; but I have never been able to find any plantation in England, and only very small areas in Scotland, Wales, and Ireland, where spruce, which stood close enough to kill the branches, have attained a considerable size, or where a spruce plantation has been a really profitable investment.

My own experience is, that land where spruce may be well grown is fit to produce a much more valuable timber, and that on ordinary land it will starve to death before it will clean itself from branches.   The late Mr. Philip Baylis, Deputy-Surveyor of Dean Forest, told me that spruces there, 50 to 60 ft. high, and so thickly planted that no vegetation would exist under them, still retained the dead branches to within 5 or 6 ft. from the ground; and I think that this will apply to most places in England and Scotland.

When the tree is of large size it usually becomes rotten at the heart near the ground; and the top is often broken by the wind.   Though the timber may be worth 4d. to 6d. per cubic foot for rough boarding or packing-cases, or for temporary sleepers and pit props in collieries or railways under construction, yet in quantity it is the most unsaleable wood we have.   When, as often happens, large quantities are blown down by a heavy gale, I have known cases where no one would go to the expense of cutting up and removing the trees if they had them for nothing, and the proprietor has had considerable expense in doing so without any return whatever.   When blown down, the shallow spreading roots tear up the ground for some distance round the tree and are very costly to get rid of, or if left leave the ground in a bad state for re-planting.

Where, however, the soil and climate allow the spruce to be crowded closely enough to clean itself, before it becomes rotten at heart or is blown down, spruce timber may be used for estate building purposes, if not with actual economy, yet in many cases more advantageously than by selling it.   Sixpence per foot is something like the average price, though 3d. to 4d. often has to be accepted.

On shallow and dry soils the spruce often begins to decay at the heart for some feet from the ground at the age of fifty to seventy years, and on such soils should not be planted at all.   Its spreading roots, which are extremely tough and elastic, are used in Scandinavia for the knees of boats, though rarely so utilised in England.   The tough and durable branches made into a wattled fence will last for a long period, and are the common farm fence in many parts of Norway and in the Alps.

When facilities exist for creosoting, spruce may be used for fencing and other outside work, such as sheds and outbuildings; but unless treated with some preservative it soon decays when exposed to wet and dry.

The spruce trees which produce the *bois de resonance*, used for sounding-boards in musical instruments, grow at high elevations in the Alps, the Jura, and in the Bohemian and Bavarian forests.   These are very old trees, the growth of which has been extremely slow and very uniform, the annual rings not exceeding $\frac{1}{12}$ in., and containing only a slight amount of autumn wood.   These trees are usually covered with lichens, and their selected timber sells at very high prices, as much as 9s. to 12s. per cubic foot.

Burgundy pitch is a resinous product of the spruce, well known under the name of *Burgony Pitch* and *Pix Burgundica* as long ago as 1640.   It was formerly produced in the Vosges Mountains, but now, according to Flückiger and Hanbury,[1] mainly in Finland, the Black Forest, Austria, and Switzerland. Flückiger states that at Oppenau, in Baden, the principal place of its manufacture in Germany, it is mixed with French turpentine from Bordeaux and with rosin from N. America; and the tapping of the trees in Government forests in Baden and Württemburg is now prohibited on account of the injury caused thereby to the timber. It is very generally adulterated in England, and is mainly used as an ingredient in plaisters.[1]

(H. J. E.)

PICEA OBOVATA, Siberian Spruce

Picea obovata, Ledebour, *Fl. Alt.* iii. t. 499, iv. p. 201 (1833); Trautvetter, in Middendorf, *Reise*, i. pt. ii. 87, 170 (1847); Maximowicz, *Prim. Fl. Amur.* 261 (1859); Regel, *Tent. Fl. Ussur.* 137 (1861); Herder, in *Bot. Jahrb.* xiv. 160 (1891); Willkomm, *Forstliche Flora*, 93 (1887); Kent, Veitch's *Man. Conif.* 441 (1900); Komarov, *Fl. Mansh.* i. 197 (1901); Clinton-Baker, *Illust. Conif.* ii. 42 (1909).

Picea vulgaris, Link, var. *altaica*, Teplouchoff, in *Bull. Soc. Nat. Mosc.* xli. pt. ii. 250 (1869).

Picea excelsa, Link, var. *obovata*, Schröter, in *Viertelj. Naturf. Ges. Zürich*, xliii. 138 (1898).

Pinus Abies, Pallas, *Fl. Ross.* i. 6 (1784) (not Linnæus).

Pinus obovata, Antoine, *Conif.* 96 (1840-1847).

Pinus orientalis, Ledebour, *Fl. Ross.* iii. 671 (in part) (1847-1849) (not Linnæus).

Abies obovata, Don, *ex* Loudon, *Arb. et Frut. Brit.* iv. 2329 (1838).

A tree, attaining in Russia and Siberia the dimensions of *P. excelsa*, which it resembles in habit of growth and in bark. Young branchlets reddish brown, covered with a dense minute pubescence, which is retained for several years, the older branchlets becoming greyish yellow. Buds, about $\frac{1}{5}$ in. long, conic, composed of closely appressed scales, rounded at their apices; terminal bud girt with a ring of keeled acuminate ciliate scales, and closely surrounded at the base by the uppermost leaves. Leaves, arranged as in *P. excelsa*, deep green in colour, $\frac{2}{5}$ to $\frac{3}{5}$ in. long, ending in a short point, quadrangular in section, with three to four stomatic lines on each side.

Cones $2\frac{1}{2}$ to $3\frac{1}{2}$ in. long, $1\frac{1}{4}$ to $1\frac{1}{2}$ in. in diameter when open, shining brown when ripe; scales numerous, thin, tough, flexible, longer than broad, $\frac{6}{10}$ to $\frac{7}{10}$ in. wide, and $\frac{7}{10}$ to $\frac{9}{10}$ in. long, fan-shaped, widest near the upper edge, tapering to the base on each side; upper margin thin, undulate, rounded or with a slightly projecting occasionally bifid apex; exposed part pale brown, glabrous; concealed part reddish brown, minutely pubescent; flat or slightly concave internally from side to side; bract $\frac{1}{5}$ in. long, lanceolate, narrowing to an acute denticulate apex. Seed $\frac{1}{8}$ in., brownish black; seed with narrow wing $\frac{2}{5}$ to $\frac{3}{5}$ in. long, broadest near the rounded denticulate apex.

The description of *P. obovata* given above is drawn up from specimens procured from Siberia, and from Perm in Russia, by Mr. H. Clinton-Baker, from specimens collected in Finland by Mr. M. P. Price, and from specimens which I gathered in

---

[1] *Pharmacographia*, 616 (1879).

northern Sweden, near Bräcke, and in Norway, near Trondhjem; all agreeing in the character of the cones, branchlets, buds, and leaves, and constituting, in my opinion, a species distinct from *P. excelsa*, of which *P. obovata* is generally considered to be a variety by Schröter and other modern botanists. These authorities have apparently paid no attention to the characteristic pubescence of *P. obovata*, a matter of importance, as in the genus Picea the presence or absence of pubescence on the branchlets is one of the most diagnostic features in the discrimination of the different species. The cones, moreover, are amply distinct in the two species.

P. obovata varies somewhat in the size of the cones and in the shape of their scales; and two main varieties have been distinguished, which are, however, connected by intermediate gradations. These varieties are: (*a*) the typical form described above, which is characterised by the scales of the cone being entire on margin; and (*b*) var. *fennica*.

1. Var. *fennica*, Henry.

Picea excelsa, Link, var. *fennica*, Schröter, in *Viertelj. Naturf. Ges. Zürich*, xliii. 138 (1898).
Picea excelsa, Link, var. *medioxima*, Willkomm, *Forst. Fl.* 75 (1887).
Picea vulgaris, Link, var. *uralensis*, Teplouchoff, in *Bull. Soc. Nat. Mosc.* xli. pt. ii. 250 (1869).
Pinus Abies, Linnæus, var. *fennica*, Regel, in *Gartenflora*, xii. 95 (1863).
Pinus Abies, Linnæus, var. *medioxima*, Nylander, in *Bull. Soc. Bot. France*, x. 501 (1863).
Pinus Picea medioxima, Christ, *Flore de la Suisse*, 254 (1883).
Abies medioxima, Lawson, *Pinet. Brit.* ii. 159 (1867).

Cone-scales, with their upper margins rounded and finely denticulate. Leaves dark green in colour.

According to Schröter this variety occurs sporadically in Amurland and Siberia, and is the common form in the Ural range and throughout Russian Lapland, northern Sweden, and northern Norway, occurring with less frequency in Finland, Livland, Kazan, and Poland. Solitary trees with cones similar to this variety have also been recorded from numerous stations in the mountains[1] of central Europe, from the Vosges and Jura throughout the Alps to the Carpathians and Bosnia.

2. Var. *alpestris*, Henry.

Picea alpestris, Stein, in *Gartenflora*, xxxvi. 346 (1887).
Picea excelsa, Link, var. *alpestris*, Schröter, *op. cit.* 141 (1898).
Abies excelsa alpestris, Brügger, in *J. B. Naturf. Ges. Graubundens*, xvii. 154 (1874), and xxix. 122 (1884).
Abies excelsa medioxima, Heer, in *Verh. Schw. Nat. Ges.* 1869, p. 70 (not Nylander).

Cones 3 to 5 in. long, with scales rounded and entire in upper margin. Leaves short, ½ to ⅔ in. stout, very glaucous.

Trees with a whitish grey bark, and with remarkable bluish white foliage, which have been found at high elevations (between 4400 and 6400 ft.) in a few localities in the Swiss Alps, from Landbeck in the Tyrol[2] to Engstelnalp in the Bernese Oberland, and from Lake Walen to Lake Como. These trees were first investigated by Heer and Brügger on account of the special name given to them by

[1] Cf. Christ, in *Garden and Forest*, ix. 273 (1896).
[2] Beissner, in *Mitt. Deut. Dend. Ges.* 1905, p. 143, describes trees like var. *alpestris* in the Engadine.

the peasants, *aviez selvadi*, or wild silver fir, the common spruce being known as *pign*. I have seen no specimens, but apart from the glaucous foliage, which is a trivial and inconstant character in conifers, *P. alpestris* would seem to be identical with *P. obovata*.

A vast amount of literature [1] has been written on the relationship of *P. obovata* to *P. excelsa*, the general result of which shows that a complete series of transitional forms connecting the two species may be found; but these are only met with in the regions where the two spruces come in contact—elsewhere they are quite distinct and easily recognisable. It is possible that these transitional forms are due to hybridisation; and further study by experimental sowings is needed to clear up the matter.

P. obovata is the most widely distributed of all the spruces, extending over the vast northerly region of eastern Europe and Asia, where the climate is severe in winter and continental in character. It occurs in northern Scandinavia, Lapland, Finland, northern and eastern Russia, throughout Siberia to the Sea of Ochotsk and Kamtschatka,[2] and in Manchuria. It extends far to the northward, reaching lat. 67° in the Kola peninsula, lat. 68° in the Ural range, attaining its most northerly point in Siberia on the Yenisei at lat. 69° 5', and crossing the Stanovoi mountains at lat. 64°, where it comes in contact with *P. ajanensis*. According to Komarov [3] it is abundant throughout the wooded parts of Manchuria, where it grows along the banks of rivers, either forming pure woods or scattered amidst other trees. Its eastern and southern limits in Asia are imperfectly known, but it forms great forests in the mountains of Dahuria and in the Altai and Sayan ranges. Seebohm [4] describes it as extending on the Yenisei "nearly as far north as the larch, where it is a very important tree for commercial purposes. Its wood is white, of very small specific gravity, extremely elastic; and it is said not to lose its elasticity by age. It makes the best masts for ships, and is for oars the best substitute for ash. Snow-shoes are generally made of this wood. The quality is good down to the roots, and it makes the best knees for shipbuilding."

In European Russia its southern limit is the northern edge of the Orenburg steppe; and it forms vast forests in the governments of Perm, Vologda, Ekaterinburg, Ufa, Viatka, and Kama, that are either pure or mixed with larch, *Pinus Cembra*, *Abies sibirica*, and birch. It appears to be the spruce prevalent in Finland and in the Baltic provinces; but in western Russia is mixed with *P. excelsa*, the limits between the two species being undefined, owing to the occurrence of transitional forms. Similarly in Scandinavia [5] it is the common spruce in the north, while in the south *P. excelsa* appears to be the prevalent form. Its occurrence as a sporadic tree in the mountains of central Europe, under the form described as *P. alpestris*,

---

[1] Cf. Teplouchoff, *loc. cit.* Korshinsky, in *Tentamen Fl. Ross. Orient.* 493 (1898), admits that cones like those of *P. excelsa* are never seen in eastern Russia. At the junction of the rivers Kama and Viatka the woods are said to be composed of both species. Cf. Kihlman, *Pfl. Stud. Russ. Lapland*, 143 (1890), on the variation of the spruce in Finland, Lapland, and northern Scandinavia. Dammer, in *Gard. Chron.* iv. 480 (1888), may also be consulted, as well as the numerous authorities quoted by Schröter, *op. cit.* 240 (1898).

[2] It is a doubtful native of the Kurile Isles, according to Miyabe in *Mem. Boston Soc. Nat. Hist.* iv. 261 (1894).

[3] *Flora Manshuriae*, i. 197 (1901).   [4] *Siberia in Asia*, 233 (1882).

[5] Cf. under *P. excelsa*, pp. 1347, 1348.

is peculiar; but may be explained as a remnant of the pre-glacial forests.[1] In habit, *P. obovata* is usually more columnar than *P. excelsa*; but little reliance can be placed on this character as a mark of distinction.

This species had not been introduced into England in Loudon's[2] time; and it is very doubtful if it occurs in cultivation in this country, except at Bayfordbury, where seedlings were raised in 1908 from seed brought from Siberia by Mr. C. F. H. Leslie. According to Kent,[3] "the Siberian spruce soon perishes under the stimulus of the high temperature of this country." Small trees in botanic gardens reputed to be this species appear to me to belong to the transitional form between *P. excelsa* and *P. obovata*, which has less pubescence on the branchlets. Plants raised from Finnish seed, procured from Rafn, are much slower in growth at Colesborne than common spruce.

In Germany, according to Mayr,[4] it is slower in growth than the native spruce, and is not more hardy. It appears[5] also to be equally slow in growth in the Arnold Arboretum, U.S.A. (A. H.)

PICEA ORIENTALIS, Caucasian Spruce

Picea orientalis, Carrière, *Conif.* 244 (1855); Boissier, *Fl. Orient.* v. 700 (1884); Masters, in
  Gard. Chron. xxv. 333, fig. 62 (1886), and iii. 754, fig. 101 (1888); Kent, Veitch's *Man.
  Conif.* 443 (1900); Clinton-Baker, *Illust. Conif.* ii. 44 (1909).
Pinus orientalis, Linnæus, *Sp. Pl.* 1421 (1763); Lambert,[6] *Genus Pinus*, i. t. 39, fig. *a* (1803).
Abies orientalis, Poiret, in Lamarck, *Dict.* vi. 518 (1804); Loudon, *Arb. et Fruit. Brit.* iv. 2318
  (1838).

A tree, attaining in the Caucasus 180 ft. in height and 12 ft. in girth. Bark brown, fissuring irregularly on old trees into thin scales. Young branchlets pale brown, slender, densely covered with a short pubescence, retained in the second and third years. Buds conical, acute, about $\frac{1}{8}$ in. long, brown; terminal buds girt at the base with a few keeled acuminate scales. Leaves, on lateral branches arranged as in *P. excelsa*, very short, $\frac{1}{4}$ to $\frac{2}{5}$ in. long, dark green, shining, bevelled and obtuse at the apex, quadrangular in section, with one to four lines of stomata on each of the four surfaces.

Staminate flowers, cylindrical, $\frac{1}{2}$ in. long, carmine red in colour; anther connective suborbicular, minutely denticulate.

Cones, 3 to 4 in. long, $\frac{3}{4}$ to 1 in. in diameter when closed, cylindrical but tapering to a narrow apex, violet coloured[7] when growing, brown when ripe; scales

---

[1] Christ, *Flore de la Suisse*, 197 (1883), compares the distribution of this spruce with *Pinus sylvestris*, var. *engadinensis*, the pine on the Engadine, which he considers to be identical with *Pinus lapponica*, Mayr, the form of the common pine that occurs in northern Scandinavia and Lapland. Cf. *ante*, vol. iii. 573.

[2] Cf. Loudon, *Trees and Shrubs*, 1030 (1842).   [3] Veitch's *Man. Conif.* 442 (1900).

[4] *Fremdländ. Wald- u. Parkbäume*, 333 (1906).   [5] Sargent, in *Garden and Forest*, x. 481 (1897).

[6] The cones figured by Lambert in this edition, t. 29, fig. *b*, were from China, and are possibly those of *P. ajanensis*. Lambert, in his second edition, t. 39 (1832), gives a new and coloured drawing of leaves and cones, collected by Sir Gore Ouseley near Tiflis, repeating also the figures of the cones from China.

[7] The scales of young cones are green, with a narrow carmine-coloured margin.

obovate with a cuneate claw, $\frac{1}{2}$ to $\frac{3}{8}$ in. wide, rounded entire and slightly bevelled in the upper margin; bract $\frac{1}{4}$ in. long, with a narrow claw and a rectangular lamina, truncate at the apex. Seed dark coloured, $\frac{1}{8}$ in. long, with the wing $\frac{1}{2}$ in. long; wing broadest about the middle, upper margin rounded.

This species is readily distinguished by its very short blunt leaves, and pale brown pubescent branchlets.

DISTRIBUTION

The oriental spruce is a native of Asia Minor and the Caucasus. It is widely spread in most of the mountain ranges of Asia Minor, being recorded for Troas, Mysia, Galatia, and Phrygia, where it generally occurs between 3000 and 7000 feet elevation. It is also met with in the valleys of the Antitaurus. It is, however, much more common, forming large forests, in the mountains between Trebizond and Erzerum, where it was discovered by Tournefort[1] at the beginning of the eighteenth century. In the Caucasus it is generally associated with *Abies Nordmanniana*, and occurs in Georgia between 2500 and 7500 feet. In the Lesser Caucasus its eastern limit is the meridian of Tiflis, being totally absent to the eastward and in the province of Talysch. As a rule it ascends higher than *Abies Nordmanniana*, occasionally forming the timber line at 7500 feet. The largest tree recorded by Radde,[2] measured, when felled, 184 ft. in height, with a diameter of stem of 4 ft. 1 in., and a cubic content of 925 ft.; it was 390 years old.

CULTIVATION

The species, according to Beissner,[3] was introduced into Europe in 1837, but Loudon, writing in 1838, speaks of it as not in cultivation; and it appears[4] to have come into this country in 1839. It has been in cultivation in the United States[5] since about 1850, where it has proved hardy as far north as eastern Massachusetts, and is one of the most beautiful of all the exotic conifers that have been planted in the neighbourhood of Boston. (A. H.)

REMARKABLE TREES

None of the spruces seems more generally successful in cultivation than this; and though it does not grow so fast as the common or the Sitka spruce, it is a really good ornamental tree, hardy in all parts of Great Britain, and ripening seed in most places. We have measured many specimens of from 60 to 70 ft. high and a few taller, among which the following may be mentioned:—

At Dogmersfield Park, Hants, the seat of Sir H. Mildmay, a fine tree with many cones, 78 ft. by 7 ft. 8 in. in 1907. At Strathfieldsaye a handsome specimen 76 ft. by 7 ft. 8 in. At Highnam a tree about 67 ft. by 7 ft. in 1905. At Penrhyn a tree recorded[6] as 58 ft. high in 1891, which was, when measured by me in 1906, 75 ft. by 5 ft. 10 in.

---

[1] *Voyage au Levant*, 288 (1717).   [2] *Kaukasusländern*, 223 (1899).   [3] *Nadelholzkunde*, 374 (1891).

[4] Lawson, *Pinet. Brit.* ii. 163 (1865). Loudon, in *Trees and Shrubs*, 1029 (1842) says: "Of late many plants have been raised in Knight's exotic nursery, from seeds received from Mingrelia and the neighbourhood of Tiflis."

[5] Sargent, *Silva N. Amer.* xii. 22, note (1898), and in *Garden and Forest*, 1895, p. 55.

[6] *Journ. R. Hort. Soc.* xiv. 485 (1892).

In Canon Ellacombe's garden at Bitton there is a dwarf bush of considerable age, which when covered with young cones is very ornamental. From it I have raised seedlings which grow very slowly.

In Scotland the finest tree we know of is at Durris, which in 1904 was 61 ft. by 6 ft. 3 in.

In Ireland Henry measured at Fota one which in 1903 was about 67 ft. by 6 ft.

(H. J. E.)

PICEA SCHRENKIANA, SCHRENK'S SPRUCE

Picea Schrenkiana, Fischer and Meyer, in *Bull. Acad. Sci. St. Petersb.* x. 253 (1842); Regel, in *Garten-flora*, xxvi. 69 (1877), and xxix. 49 (1880); Fedtschenko, in *Bull. Herb. Boissier*, vii. 189 (1899); Kent, Veitch's *Man. Conif.* 451 (1900); Clinton-Baker, *Illust. Conif.* ii. 48 (1909).

Picea tianschanica, Ruprecht, in *Mém. Acad. Sci. St. Petersb.* xiv. No. 3, p. 72 (1870).

Picea obovata, Ledebour, var. *Schrenkiana*, Masters, in *Journ. Linn. Soc.* (*Bot.*) xviii. 506 (1881).

Pinus Schrenkiana, Antoine, *Conif.* 97 (1840-1847).

Pinus obovata, Antoine, var. *Schrenkiana*, Parlatore, in De Candolle, *Prod.* xvi. 2, p. 415 (1868).

Pinus orientalis, Linnæus, var. *longifolia*, Ledebour, *Fl. Ross.* iii. 671 (1847).

Abies Schrenkiana, Lindley and Gordon, in *Journ. Hort. Soc. Lond.* v. 212 (1850).

A large tree, attaining in Turkestan the dimensions of *P. obovata*. Young branchlets ashy grey, stout, glabrous. Buds dome-shaped or sub-globose, $\frac{1}{3}$ in. in length, rounded at the apex, light brown, with scarious scales; terminal bud girt with a ring of acuminate keeled pubescent ciliate scales and closely surrounded at the base by the uppermost leaves of the branchlet.

Leaves in an imperfect radial arrangement, dense and pointing forwards on the upper side of the branchlet, spreading with a few leaves pointing forwards and not truly pectinate on the lower side of the branchlet; $\frac{3}{4}$ to $1\frac{1}{4}$ in. long, straight or curved, rigid, gradually tapering at the distal end to a long fine sharp-pointed apex[1]; obscurely quadrangular in section, with three to four lines of stomata on each of the four sides.

Cones, 3 to 4 in. long, cylindrical, narrowing towards the obtuse apex, shining dark brown when ripe; scales numerous, closely imbricated, longer than broad, about $\frac{1}{2}$ in. wide, obovate-cuneate, with the upper exposed part thin and glabrous, concealed part thicker and minutely pubescent; upper margin rounded, entire, undulate; bract $\frac{1}{6}$ in. long, ovate. Seed, light brown, $\frac{1}{6}$ in. long; seed with wing $\frac{1}{2}$ in. long; wing narrow, widest near the rounded apex.

This species was discovered in 1840 by Schrenk, and is widely distributed in Central Asia, occurring mainly in the Alatau mountains and in the Thianshan[2] range in Turkestan, where, according to Fedtschenko, it forms vast forests, now rapidly disappearing, as far south as lat. 41°, at 4000 to 8000 ft. altitude towards the north, and at 8000 to 10,000 ft. towards the south. It does not appear to

---

[1] In wild specimens from old trees, the leaves end in a short acute callous tip.

[2] Both Regel and Komarov agree that the spruce in the Thianshan range, considered by Ruprecht to be a distinct species (*P. tianschanica*), is identical with *P. Schrenkiana* in the Alatau range.

occur in the mountains uniting the Thianshan range with the Pamirs; and its western limit is probably the Alexandrovoski mountains in Russian Turkestan. Its eastern limit is not as yet clearly known;[1] but Przewalski found extensive woods of it, not only in the Thianshan range, but also in the upper course of the Yellow River in Mongolia, near Lake Kokonor, and in the adjoining Nan-Shan range.[2]

Mr. M. P. Price informs us that its most northerly point appears to be in the Barluk mountains, lat. 46°, where there are a few scattered forests in the higher valleys. He observed this tree at 9200 feet altitude in the pass between the valley of the river Baratala and the plateau of Lake Sairam. It bears the greatest extremes of heat and cold in these regions. Most of the trees, which he saw, scarcely exceeded 50 to 70 ft. in height and 7 to 8 ft. in girth. On a section 2 ft. 10 in. in diameter from the base of a tree, which had grown in the vicinity of Lake Issik Kul and was preserved in the museum at Vernoe, he counted 296 annual rings. The wood is used for building houses in Russian Turkestan; but is of little economic importance on account of the inaccessibility of the forests.

In the eastern part of the Thianshan range, where the climate is very severe, and the thermometer sinks at least 7° F. below freezing every night during summer, *P. Schrenkiana*, nevertheless, forms open woods at about 8000 ft. elevation, which are remarkable for the peculiar narrow columnar form of the trees. This is well shown by two photographs, taken by Baron von Dungern, which are reproduced in *Mitt. Deut. Dend. Ges.* 1910, pp. 227, 229. He explains the cypress-like habit as due to the fact that the shoots of the lateral branches are almost invariably frozen, soon after their production in early summer; whilst those of the leading branches, which are later in the season in emerging from the bud, escape destruction by the severe frosts.

This species was distributed by the St. Petersburg Botanic Garden after its re-discovery in 1877 by Regel in Turkestan. It has never become common in cultivation. There are two trees at Kew, 8 and 10 ft. high, obtained from Messrs. Veitch in 1882; and smaller specimens at Bayfordbury and in other private collections. It appears to be hardy, though slow in growth, and is very distinct in appearance, most of the branches being rigid and ascending. (A. H.)

---

[1] The spruce collected in Kansu, in north-western China, by Futterer and Holderer, identified with *P. Schrenkiana* by Diels, *Flora von Central-China*, 217 (1901); and another, collected by Bretschneider, near Peking, similarly identified by Masters, in *Journ. Linn. Soc. (Bot.)* xxvi. 554 (1902), appear to be identical with *Picea Mastersii*, Mayr, *Fremdländ. Wald- u. Parkbäume*, 328, figs. 105-107 (1906).

[2] Cf. Köppen, *Holzgewächse Europ. Russlands*, ii. 538 (1889). Regel, in *Act. Hort. Petrop.* vi. 485 (1880), states that it grows not only in the high mountains, but also along the rivers Baratala, Kash, and Yuldus.

PICEA SMITHIANA, Western Himalayan or Morinda Spruce

Picea Smithiana, Boissier, *Fl. Orient.* v. 700 (1884); Kent, Veitch's *Man. Conif.* 454 (1900).
Picea Morinda, Link, in *Linnœa*, xv. 522 (1841); Masters, in *Gard. Chron.* xxiv. 393, fig. 85 (1885);
   Hooker, *Flora Br. India*, v. 653 (1888) (in part); Gamble, *Indian Timbers*, 716 (1902);
   Brandis, *Indian Trees*, 692 (1906); Clinton-Baker, *Illust. Conif.* ii. 40 (1909).
Picea Khutrow, Carrière, *Conif.* 258 (1855).
Pinus Smithiana, Wallich, *Pl. Asiat. Rar.* iii. 24, t. 246 (1832).
Pinus Khutrow, Royle, *Illust. Him. Plants*, 353, t. 84 (1839).
Abies Smithiana, Lindley, in. *Penny Cycl.* i. 31 (1833); Loudon, *Arb. et Frut. Brit.* iv. 2317 (1838).
Abies Khutrow, Loudon, *Trees and Shrubs*, 1032 (1842).
Abies Morinda, Nelson (Senilis), *Pinaceœ*, 49 (1866).

A tree, attaining in the Himalayas over 200 ft. in height and 20 ft. in girth. Bark greyish brown, divided by shallow fissures into small rounded or quadrangular scales. Young branchlets grey, shining, glabrous. Buds about $\frac{1}{2}$ in. long, spindle-shaped or ovoid, acute at the apex, brownish, resinous; scales numerous, densely imbricated, rounded at the apex; terminal bud girt at the base with a ring of acuminate keeled scales. Lateral branches always pendulous, with the leaves radially arranged and directed outwards and towards the apex of the branchlet at an acute angle. Leaves long and slender, about $1\frac{1}{2}$ in. long and $\frac{1}{25}$ in. broad, incurved, tapering towards the apex, which ends in a slender cartilaginous point; obscurely 4-angled, with about two lines of stomata on each of the four sides.

Staminate flowers, about 1 in. long and $\frac{1}{2}$ in. in diameter, cylindrical, obtuse, light yellow; anther connective orbicular, crenate.

Cones, 4 to 6 in. long, $1\frac{1}{2}$ to 2 in. in diameter, cylindrical, narrowed towards the base, obtuse at the apex; bright green and smooth when growing; shining brown when mature; scales about an inch wide, broadly obovate from a cuneate base, smooth, convex, rounded and entire in margin; bract obsolete. Seed dark brown, $\frac{1}{4}$ in. long, with the wing $\frac{3}{4}$ in. long; wing spatulate, broadest near the truncate denticulate apex.

P. Smithiana occurs throughout the western Himalayas, between 7000 and 11,000 ft. elevation, being common from Garhwal to Kashmir, and also occurring in Gilgit, Chitral, and Kafiristan. It extends westwards to Afghanistan, where Aitchison found it in the Kuram and Hariab district, between 8000 and 11,000 ft., occasionally extending as high as 12,000 ft., where it struggles for existence with *Pinus excelsa*. According to Gamble, it is a very fine tree in the Himalayas, often attaining a greater height than the deodar, but probably never equalling the latter in girth. Large trees measured near Mundali in Jaunsar were 175 to 215 ft. in length and 19 to 23 ft. in girth.[1] It forms mixed forests with *Abies Pindrow*, which cover mainly the northern and western slopes of the mountains, usually between 7500 and 8500 ft. In these forests the spruce is more common on the drier ridges, the silver fir growing in the moister ravines. *P. Smithiana* also forms mixed forests

[1] I am informed by Sir G. Watt that a tree, recorded by Sir E. Buck, near Nagkunda, measured no less than 250 ft. high. Cf. Frontispiece of Vol. V.—H. J. E.

with the deodar. Grown in dense forest the stems are often free from branches to a great height, crowned by a conical pyramid of foliage with pendulous branches. In this condition, it produces seed at intervals of three or four years, and in small quantity. The rank undergrowth consists of *Strobilanthes*, small bamboos, raspberries, balsams, and other plants, which render natural reproduction of seedlings rare and difficult. Clear cutting and artificial regeneration have been found to be the most successful modes of treating these forests. This spruce is attacked in the Himalayas by the aphis, *Chermes abietis*, which is common on the European spruce, and produces cone-like excrescences on the twigs. A fungus, *Peridermium incarcerans*, Cooke, often occurs as curious tassel-like orange bunches on the branchlets. The leaves are attacked by another fungus, *Æcidium Thomsoni*.[1] (A. H.)

CULTIVATION

P. Smithiana was introduced into cultivation in 1818 by Dr. Govan of Cupar, who gave the seed to the Earl of Hopetoun, from which the first trees were raised at Hopetoun House, near Edinburgh. It is a thriving tree in many parts of the British Isles; and though the young shoots are liable to be nipped by frost, this does not seem to do the tree permanent injury. It does not, however, seem to succeed on limestone soil.

The tallest specimen[2] I have seen in England is at Melbury, where, in 1906, I measured one 85 to 90 ft. high and 8 ft. 10 in. in girth. (Plate 345.)

At Carclew a tree was reported[3] in 1891 as 80 ft. high, but when I measured it in 1905 it was 86 ft. by 7 ft. 9 in. At Pencarrow Mr. Bartlett measured a perfect specimen planted by Sir W. Molesworth about 1850, which in 1907 was 57 ft. by 6 ft. 7 in. At Bicton in 1902 I measured one 65 ft. by 6 ft. 9 in. At Redleaf, Kent, in 1907 a tree 75 ft. by 9 ft. had many cones on the lower branches, which rested on the ground. At Walcot there is a fine tree 60 ft. by 5½ ft.

At Barton, Bury St. Edmunds, there are two fine trees, one of which was 84 ft. high by 7 ft. 1 in. in girth in 1904, the other 77 ft. by an inch less in girth. These trees were raised from seeds sent by Lady Napier, to whom they had been given by Wallich, and the seedlings were planted out in 1843. The trees were not injured in the least by the severe winter of 1860-1861, and commenced to bear cones for some years before 1869, having a very abundant crop in that year.[4] A tree at Hardwicke House, Suffolk, planted later than those at Barton, was measured by Sir Hugh Beevor in 1904 as 73 ft. by 7 ft.

In Wales the largest I have seen, a tree at Margam Park, was 81 ft. by 6½ ft. in 1907.

In Scotland there are many good specimens, of which those at Hopetoun are the oldest, having been raised from seed sent to the Earl of Hopetoun by

---

[1] Described and figured by Berkeley in *Gard. Chron.* 1852, p. 627.

[2] The *Picea Smithiana* reported in *Gard. Chron.* 1869, p. 713, to be growing at Shelsley Walsh in the Teme Valley in Worcestershire, is *P. excelsa*. Cf. p. 1341.

[3] *Journ. R. Hort. Soc.* xiv. 488 (1892).   [4] Bunbury, *Arboretum Notes*, 134.

Dr. Govan in 1818. When I saw them in 1904 the best of these was about 70 ft. by 8½ ft. Fowler[1] states that the two trees at Hopetoun House were planted in their present position in 1824, one being a seedling, the other a grafted plant worked on the common spruce, four feet above the ground. In 1871 the graft had outgrown the stock all round for 2 to 3 inches. The seedling tree in that year was 60 ft. high by 7 ft. in girth at four feet from the ground, the grafted tree being scarcely so tall. Mr. T. Hay, gardener at Hopetoun, remeasured these trees in January 1911, and informed me that the grafted tree is still in fair condition, and measures 70 ft. high. Its girth below the graft is 6 ft. 2 in., and above it 7 ft. 2 in. The seedling tree is more healthy and measures 75 ft. by 8 ft. 8 in. at 4 ft. from the ground.

At Smeaton Hepburn, a tree, planted in 1840, was measured by Henry in 1905 as 67 ft. by 6 ft. 5 in.

In Ireland this species thrives remarkably well, and there are many fine specimens. At Woodstock, Kilkenny, in 1909 I measured a tree 72 ft. by 8½ ft. At Mount Shannon near Limerick, a tree measured, in 1905, 69 ft. by 8½ ft. in girth. At Fota, Queenstown, there is a fine tree, which was, in 1903, 63 ft. by 8 ft. At Glenstal, Co. Limerick, in the same year, a tree was 11½ ft. in girth. with an estimated height of 70 ft. At Bessborough in Co. Kilkenny, a tree, which was figured in the *Gardeners' Chronicle*, May 21, 1904, is, we are informed by Viscount Duncannon, 60 ft. high by 6 ft. 9 in. in girth. Another at Emo Park, Portarlington, was 60 ft. by 8 ft. in 1907; and one at Coollattin was 59 ft. by 4 ft. 8 in. in 1906.

In the United States,[2] the tree is too tender for the climate of Boston, and does not do well even at Washington. There are no large trees of this species in the United States.

TIMBER

According to Gamble, the rate of growth in India is fairly fast, averaging about 11 rings per inch of radius, or 125 years to a girth of 6 ft. The wood is similar to that of the European spruce, and affords excellent planking for floors, walls, and ceilings. It is used for shingles, for packing cases, for building huts, for water-troughs, etc. In some places it is utilised for making tea boxes. It averages in weight 30 lbs. to 32 lbs. per cubic ft. The bark was formerly used extensively by the shepherds for roofing their huts, but this practice has been stopped in the Government forests. On account of the expense of transport, it is never likely to be exported.

(H. J. E.)

---

[1] In *Gard. Chron.* 1872, p. 76. [2] *Garden and Forest*, 1893, p. 14, and 1897, p. 482.

PICEA GLEHNII

Picea Glehnii, Masters, in *Gard. Chron.* xiii. 300, fig. 54 (1880), and in *Journ. Linn. Soc. (Bot.)*, xviii. 512, fig. 13 (1881); Mayr, *Abiet. Jap. Reiches*, 56, 102, t. 4, fig. 11 (1890), and *Fremdländ. Wald- u. Parkbäume*, 327 (1906); Kent, Veitch's *Man. Conif.* 437 (1900); Shirasawa, *Icon. Ess. Forest. Japon*, ii. t. 3, figs. 19-42 (1907); Clinton-Baker, *Illust. Conif.* ii. 39 (1909).

Abies Glehnii, Schmidt, in *Mém. Acad. Imp. Sc. St. Pétersb.* xii. 176, t. 4 (1868).

A tree, attaining in Yezo over 100 ft. in height. Bark different from any of the other spruces, reddish in colour, and fissuring into broad thin loose plates. Young branchlets slender, reddish, with dense short pubescence in the furrows between the pulvini, not spreading over the surface of the latter. Buds ovoid, brown, $\frac{1}{8}$ in. long, composed of a few glabrous scales; terminal buds girt with a ring of scales ending in long subulate points. Leaves arranged on the branchlets as in *P. excelsa*, $\frac{1}{3}$ to $\frac{1}{2}$ in. long, slender, ending in a short cartilaginous point; rhombic in section, with about two stomatic lines on each of the two upper sides, and a single line on each of the lower sides.

Cones, about 2 in. long by 1 in. in diameter when closed; violet, with a red edge to the scales when growing, shining brown when ripe; cylindrical, with an obtuse narrowed apex; scales, when ripe, spreading from the axis at a right angle, suborbicular, with a cuneate claw, about $\frac{1}{2}$ in. wide, with the thin upper margin entire, slightly erose, or faintly denticulate; bract spatulate, $\frac{1}{5}$ in. long, denticulate at the apex. Seed blackish, $\frac{1}{6}$ in. long, with wing $\frac{2}{5}$ in. long; wing broadest about the middle, rounded at the apex, outer margin denticulate.

This species is readily distinguishable from the other short-leaved spruces by the reddish branchlets, with the pubescence confined to the furrows. It resembles *P. orientalis* in the colour of the foliage, but is very distinct in the terminal buds, which have a ring of subulate scales, similar to *P. nigra* and *P. rubra*.

DISTRIBUTION

P. Glehnii[1] was discovered in Saghalien in 1861 by Glehn, the comrade of F. Schmidt on the expedition sent out by the Russian Geographical Society to Eastern Asia. It was subsequently found in Yezo by Maries in 1877.

In Saghalien, this species is confined to the southern half of the island, where it grows on the plains and in the valleys, never attaining, according to Schmidt, a great size, being seldom over a foot in girth. According to Mayr, it is probably absent from the Kurile Isles,[2] as it was not noticed by him on Shikotan; according to Komarov,[3] it does not occur in Russian Manchuria.   (A. H.)

---

[1] The Formosan spruce, identified with *P. Glehnii* by Matsumura in *Tokyo Bot. Mag.* xv. 141 (1901), is quite distinct, and has been named *P. morrisonicola*, Hayata, in *Journ. Coll. Sci. Tokyo*, xxv. 220 (1908).

[2] Miyabe does not include it in his Flora of the Kuriles.   [3] *Floræ Manshuriæ*, i. 200 (1901).

P. Glehnii attains its maximum development in Yezo, where, according to Mayr, it is much commoner in the west of the island than in the east.   In western Yezo it forms mixed forests in company with *P. ajanensis*, chiefly on the cooler parts of the mountains, the trees reaching on an average nearly 120 feet in height. Mayr mentions a peculiar forest of this species, which occurs on the volcanic Iwo-san (1500 feet elevation) east of Lake Kucharro.   In the eastern part of the island, it forms pure forests in the river valleys in swampy situations, which are often several hundreds of acres in extent; but the trees are of no great size, averaging only 80 ft. in height.   This species is known to the Japanese as *Shinko matsu* or *Aka-eso*.

According to Miyabe, it is rare near Sapporo and only found at high elevations mixed with *P. ajanensis*.   Near Lake Shikotsu at 1500 feet elevation, I found it much less abundant than *P. ajanensis*, and could not procure any fruiting specimens. A self-sown seedling which I brought from here is growing very slowly at Coles-borne and is now only 1 foot high.

I could not learn whether the wood of the tree is distinguished from that of the common Yezo spruce.   Some very broad clean pieces which I saw in the saw-mill at Sunagawa had a close grain and a shiny satiny surface when planed, making it suitable for interior work where strength is not required.

CULTIVATION

According to Beissner [1] seeds of this species arrived in Germany before 1891, from which young plants were raised.   It is scarcely known in cultivation in England. There are young plants at Kew, about 2 ft. high, which are thriving; and small specimens at Bayfordbury and Brickendon Grange, Herts, and in the Cambridge Botanic Garden.   It is too soon as yet to form any opinion as to the suitability of this species to our climate; but I do not expect that it will attain any size.

(H. J. E.)

PICEA POLITA

Picea polita, Carrière, *Conif.* 256 (1855); Masters in *Gard. Chron.* xiii. 233, fig. 44 (1880), and
   Journ. Linn. Soc. (Bot.) xviii. 507, pl. 19 (1881); Mayr, *Abiet. Jap. Reiches*, 46, t. 3, f. 7
   (1890), and *Fremdländ. Wald- u. Parkbäume*, 335 (1906); Kent, Veitch's *Man. Conif.* 446
   (1900); Shirasawa, *Icon. Ess. Forest. Japon*, ii. t. 2, figs. 18-29 (1907); Clinton-Baker, *Illust.
   Conif.* ii. 45 (1909).
Picea Torano, Koehne, *Deutsche Dendrologie*, 22 (1893).
Abies Torano,[2] Siebold, in *Verhand. Batav. Genoot. Konst. Wet.* xii. 12 (1830).
Abies polita, Siebold et Zuccarini, *Flor. Jap.* ii. 20, t. 111 (1842).
Pinus polita, Antoine, *Conif.* 95 (1840-1847).

A tree, occasionally attaining in Japan 120 ft. in height, but usually considerably smaller.   Bark fissuring into small scales, exposing the yellowish brown cortex

---

[1] *Nadelholzkunde*, 377 (1891).
[2] This specific name is uncertain, as it was unaccompanied by any description, and cannot be adopted.

beneath. Buds ovoid, acute at the apex, up to $\frac{5}{8}$ in. long, shining reddish brown; scales closely imbricated, ovate, rounded at the apex. Young branchlets stout, glabrous, shining, pale yellow. Leaves on lateral branchlets in an imperfect radial arrangement, all directed outwards with their tips curving upwards; about $\frac{3}{4}$ to $\frac{7}{8}$ in. long, $\frac{1}{16}$ in. wide, very rigid, stout, curved, ending in a sharp spine-like point; compressed rhomboidal in section, with 4 to 6 lines of stomata on each of the four surfaces.

Cones, about 3 to 4 in. long, $1\frac{1}{2}$ in. in diameter when closed, yellowish green when growing, shining chestnut brown when mature; ovoid-cylindrical, obtuse at the apex; scales obovate, with a cuneate base, about $\frac{9}{10}$ in. wide; upper margin rounded, with a few irregular denticulations; bract oblong, $\frac{1}{4}$ in. long, slightly narrowed at the denticulate apex. Seed mottled grey, about $\frac{1}{8}$ in. long, with wing $\frac{7}{8}$ in. long; wing broadest near the truncate denticulate apex.

The very rigid sickle-shaped leaves, ending in prickly spines, and arranged radially on the branchlets, are unlike those of any other spruce.

P. polita is confined to the main island of Japan, having nearly the same distribution as *P. hondoensis* and *P. bicolor*, extending from about lat. $35\frac{1}{2}°$ to lat. $38°$, and not reaching the extreme north of the island. It is found in warmer situations than the other two spruces, and, unlike them, never forms pure woods. It always occurs as isolated trees or in small groups, scattered through the broad-leaved forest. It is the tallest of the three, the largest specimens seen by Mayr being nearly 120 ft. high; and is a much rarer tree, of no economic importance in Japan, where it is known as *hari-momi*.

This species was introduced into cultivation by J. Gould Veitch in 1861, and is perfectly hardy; but it has nowhere attained considerable dimensions. Kent states that the best specimens occur in Devon and Cornwall; but the largest which we have seen is one at Highnam, Gloucester, 30 ft. by 2 ft. in 1910. There is also a healthy specimen at Bayfordbury, planted in 1879, which has borne cones; and another at Hatfield, very thriving. A tree at the Heatherside Nursery, Farnborough, about 20 ft. high, bore cones in 1909. There are two good young trees at Castle Kennedy. A fine specimen at Castlewellan, planted in 1884, was about 25 ft. high in 1907.

According to Mayr this species, with *P. bicolor* and *P. pungens*, are the latest to grow in Germany, not opening their buds until June. It is much injured by squirrels, and will probably be of no economic value, either on the Continent or in England.

(A. H.)

PICEA BICOLOR

Picea bicolor, Mayr, *Abiet. Jap. Reich.* 49, t. 3, fig. 8 (1890), and *Fremdländ. Wald- u. Parkbäume,* 323 (1906); Shirasawa, *Icon. Ess. Forest. Japon,* i. text 19, t. 4, figs. 1-14 (1900).

Picea Alcockiana, Carrière, *Conif.* 343 (1867); Masters, in *Gard. Chron.* xiii. 212, figs. 41, 43 (1880), and in *Journ. Linn. Soc. (Bot.)* xviii. 508, figs. 7-9 (1881); Hennings, in *Gartenflora,* xxxviii. 216, fig. 40 (1889); Kent, Veitch's *Man. Conif.* 429 (1900); Henry, in *Trees of Great Britain,* i. 89, 90 (1906).

Picea japonica,[1] Regel, *Index Sem. Hort. Petrop.* 33 (1865).

Picea acicularis, Beissner, *Nadelholzkunde,* 380 (1891).

Abies bicolor, Maximowicz, in *Mél. Biol.* vi. 24 (1866).

Abies acicularis, Maximowicz, in *Index Sem. Hort. Petrop.* 74 (1868).

Abies Alcockiana, Gordon, *Pinetum,* 4 (1875) (not Lindley).

Pinus Alcoquiana, Parlatore, in De Candolle, *Prod.* xvi. 2, p. 417 (1868).

A tree, attaining in Japan 80 ft. in height. Bark greyish brown, fissuring into small scales. Young branchlets yellowish, glabrous on the lateral branches, but pubescent in the furrows between the pulvini on strong leading shoots; older branchlets shining reddish brown. Buds, about ⅕ in. long, conic, rounded at the apex, without resin, and with few scales, scarious in margin. Leaves, on lateral branches arranged as in *P. excelsa,* about ¾ in. long, rigid, curved, ending in a short cartilaginous point, rhombic in section, with two conspicuous white stomatic bands on the upper two sides, each of 5 or 6 lines, and two bands of about 2 lines each on the two lower green sides.

Cones, averaging 3½ in. long and 1 in. in diameter when closed; bluish red with green margins to the scales when growing, brownish when mature; ovoid-cylindrical: scales obovate with a cuneate base, about ⅗ in. broad, thin and faintly denticulate in the upper rounded margin; bract ⅕ in. long, spatulate, with a slightly expanded denticulate lamina. Seed, ⅕ to ¼ in. long, brown; seed with wing ⅗ in. long; wing widest about the middle, rounded and faintly denticulate at the apex.

This species, as its name implies, differs from the other quadrangular-leaved spruces, in the conspicuous white broad stomatic bands on the upper surface of the leaf, contrasting with the green lower surface, and in this respect it simulates the flat-leaved spruces, and has been confused[2] with *P. hondoensis* and *P. ajanensis.* The leaves of the latter are flat and not rhombic in section, and are devoid of the faint stomatic lines on their lower surface, which are readily seen in *P. bicolor.*

History

This species was discovered in 1860 on Fujiyama by J. G. Veitch, who collected cones of it, unfortunately mixed with twigs of *P. hondoensis.* Lindley, in 1861, described a mixture of the two species, and his name, *Abies Alcoquiana,* Veitch,[3]

---

[1] A name without any description. It is identified in *Index Sem. Hort. Petrop.* 3 (1866) with *Abies bicolor,* Maxim. Seeds were sent from Japan in 1865 by Tschonoski.   [2] Cf. Vol. I. p. 90.

[3] *Ex* Lindley, in *Gard. Chron.* 1861, p. 23. Lindley's description comprises the leaves of *P. hondoensis* and the cones of *P. bicolor.* The type specimen, in which both these are mixed in one packet, is in the herbarium at Cambridge.

cannot stand. Maximowicz gave a correct description of the species under the name *Abies bicolor* in 1866.

As explained in our article[1] on *P. hondoensis*, seeds of both species were early distributed as *P. Alcockiana* ; and in gardens most trees named *P. Alcockiana* are in reality *P. hondoensis*.

This species was introduced into the St. Petersburg Botanic Garden by seeds sent from Japan in 1868 by Tschonoski under the name *Abies acicularis*,[2] Maximowicz.

DISTRIBUTION

This species occurs only in the main island of Japan, where, like *P. hondoensis* and *P. polita*, it is confined to the central ranges between lat. $35\frac{1}{2}°$ and lat. 38°. It forms part of the coniferous forest, which covers these mountains at varying altitudes from south to north, usually above the zone of broad-leaved trees ; but occasionally scattered trees are met with in the upper limits of this zone. Mayr never saw any trees over 80 ft. in height, though he thinks that it occasionally attains greater dimensions.

This species is rare in collections, the largest we have seen being at Kew, where there are two trees, 25 and 30 ft. high, one of which bore cones in 1900. There are also specimens at Westonbirt, Pencarrow, Murthly, Castle Kennedy, and Glasnevin. Mr. H. Clinton Baker collected cones from the tree at Pencarrow in August 1908 ; and I saw at Castlewellan in 1907 a tree about 20 ft. high bearing cones.

The tree at Blackford Park, Edinburgh, mentioned by Kent, was planted about 1882-1884, and measured 20 ft. by 1 ft. 7 in. in 1906. The gardener, Mr. Small, states that it is late in starting into growth in the spring, and in consequence escapes late frosts.

Probably the finest tree in cultivation is growing in Mr. Hunnewell's pinetum at Wellesley, Mass., U.S.A. It bears cones freely, some of which I gathered in 1906, when the tree measured about 36 ft. high by 3 ft. in girth.[3] (A. H.)

[1] Vol. I. p. 90.
[2] Young plants with slender sharp-pointed needles were distributed under this name.
[3] Cf. Sargent, *Pinetum at Wellesley*, 1905, p. 11.

PICEA MAXIMOWICZII

Picea Maximowiczii, Regel, in *Index Sem. Hort. Petrop.* 33 (1865); Carrière, *Conif.* 347 (1867);
   Masters, in *Gard. Chron.* xiii. 363 (1880), and *Journ. Linn. Soc.* (*Bot.*) xviii. 507 (1881);
   Mayr, *Abiet. Jap. Reiches*, 98 (1890).
Picea obovata, Ledebour, var. *japonica*, Beissner, *Nadelholzkunde*, 370 (1891).
Picea Tschonoskii, Mayr,[1] *Fremdländ. Wald- u. Parkbäume*, 339 (1906).
Abies obovata, Loudon, var. *japonica*, Maximowicz, in *Index Sem. Hort. Petrop.* 1 and 3 (1866);
   Franchet, *Enum. Pl. Jap.* i. 466 (1875).
Abies Maximowiczii, Neumann, *Cat.* 1865, *ex* Parlatore, in De Candolle, *Prod.* xvi. 2, p. 431 (1868);
   Veitch, *Man. Conif.* 80 (1881).

A small tree. Young branchlets reddish brown, glabrous, with the apices of the pulvini all directed outwards and forwards. Buds about $\frac{1}{5}$ in. long, ovoid, acute, with glabrous rounded resinous scales. Leaves on lateral branches radially spreading on all sides at nearly a right angle to the branchlet, but with their tips pointing slightly forwards; $\frac{3}{8}$ to $\frac{1}{2}$ in. long, rigid, tapering near the apex which is tipped with a short blunt point; green, quadrangular in section, with three to five stomatic lines on each surface; resin-canals two, lateral, close to the epidermis.

Cones, $1\frac{3}{4}$ to 2 in. long, 1 in. in diameter when open, shining brown when ripe, cylindrical, but tapering at both ends: scales numerous, obovate with a cuneate claw, $\frac{1}{2}$ in. wide; rounded, entire, and bevelled in the upper margin: glabrous in the exposed part, elsewhere covered with a minute reddish pubescence: bract about $\frac{1}{5}$ in. long, oblong, with a rounded faintly denticulate apex. Seeds, not extending to the upper and lateral margins of the scale, $\frac{1}{6}$ in. long, dark brown mottled with lighter streaks; seed with wing $\frac{1}{2}$ in. long; wing widest near the upper rounded denticulate margin.

This species is readily distinguishable by its short leaves radially arranged, and its resinous buds. At Kew it produces new shoots a month earlier than *P. bicolor*.

This spruce is a native of Japan, where it was collected in 1864 on Mt. Fujiyama by Tschonoski,[2] a young Japanese collector in the employment of Maximowicz. One of the original specimens from this locality is preserved at Kew, where there is also an imperfect specimen,[3] collected in the same year in the province of Senano by Tschonoski, which was recognised by Maximowicz to be the same species.[4] It appears to be very rare, and has not since been found by Japanese botanists. Maximowicz considered it to be a variety of *P. obovata*, from which it is clearly distinct; but it is rather related to *P. bicolor*, though differing much in foliage and in cones.

[1] Mayr erroneously considered that the tree cultivated as *P. Maximowiczii* was different from Tschonoski's Fujiyama specimen. He identified the latter with *P. bicolor*, and proposed a new name, *P. Tschonoskii*, for the former.

[2] Maximowicz, in *Rhamn. As. Or.* 17 (1866), gave an account of Tschonoski, who was a Japanese and not a Russian as some authors have supposed. He gathered about 800 species of Japanese plants, and sent seeds of many kinds to St. Petersburg.

[3] Consisting of a cone and a single leaf. The cones on the Grignon tree, about 2 in. long, are intermediate in size between those of the Fujiyama tree (which are $1\frac{3}{4}$ in. long) and those of the Senano specimen (about $2\frac{1}{2}$ in. long).

[4] The Senano specimen is labelled *Abies obovata*, Loudon, var. *japonica*, Maximowicz; and the Fujiyama specimen is named *Picea Maximowiczii*, Regel.

Seeds sent to St. Petersburg by Tschonoski in 1865 were distributed by Regel to various botanic gardens on the Continent. The best specimen that I have seen is a tree with ascending branches at the Agricultural School of Grignon near Paris, which is about 30 ft. in height by 19 in. in girth; but M. Hickel[1] tells me that there are still finer trees elsewhere in France.

A smaller tree in the Arnold Arboretum, U.S.A., also bears cones, smaller in size than those on the tree at Grignon. Another in Mr. Hunnewell's pinetum at Wellesley, Mass., was 11 ft. high in 1905. There is also a small specimen[2] in the spruce collection at Kew, a bush about 4 ft. high; and two trees at Handcross Park, Sussex, the taller of which was 32 ft. by 2 ft. 5 in. in 1911. These were planted about thirty years ago, and have not as yet borne cones. (A. H.)

PICEA NIGRA, BLACK SPRUCE

Picea nigra, Link, *Handb*. ii. 478 (1831); Kent, Veitch's *Man. Conif*. 438 (1900); Clinton-Baker, *Illust. Conif*. ii. 41 (1909).

Picea Mariana, Britton, Sterns, and Poggenburg, *Cat. Pl. N. York*, 71 (1888); Sargent, *Silva N. Amer*. xii. 28, t. 596 (1898), and *Trees N. Amer*. 39 (1905).

Picea brevifolia,[3] Peck, *Spruces of the Adirondacks*, 13 (1897), and in *Bull. Torrey Bot. Club*, xxvii. 409 (1900).

Abies Mariana, Miller, *Dict*. ed. 8, No. 5 (1768).

Abies nigra, Du Roi, *Harbk. Baumz*. ii. 182 (1800); Loudon, *Arb. et. Frut. Brit*. iv. 2312 (1838).

Abies denticulata, Michaux, *Fl. Bor. Amer*. ii. 206 (1803).

Pinus Mariana, Du Roi, *Obs. Bot*. 38 (1771).

Pinus nigra, Solander, in Aiton, *Hort. Kew*. iii. 370 (1789).

A tree, attaining in America 100 ft. in height and 9 ft. in girth, but usually much smaller. Bark brownish, fissuring into irregular thin appressed scales. Buds small, ovoid, acute; the terminal buds surrounded at the base by ciliate pubescent scales with conspicuous long subulate points. Young branchlets brownish, with dense short erect glandular pubescence, retained on the dark-coloured branchlets of the second year. Leaves, arranged on lateral branches as in the European spruce, about ½ in. long, bluish or glaucous green, slightly incurved, ending in a short cartilaginous point, quadrangular in section, with four lines of stomata on each of the two sides turned towards the branchlets, and with one to two lines on each of the other sides.

Cones persistent on the branches for several years, ovoid, acute at the apex, ¾ to 1½ in. long, dark purple when growing, dull brown when ripe; scales rigid, woody, pubescent, about ⅖ in. wide, rounded or rarely pointed at the apex, denticulate

---

[1] M. Hickel informs me that the older trees in France, which were planted about 1868, were originally raised in Thibaut and Keteleer's nursery at Sceaux, from seed given them by Carrière, which he received from Regel. Of late years this spruce has been propagated by grafting.

[2] This is perhaps the same shrub as that from which a specimen in the Kew Arboretum herbarium was taken in 1882, labelled "low bush, 1 to 2 ft. (rounded). Pinetum, Aug. 3, 1882. J. D. Hooker." The low stature of the shrub at Kew indicates probably an alpine origin for the seed from which it was raised.

[3] This is the ordinary stunted form of *P. nigra*, growing on swamps and exposed mountain summits, and is not distinguishable even as a variety by Sargent, or by Britton and Shafer, *N. Amer. Trees*, 57 (1908); Rehder, in Bailey, *Cyc. Am. Hort.* iii. 1334, fig. 1794 (1901), and in *Rhodora*, ix. 109 (1907), has distinguished it as var. *brevifolia*.

in margin. Seeds dark brown, about $\frac{1}{8}$ in. long, with pale brown wings broadest above the middle and very oblique at the apex.

Dwarf and fastigiate forms,[1] and varieties in which the foliage is variegated with white or golden yellow in colour are mentioned by Beissner.

Var. *Doumetii*, Carrière, *Conif.* 242 (1855). This variety was first noticed about 1835 in the garden of the Château de Baleine[2] near Moulins in France. It is a small tree or large shrub, with short numerous branches, forming a dense conical pyramid of foliage. The leaves are very crowded, thin and sharp-pointed. As seen at Kew this variety is very distinct in appearance.

There are remarkable black spruces[3] in the Wilhelmshöhe and Karslane parks at Cassel in Germany, which are pyramidal in habit and bluish in foliage. Self-layering occurs, and numerous colonies of young plants are produced round the parent trees.

DISTRIBUTION

The black spruce is widely spread throughout the Dominion of Canada, occurring as far north as Labrador on the Atlantic coast, and reaching lat. 65° in the valley of the Mackenzie River, whence, crossing the Rocky Mountains, it spreads in the interior of Alaska to the valley of the White River.[4] Farther south, it is restricted to the eastern side of the Rocky Mountains, extending throughout Alberta, Assiniboia, northern Saskatchewan, and northern Manitoba (where it attains its largest size) to central Wisconsin and Michigan. It is common in Newfoundland and all the eastern provinces of Canada, except southern Ontario; and spreads in the north-eastern United States to Pennsylvania, reaching its most southerly point in the Alleghanies in southern Virginia.

Towards the northerly part of its range it is abundant, and grows on well-drained alluvial soils and on the stony slopes of barren hills; while towards the south it is almost entirely restricted to bogs and swamps. Mr. H. E. Ayres in *Garden and Forest*, vii. 504, fig. 80 (1894), describes and figures it under these conditions in Minnesota, as the "Muskeag" spruce, this being the name by which the sphagnum bogs so common in North America are known. He states that in these swamps the trees grow slowly to a height of 60 ft. with very drooping branches, the trunks never exceeding about 10 in. in diameter. The cones are densely crowded at the summit of the tree, and are sometimes produced on trees only 3 ft. high.

(A. H.)

CULTIVATION

The black spruce was introduced[5] into England by Bishop Compton about 1700; but Sir W. Watson, who gave a list of the principal trees which he found in the Fulham Palace gardens in 1751, does not include it.

[1] For *Picea nigra*, var. *virgata*, Rehder, see under *P. rubra*, p. 1378.
[2] When I visited this place in 1909, I found that the original specimen, a tree about 30 ft. high, was dead; but two plants grown from its layers are now about 15 ft. high, with pointed tops; while others, which were raised from cuttings, form dense dwarf bushes.—(H. J. E.)
[3] See *Gard. Chron.* xi. 81, *Suppl. Illust.* (1892). The black spruce appears to layer frequently; and Loudon figures an instance which was noticed in 1828 at Braco Castle, Perthshire.
[4] Cf. Sargent, *Silva N. Amer.* xiv. 106 (1902).   [5] Aiton, *Hort. Kew.* iii. 371 (1789).

It is common in cultivation and usually sold in nurseries under the name of blue spruce, but it never attains large dimensions and is of no economic importance. It has been recommended for planting in boggy and marshy situations, but is always much surpassed in growth by *Picea sitchensis*, and seems to be a short-lived tree in this country.

One of the best specimens we know of is the one figured in Plate 346, which grows on the north edge of a plantation of common spruce at Lyde, near Colesborne, on my property. This tree has been favoured by a moist clay soil, a sheltered position and a cold damp climate; and has attained at about fifty-five years old a height of 56 ft. with a girth of 2 ft. 10 in. As the figure (Plate 346) shows, it has become self-layered under the shade of a hedge, which was cut away to show it; and one of the lower branches has already attained half the height of the parent stem. Though it has not increased much in the last ten years, this tree is in good health, but several others, planted at the same time on dry land, are not half the size and dead or dying. As usual in England it bears cones in abundance near the top of the tree. I have seen a tree at Woburn about 60 ft. by 4½ ft.; and there is one at Merton which was about 40 ft. by 5 ft. 10 in. in 1905.

In the west of Scotland it grows well, but so far as we have seen never attains a large size; the tallest recorded[1] in Scotland in 1891 was 46 ft. by 3 ft. 5 in. at Mount Stuart. Of the numerous trees planted in two groups in 1832 at Keillour, Perthshire, at the lower end of a peat-bog, Henry only found a few surviving in 1904, none exceeding 40 ft. in height. At Dawyck, a tree was 37 ft. by 2½ ft. in 1911.

In Ireland the best we have seen was measured by Henry at Fota in 1903, when it was 60 ft. by 4 ft. 10 in. (H. J. E.)

PICEA RUBRA, RED SPRUCE

Picea rubra, Link, in *Linnæa*, xv. 521 (1841) (not Dietrich[2]); Gorrie, in *Trans. Bot. Soc. Edin.* x. 353 (1870); Kent, Veitch's *Man. Conif.* 450 (1900); Sargent, in *Bot. Gaz.* xliv. 226 (1907); Clinton-Baker, *Illust. Conif.* ii. 47 (1909).

Picea nigra, Link, var. *rubra*, Engelmann, in *Gard. Chron.* xi. 334 (1879).

Picea rubens, Sargent, *Silva N. Amer.* xii. 33, t. 597 (1898), and *Trees N. Amer.* 41 (1905).

Pinus rubra, Lambert, *Genus Pinus*, i. 43 (1803) (not Miller[3]).

Picea acutissima, Jack, in *Garden and Forest*, x. 63 (1897).

Abies rubra, Poiret, in Lamarck, *Dict.* vi. 520 (1804); Loudon, *Arb. et Frut. Brit.* iv. 2316 (1838).

A tree, attaining in America 100 ft. in height and 9 ft. in girth. Bark, branchlets, and buds, similar to *P. nigra*. Leaves yellowish or dark green, not glaucous, about ⅝ in. long, incurved, acute or rounded at the apex, quadrangular in section, marked on each of the two upper sides by about four stomatic lines, and on each of the two lower sides by two to three stomatic lines.

---

[1] *Journ. Roy. Hort. Soc.* xiv. 506 (1892).

[2] *Picea rubra*, Dietrich, *Fl. Berol.* ii. 795 (1824) is the common European spruce, *Picea excelsa*.

[3] *Pinus rubra*, Miller, *Gard. Dict.* No. 3 (1795) is the common European pine, *Pinus sylvestris*.

Cones, ovoid-oblong, $1\frac{1}{4}$ to 2 in. long, green or purplish green when growing, shining reddish brown when mature, usually falling in the second summer : scales $\frac{2}{5}$ in. broad, rounded entire or denticulate at the thin upper margin : bract inconspicuous, oblanceolate, about $\frac{1}{6}$ in. long. Seeds mottled dark brown, about $\frac{1}{8}$ in. long, with wings broadest near the rounded apex, the total length of seed and wing being about $\frac{2}{5}$ in.

VARIETIES

1. *Picea australis*, Small, *Flora S.E. United States*, 30 (1903), is probably a variety [1] of the red spruce, which differs in bearing small cones that are said to fall directly after shedding their seed. Large trees of this kind, attaining 130 ft. in height, are reported to occur on the summits and rocky slopes of mountains in Virginia and North Carolina.

2. A solitary red spruce,[2] with snake-like branches, similar in habit to *P. excelsa*, var. *virgata*, was discovered in 1892, near Williamstown, in north-western Massachusetts. From it young plants were raised by grafts in the Arnold Arboretum.

DISTRIBUTION

The red spruce has a much more southerly distribution than the black spruce,[3] and does not extend farther north than Prince Edward Island and Nova Scotia. It is widely spread in New England,[4] through Maine, New Hampshire, Vermont, and northern Massachusetts ; but is not known in Rhode Island and Connecticut. In New York, especially in the Adirondacks, it forms extensive forests ; and extends through the Alleghany Mountains southward through Pennsylvania and West Virginia to the high peaks of North Carolina. Pinchot[5] has given a complete account of this species, which provides the only merchantable spruce timber in the United States ; and states that it is remarkable for its tolerance of dense shade and its capacity for recovering after years of suppression. In the Adirondacks, it ascends to 4500 ft., and is often seen on steep southern slopes ; but elsewhere is mixed with balsam fir, hemlock, white pine, birch, maples, and beech.   (A. H.)

CULTIVATION

The red spruce was first accurately described and figured by Lambert from a tree growing in England, which was said to have been brought from Newfoundland. According to Aiton,[6] it was cultivated near London by Miller before 1755 ; but it is doubtful if this tree was distinguished from the black spruce at that date.

P. rubra is rare in collections, the only large trees which we have seen being

---

[1] Cf. Britton and Shafer, *North American Trees*, 58 (1908).

[2] Figured in *Garden and Forest*, viii. 45, fig. 7 (1895). This is identical with *Picea nigra*, Link, var. *virgata*, Rehder, in Bailey, *Cycl. Am. Hort.* iii. 1334 (1901), corrected to *P. rubra forma virgata*, Rehder, in *Rhodora*, ix. 110 (1907). Cf. *Mitt. Deut. Dend. Ges.* 1907, p. 116.

[3] Keiler, *Our Native Trees*, 470 (1907) says : "Black spruce is a tree of the far north, existing but precariously south of the northern boundary of the United States ; while red spruce is an Appalachian tree, attaining its greatest dimensions in northern New Hampshire and Pennsylvania."   [4] Dame and Brooks, *Trees of New England*, 14 (1902).

[5] *The Adirondack Spruce*, pp. 1-157 (1898).   [6] *Hort. Kew.* v. 319 (1813).

one at Stanage Park, Herefordshire, which measured in 1911, 72 ft. by 5 ft. 9 in. ; and another at Merton, Norfolk, 63 ft. high and 4 ft. 7 in. in girth in 1909.

The large trees reputed[1] to be of this species at Dropmore are undoubtedly *P. excelsa*.

The only place I know where this tree has been planted in quantity is on the drive of the Rhinefield enclosure in the New Forest, where there are a number of red and white spruce along the south end of the main avenue. The largest of the former was on the north side of the first cross ride on the west and measured 40 ft. by 4 ft. in 1906 ; most of the trees had old and new cones, low down on the ends of the branches, from which I have raised seedlings. The largest of the white spruce at the corner of the second cross ride on the west was 56 ft. by 5 ft. 4 in., and I found cones on only one of the trees.

Some small trees sent from America grow very slowly at Colesborne.

In 1870 Gorrie[2] found a few trees of this species, about 12 to 18 ft. in height, and bearing cones, which were growing on the railway bank near Tynehead in Midlothian at 800 ft. elevation. They had been raised about fifteen years previously from seed obtained in Newfoundland. Some of the seedlings which had been planted two or three miles off in a dry heavy soil had dwindled and died. Dr. A. W. Borthwick visited this place in 1906, and sent me specimens from these trees, from the cones of which I have raised seedlings. The trees are now about 35 ft. in height, growing mixed with common spruce, but not so large as white spruce at the same place. Gorrie also reported in the same year trees about 15 to 20 ft. high growing in Dunmore Park, near Stirling. We have not been able to discover whether these are living.

At Avondale, in Ireland, there is an experimental plot, covering about two acres, which was planted in 1907 with red spruce, mixed with a small proportion of white and black spruce. The red spruce in this plot is extremely thriving, being about 6 ft. high in January 1912, and exceeding in vigour a plot of European spruce beside it.

In France the species seems to grow remarkably well at Les Barres, according to Pardé, who figures[3] a group of three trees about 45 ft. by 4 ft. There are others even larger planted in 1832 which have produced several natural seedlings. Beissner says that there are fine specimens in the Karls-aue at Cassel, at Herrenhausen in Hanover, and at Worlitz.   (H. J. E.)

In 1908 I visited in the Hertogenwald in Belgium, a plantation of red spruce consisting of five acres in two separate plots at an elevation of 1250 ft. The soil here is a loamy clay, on which the European spruce thrives remarkably well. The plots had been accurately measured in 1907, when the trees were fifty-five years old with the following results :—

Number of trees per acre, 950.

Total volume in the round per acre, 3265 cubic ft.

Annual increment, about 60 cubic ft. per acre.

[1] Kent, Veitch's *Man. Conif.* 451, note (1900).   [2] In *Trans. Bot. Soc. Edin.* x. 353 (1870).
[3] *Arb. Nat. des Barres*, 102, pl. 46 (1906).

The two best trees were, in 1908, 50 ft. high and 10 in. in diameter at five feet from the ground. The plantation had not been properly thinned at an early period.

The European spruce in the same district, at a higher elevation, about 1800 ft., averaged at forty-five years old 445 trees per acre, with an annual increment of 136 cubic ft. per acre; and the best trees were 60 ft. high by 12½ in. in diameter.

The red spruce had a redder and more scaly bark than the common spruce; and was more densely clothed with foliage, the improvement of the soil due to the decay of the fallen leaves being well marked. The trees bear cones about every two years; and I noticed several seedlings in the open ground adjoining the plantation. One of the trees had a sucker from its roots about 3 ft. high. (A. H.)

PICEA ALBA, White Spruce

Picea alba,[1] Link, *Handb.* ii. 478 (1831), and in *Linnæa*, xv. 519 (1841); Kent, Veitch's *Man. Conif.* 427 (1900); Clinton-Baker, *Illust. Conif.* ii. 34 (1909).

Picea canadensis, Britton, Sterns, and Poggenburg, *Cat. Pl. N. York*, 71 (1888); Sargent, *Silva N. Amer.* xii. 37, t. 598 (1898), and *Trees N. Amer.* 42 (1905).

Picea laxa, Sargent, in *Garden and Forest*, ii. 496 (1888); Jack, in *Garden and Forest*, x. 63 (1897).

Abies canadensis, Miller, *Dict.* 8th ed. No. 4 (1768).

Abies alba, Michaux, *Fl. Bor. Amer.* ii. 207 (1803) (not Miller); Loudon, *Arb. et Frut. Brit.* iv. 2310 (1838).

Abies curvifolia, Salisbury, in *Trans. Linn. Soc.* viii. 315 (1807).

Abies laxa, Koch, *Dendrologie*, ii. 2. p. 243 (1873).

Pinus canadensis, Du Roi, *Obst. Bot.* 38 (1771) (not Linnæus).

Pinus laxa, Ehrhart, *Beiträge*, iii. 24 (1788).

Pinus alba, Solander, in Aiton, *Hort. Kew.* iii. 371 (1789); Lambert, *Genus Pinus*, i. 39, t. 26 (1803).

A tree, attaining in America 70 to 100 ft. in height and 6 to 8 ft. in girth. Bark ¼ to ½ in. thick, with thin greyish plate-like scales. Young branchlets slender, glabrous, glaucous; becoming greyish or pale brown in the second year. Buds, ¼ in. long, ovoid, rounded or obtuse at the apex; with glabrous scales, non-ciliate, rounded and bifid at the tip, and usually loosely imbricated. Leaves on lateral branches arranged as in the common spruce, usually with a disagreeable odour[2] when bruised, bluish,

---

[1] The oldest specific name (*canadensis* of Miller) for this species is not available, as it was previously used by Linnæus for the eastern hemlock, his *Pinus canadensis* being *Tsuga canadensis*. Moreover, Jack, in *Garden and Forest*, x. 63 (1897), gives some reasons for supposing that Miller indicated the red spruce by his name *Abies canadensis*; and on this account Jack proposes the name *Picea canadensis* for the red spruce, and *Picea laxa* for the white spruce. The latter name is based on Ehrhart's *Pinus laxa*, which is earlier than Solander's *Pinus alba*. Voss, in *Mitt. Deut. Dend. Ges.* 1907, p. 93, proposes *Picea glauca* for the white spruce, based on *Pinus glauca*, Moench, which is earlier than any name except Miller's, but was applied to the glaucous variety. In the midst of the confusion, in which the nomenclature of the American black, white, and red spruces is involved, it is most convenient to adopt for them the names *Picea nigra*, *Picea alba*, and *Picea rubra*, which were first used in combination by Link, and which have been in common use for a great number of years. Moreover, these names are unambiguous, as they have always been applied in each case to the same species.

[2] Hence the name cat or skunk spruce often given to the tree in America. Usually the odour is only perceived when the leaves are bruised, but in certain states of the air it can be detected at some distance from the tree. Cf. *Garden and Forest*, x. 63 (1897).

about $\frac{1}{2}$ in. long, incurved, ending in a rounded or acute (not acuminate) cartilaginous tip; quadrangular in section, with three to four rows of stomata on each side.

Cones, deciduous in the autumn or winter of the first year after the escape of the seeds, sessile or shortly stalked; slender, cylindrical but tapering at both ends, about 2 in. long and $\frac{1}{2}$ in. in diameter, green when growing, shining pale brown when ripe: scales few, loosely imbricated, thin and flexible, so that the cone can be easily crushed by the hand, orbicular or oval, $\frac{1}{3}$ in. broad, rounded or truncate at the entire anterior margin: bract about $\frac{1}{8}$ in. long, oblong with a slightly enlarged ovate denticulate lamina. Seed, $\frac{1}{8}$ in. long, brown, partly embraced by the inflexed margins of the base of the narrow pale wing, which is broadest near the rounded denticulate apex; seed with wing, $\frac{3}{8}$ in. long.

The three American species are often confused, though they have been clearly recognised by botanists in Europe since Lambert's time. In America the younger Michaux and Asa Gray united *P. rubra* with *P. nigra*; but all modern American botanists and foresters keep the three species distinct. The best account of their history is given by Dr. G. Lawson of Halifax, Nova Scotia, in *Proc. Canad. Institute*, 1887, pp. 169-179. Formerly the white spruce was considered to be a native of the Rocky Mountains, but the tree inhabiting Alberta, British Columbia, and Montana is now considered to be distinct, and has been named *P. albertiana*.

P. alba is readily distinguished by its bluish disagreeably smelling foliage and glabrous branchlets, and cannot be confused with *P. nigra* and *P. rubra*, which have pubescent branchlets and peculiar buds with long subulate scales. The cones of the white spruce are easily crushed by the hand on account of their thin flexible scales, and are very different in shape from those of the other two species, which have firm rigid scales.[1]

VARIETIES

1. Var. *arctica*, Kurz, in *Bot. Jahrb.* xix. 425 (1895).

> *Abies arctica*, Murray, in *Journ. Bot.* v. 253, t. 69 (1867).
> *Pinus alba*, var. *arctica*, Parlatore, in De Candolle, *Prod.* xvi. 2, p. 414 (1868).

Towards the northern limit of its area the white spruce has thicker leaves and smaller cones, with more concave scales and bracts slightly different in shape. This form was first collected by Seemann in north-western Alaska. According to Sargent[2] the branchlets of the white spruce in the interior of Alaska are sometimes slightly pubescent, and in all probability this variety is a connecting link between *P. alba* and *P. albertiana*.

A few peculiar forms have arisen in cultivation :—

2. Var. *nana*, Loudon. A round compact bush, rarely exceeding 6 ft. in height.

3. Beissner mentions fastigiate, pendulous, and variegated forms, which we have not seen in England.

---

[1] Trelease, in *Bot. Gaz.* xxix. 196 (1900) describes remarkable burrs, almost globose in shape and covered with smooth bark, which are occasionally seen on the trunk and branches of the white spruce in the United States.

[2] *Silva N. Amer.* xii. 38, note (1898).

4. Var. *cœrulea*, Carrière, *Conif.* 320 (1867).

> *Pinus glauca*, Moench, *Bäume Weiss.* 73 (1785).
> *Abies rubra cœrulea*, Loudon, *Arb. et Frut. Brit.* iv. 2316 (1838).
> *Abies cœrulea*, Forbes, *Pin. Woburn.* 99 (1839).
> *Picea cœrulea*, Link, in *Linnœa*, xv. 522 (1841).
> *Pinus rubra violacea*, Endlicher, *Syn. Conif.* 114 (1847).
> *Picea canadensis glauca*, Sudworth, in *U.S. Forestry Bulletin*, No. 14, p. 37 (1897).

A small tree of dense pyramidal habit, with very glaucous leaves closely pressed against the branchlets. This variety, which according to Carrière frequently arises in the seed-bed, appears to have been known since 1785, and is unquestionably a form of *P. alba*, though it has been by various authorities ascribed to *P. rubra*.

(A. H.)

DISTRIBUTION

The white spruce is a native of eastern Canada and the northern part of the United States, extending southward to the Black Hills of Dakota, the northern parts of Minnesota, Wisconsin, and Michigan, New York, Vermont, northern New Hampshire, and the coast of Maine as far south as Casco Bay. It is recorded[1] for a few stations in Massachusetts, its most southerly limit.

Its westerly distribution in the Dominion of Canada is uncertain; but according to Dr. Lawson, the white spruce is essentially a maritime species, growing along the Atlantic and northern coasts of Canada, and extending by way of the St. Lawrence to the great lakes, as far as Lake Winnipeg. It is common in Newfoundland, Nova Scotia, and New Brunswick, and on the streams which flow from the north into the St. Lawrence, ranging westward through Ontario to the treeless plains of Manitoba, where it occupies sandhills and the dry slopes of river banks. In Labrador it is widely but not generally distributed, growing in the south in well-watered valleys and ascending rocky hills to elevations of 2000 ft. West of Hudson Bay it grows to a large size on river terraces to the borders of the barren lands; and its stems choke the mouths of every arctic American river, strewing the shores with driftwood and testifying to its abundance on their shifting banks.[2]

CULTIVATION

The white spruce was first described by Miller in 1731, and is said by Loudon to have been introduced into England by Bishop Compton in 1700.

Though the name is often found in nursery catalogues and it has no doubt been planted in many places, yet it is nowhere in England so far as we have seen of any special value, either as a timber or an ornamental tree. In some parts of Denmark, however, it has been largely planted as a shelter tree on poor sandy land, in alternate rows with *Pinus montana*, as it is found to grow on poorer soil and to bear salt sea

---

[1] Dame and Brooks, *Trees of New England*, 17 (1902).

[2] E. T. Seaton, *Arctic Prairies*, 329 (1912), measured a tree near Fort McKay 118 ft. high. A log here, 84 ft. long, was 22 in. in diameter at the butt and 15 in. at the small end. At tree limit on the eastern shore of Artillery Lake, a tree, 8 ft. high and 1 ft. in diameter at the butt, showed 300 annual rings.

wind better than common spruce. From what I saw, however, during our visit to Denmark in 1908,[1] it is not likely to become a timber tree of any value here.

In a paper on the "Reclamation of Moors in Belgium,"[2] Mr. A. P. Grenfell says that it forms an excellent mixture with common spruce on poor soils, and in exposed situations in that country, and that it is more windfirm than common spruce.

The white spruce is extremely hardy, and will thrive in exposed situations on high hills, where the common spruce succumbs to the continued action of cold winds in spring. Annand[3] gives an instance of its success on poor peaty soil at a high elevation in the north of Scotland, and recommends it for planting as a wind-break. He tells us that at Carragill in Cumberland, where it has been planted in perhaps the most exposed situation in England, it continues to grow as a low tree between 1600 and 2000 feet, where there is practically no soil, and above the zone in which the common spruce can exist. He considers it specially valuable on wet soils; and says that it has been planted for shelter to a considerable extent in the Moorfoot hills in Midlothian, and in hilly districts in Peeblesshire, Aberdeen-shire, and Caithness. At Durris,[4] however, *P. sitchensis* far surpasses it in growth in such conditions, and appears to be equally hardy. In the Hertogenwald in Belgium *P. alba* has been planted with some success in the wettest parts of the peat mosses at high altitudes.

At Weston Hall, Staffordshire, on good alluvial soil, a plantation was made of the common spruce in 1868, amongst which are scattered a few *P. alba*. The best of the white spruce was 45 ft. by 2 ft. 9 in. in 1909, while the European spruce averaged 60 ft. by 5 ft. 3 in.

The tallest white spruce in England is probably one at Woburn, which measured in 1909 72 ft. by 4 ft. 6 in. There are two fine trees at Powis Castle, the best of which measured 56 ft. by 5 ft. 3 in. in 1906. At Eastnor, a white spruce measures 46 ft. by 2 ft. 10 in.

In the Keillour pinetum, Perthshire, which was planted in 1832, the best *P. alba* measured 52 ft. by $5\frac{1}{2}$ ft. in 1905. In this poor boggy soil, the West American conifers much surpass both *P. alba* and *P. nigra* in growth, the growth of *P. sitchensis* and *Abies grandis* being astonishing.[5] Kent mentions a tree, 45 ft. high, growing on light loam at Dolphinton, Lanarkshire.

At Fota, a white spruce was 45 ft. by 5 ft. in 1907.

Timber of the Black, Red, and White Spruces

In the United States, only the red and white spruce yield merchantable timber, the black spruce never attaining large enough dimensions. In Canada the red spruce is never mentioned by foresters or lumbermen, and only the white and black spruces are said to produce lumber. According to Dr. Lawson, the black spruce is

---

[1] *Quarterly Journal of Forestry*, iii. 75 (1909).   [2] *Ibid.* ii. 273 (1908).

[3] In *Trans. Roy. Scot. Arb. Soc.* xvi. 473 (1901).

[4] Cf. Crozier's account in *Trans. Roy. Scot. Arb. Soc.* xxiii. 7-16, plate 1 (1910), and in our Vol. I. p. 95.

[5] Cf. our account of the pinetum at Keillour, in Vol. I. p. 96. Complete details of the original planting operations in 1832 at Keillour are given in *Proc. Hort. Soc.* iii. 297 (1863).

famed amongst the Canadian lumbermen as a tree yielding sound, strong, and lasting timber; while red spruce produces softer wood, less durable "under exposure to the open air, as is known from experience; every year the red spruce poles have to be replaced more frequently than the black in fences.[1]"

Langelier[2] states that the black spruce is the prevailing coniferous tree in northern Quebec, where the forests are estimated to be capable of supplying 400,000,000 cords of pulp wood. White spruce is less abundant in this region, but attains a larger size, and is utilised for lumber, only the tops being converted into pulp wood. In the southern section of the Abitibi territory white spruce attains splendid dimensions over an area of 15,000,000 acres, and Mr. H. O'Sullivan has seen trees over 100 ft. in height and 20 in. in diameter. Dr. Bell is quoted as saying that "white spruce is perhaps the most valuable tree of the district. It grows to a great size everywhere along the rivers and lakes, where it often girths upwards of 6 ft. The timber is sound; as a rule the trunks run to a great height, and in every respect the white spruce ranks among the very best timber for the manufacture of first-class saw-logs."

J. M. Macoun, in *Forest Wealth of Canada*, says that the wood of the three species is not separated commercially, and that they are used for the same purposes. The black spruce is perhaps the best suited for masts or spars. Of the white spruce he says that the wood is tougher, stronger, and more elastic than that of pine, and is very largely used as lumber, and for railway ties, fence-posts, piles, and telegraph poles.

The wood of these Canadian spruces now supplies the greater part of the material used for pulp-making, which has recently become one of the great industries of Canada. According to a paper on *Pulp Wood in Canada*, by George Johnson, which was printed for the Minister of Agriculture at Ottawa in 1904, no less than 15,000,000 to 20,000,000 dollars are now invested in this manufacture; and as it is estimated that no less than 450,000,000 acres of land in Canada are covered more or less densely with spruce which reproduces itself very rapidly when cut, there is no risk of the supply failing. Great Britain and the United States are said to consume about 900,000 tons of pulp wood annually, the product of about 90,000 acres.

The black spruce is considered better than the white for this purpose and grows mostly on the hills and rocky ground, whilst white spruce loves valleys, where there is more soil.

To show the rapid increase in the value of these timber lands it is stated that in 1892 spruce limits were sold in the province of Quebec as low as eight dollars per square mile, whereas in 1899 similar limits realised 150 dollars per mile, and the price has risen higher lately.

English papermakers are said to have found out that Canadian spruce pulp makes a stronger and better newspaper than Scandinavian pulp; and the immense water-power of the Dominion makes both the transport of the logs and the manufacture cheaper than in most parts of Europe.   (H. J. E.)

---

[1] *Proc. Canad. Inst.* 1887, p. 169.   [2] *Report 6th Meeting Canada Forestry Association,* 1905, p. 65.

PICEA ALBERTIANA, Alberta White Spruce

Picea albertiana, Stewardson Brown, in *Torreya*, vii. 126 (1907); Rehder, in *Mitt. Deut. Dend. Ges.*, 1907, p. 69; Britton and Shafer, *N. Amer. Trees*, 58 (1908).
Picea columbiana, Rydberg, in *Mem. New York Bot. Garden*, i. 11 (1900) (not Lemmon[1]); M. E. Jones, in *Montana Botany Notes*, 10 (1910).
Picea alba, Mayr, *Fremdländ. Wald- u. Parkbäume*, 319 (in part), fig. 101 (1906) (not Link).

A tree, attaining in western North America 160 feet in height. Bark thin, greyish white, scaling off in small quadrangular plates, furrowed at the base of old trunks. Young branchlets greyish or light yellow; yellow or orange in the second year; glabrous or more usually with a minute pubescence on the pegs (from which the leaves arise), which is occasionally scattered over the rest of the surface of the pulvini. Buds about ⅛ in. long, ovoid, slightly resinous, with scarious scales rounded and entire in margin; terminal buds girt at the base with acuminate ciliate keeled scales. Leaves bluish green, in an imperfect radial arrangement on the lateral branches, but more crowded on the upper side of the branchlets; ½ to 1 in. long, soft or rigid, curved, ending in a short point, quadrangular in section, with three to five stomatic lines on each side.

Cones, 1 to 2¼ in. long, cylindrical, obtuse at the apex, sessile, about 1 in. wide when open, shining light brown when ripe: scales numerous, thin, and flexible, fan-shaped, wide, and rounded anteriorly, cuneate on the sides, flatter than those of *P. alba*; upper margin thin, undulate, or faintly denticulate; light brown and glabrous on the exposed part, minutely pubescent and reddish brown on the concealed part: bract ⅛ in. long, with an oblong claw, slightly expanded into a denticulate lamina, which is either emarginate or rounded at the apex. Seed ⅛ in. long, mottled dark brown; seed with wing ½ in. long; wing contracted just above the seed, widest in the upper two-thirds, ending in an oblique denticulate apex.

This species is very variable in the amount of pubescence on the branchlets, which is occasionally absent both in specimens from Montana[2] and from Alberta. The cones are also variable in size, and in the faint denticulation of the margin of the scale. It is most closely allied to *P. alba*, of which it may be considered the Rocky Mountain form. In *P. alba* the branchlets are always perfectly glabrous, with less prominent pulvini; and the leaves are differently arranged in the two species. The buds of *P. alba* are non-resinous, with scales emarginate or two-lobed, and not entire as in *P. albertiana*. The cones of *P. alba* are less rigid, being easily crushed by the hand, and have very fragile scales, entire in margin, more concave internally from side to side, and more reddish brown in colour than those of *P. albertiana*. The seeds are similar in the two species, but those of *P. alba* have shorter wings.

[1] *P. columbiana*, Lemmon, is imperfectly described, and is referred by Sargent and by Britton to *P. Engelmanni*. Lemmon's description may have partly included *P. albertiana*; but the latter name, being quite certain, must stand for the species now treated here. Cf. p. 1388.

[2] Three trees growing together in a clump at 3300 ft. altitude, near Belton in Montana, which I examined in 1906, were precisely alike in size, bark, and habit. One had perfectly glabrous branchlets, silvery leaves, and large cones. Another had very pubescent branchlets, green leaves, and small cones. The third was intermediate.

This species, the exact distribution of which has not yet been clearly defined, ranges from Wyoming[1] and western Montana northward to Alberta and British Columbia. It occurs in the Rocky Mountains at lower elevations than *P. Engelmanni*, extending from 3000 to 5000 ft. The type specimen was collected at Bankhead, Alberta, by Stewardson Brown; and I have received from Macoun specimens from the neighbourhood of Banff, in the same province. Rehder states that this spruce near Banff occasionally attains 160 ft. in height, and forms extensive forests, in one of which he took a fine photograph, which shows well the habit of the tree, and is reproduced by Mayr[2] in his article on the white spruce. *P. albertiana* is the white spruce referred to by Sargent[3] as a native of "the Rocky Mountains of Alberta, British Columbia, and northern Montana, where it lines the banks of streams and lakes up to 5000 ft. elevation, attaining a large size, and sending up tall spire-like heads of dark foliage."

In Montana this spruce is not found on the east side of the continental divide, but is common in the Flathead[4] region, where it forms a low tree in marshy situations; but on moist alluvial soil, in mixture with the Douglas fir, western larch, and *Thuya plicata*, it attains large dimensions. It usually occurs in small groups in these mixed forests, occupying the moister ground, and bearing considerable shade. The largest tree which I measured, growing near Nyack on the Northern Pacific railway, was 150 ft. by 10 ft. A tree 114 ft. by 4 ft. 9 in. showed, when cut down, 114 annual rings; another, 15 in. in diameter, showed 160 rings, the bark being only $\frac{1}{4}$ in. thick.

It is possible that the trees referred to *P. Engelmanni*, in Idaho, Washington, and Oregon, may wholly or in part belong to *P. albertiana*; and a further study of the spruces in western America is desirable, as the variability in *P. albertiana* points possibly to hybridisation with Engelmann's spruce.

This spruce is the finest species in North America, except *P. sitchensis*, and is worth a trial as an ornamental tree. It was introduced into England by Elwes, who received seeds from Mr. J. M. Macoun of Ottawa in 1906, which have produced plants, the largest of which in 1912 were about 18 in. high, and which have been distributed to several places in England and Scotland. According to Rehder, it was sent by Baron von Fürstenberg to Germany in 1907.　　　　(A. H.)

---

[1] Britton and Shafer, *N. Amer. Trees*, 58 (1908), give Wyoming as a habitat; but I have seen no specimens.

[2] *Fremdländ. Wald- u. Parkbäume*, fig. 101 (1906). This photograph is also reproduced in Möller's *Deut. Gärtn. Zeit.* 1905, p. 117.

[3] *Silva N. Amer.* xii. 39 (1898).

[4] The spruce described as *P. Engelmanni* by Whitford, in *Bot. Gaz.* xxxix. 196 (1905).

PICEA ENGELMANNI, ENGELMANN'S SPRUCE

Picea Engelmanni, Engelmann, in *Trans. St. Louis Acad.* ii. 212 (1863), and in *Gard. Chron.* 1863, p. 1035; Sargent, *Silva N. Amer.* xii. 43, t. 599 (1898), and *Trees N. Amer.* 43 (1905); Kent, Veitch's *Man. Conif.* 431 (1900); Britton, *N. Amer. Trees*, 59 (1908); Clinton-Baker, *Illust. Conif.* ii. 37 (1909).

Picea columbiana, Lemmon, in *Garden and Forest*, x. 183 (1897).

Picea pseudopungens, Dieck,[1] *Verkaufs-Verzeichniss Zöschen*, 28 (1904).

Abies Engelmanni, Parry, in *Trans. St. Louis Acad.* ii. 122 (1863).

Abies commutata, Murray, in *Gard. Chron.* iii. 106 (1875).

Pinus commutata, Parlatore, in De Candolle, *Prod.* xvi. 2, p. 417 (1868).

A tree with disagreeably smelling foliage, attaining in America 150 ft. in height and 15 ft. in girth, though usually considerably smaller. Bark reddish, exuding resin, with thin loose scales. Young branchlets greyish yellow, with a sparse minute erect glandular pubescence. Buds conical, about $\frac{1}{5}$ in. long, obtuse at the apex; the terminal bud closely surrounded at the base by the uppermost leaves; scales scarious, rounded, without resin.

Leaves, arranged on lateral branches as in *P. excelsa*, $\frac{7}{8}$ to 1 in. long, soft and flexible, tapering towards the apex, which ends in a sharp point; bluish green in colour, with a cat-like odour when bruised; quadrangular in section, with four to five stomatic lines on each side.

Cones horizontal at first, ultimately pendulous, sessile, green tinged with scarlet when growing, light brown when ripe, cylindrical but narrowed at both ends; very variable in size, $1\frac{1}{2}$ to 3 in. long: scales numerous, thin and flexible, rhombic or ovate, minutely pubescent in the lower half, longer than broad, $\frac{2}{5}$ in. wide, with their upper margin truncate or rounded and lacerate: bract $\frac{1}{4}$ in. long, with an oblong claw, and an oval expanded denticulate lamina. Seed about $\frac{1}{10}$ in. long, dark brown; seed with wing $\frac{3}{10}$ in. long; wing broadest near the rounded oblique faintly denticulate apex.

This species resembles *P. alba* in the peculiar odour of the leaves, but is readily distinguishable by the very sparse minute pubescence on the branchlets.

VARIETIES

1. Forms with glaucous or silvery foliage appear in the seed-bed, and are known as var. *glauca* and var. *argentea*. The tree, however, usually cultivated under the name *P. Engelmanni glauca* is a form of *P. pungens*, agreeing with the latter species in having glabrous branchlets and peculiar buds with reflexed scales.

2. Var. *microphylla*, Hesse. A dwarf form with short leaves, mentioned by Beissner, *Nadelholzkunde*, 345 (1891).

3. Var. *Fendleri*, Henry (var. *nova*).

At Kew one of the tallest spruce trees has long been labelled *P. Engelmanni*, but differs remarkably from that species in habit. Young branchlets pendulous,

---

[1] Dieck, *Neuh. Offert. Zöschen*, 1892, p. 38, mentions *Abies Engelmanni*, var. *pseudopungens*, as a supposed new variety, raised from seed collected by Purpus north of Lytton in British Columbia.

brownish yellow, densely covered with minute erect glandular pubescence, retained for several years. Leaves disagreeable in odour when bruised, radially spreading from the branches equally on all sides, long and slender, about $1\frac{1}{8}$ in. long, bluish in tint, sharp pointed, quadrangular in section, with four stomatic lines on each of the inner two sides, and two lines on each of the outer two sides. This tree has never borne either flowers or cones.

In the pendulous branches and the radial arrangement of the leaves it resembles *P. Smithiana*; but differs entirely from this species in branchlets and buds. The tree at Kew, the history of which cannot be ascertained, now measures 40 ft. high by 3 ft. in girth.

It is probably identical with a specimen, preserved in the Kew Herbarium, which was gathered by Fendler in 1847 in New Mexico. This specimen, which bears no cones, has similar branchlets and buds, and leaves similar in length and stomatic lines, but somewhat stouter. Engelmann has marked this specimen in pencil: "vigorous long-leaved form, young tree"; but in all probability it is a distinct species. Until the tree at Kew bears cones, the identification must remain doubtful. The seed was perhaps sent home by Roezl, who collected in this region.

DISTRIBUTION

Engelmann's spruce is an alpine tree, widely distributed in western North America, extending in the Rocky Mountains from Alberta[1] to southern New Mexico and Arizona, and westwards to the Selkirk and Cascade Mountains of British Columbia, Washington, and Oregon. Towards the south it occurs at 8000 to 11,500 feet elevation, while farther north it descends to 5000 feet. It attains its largest size and forms a great part of the forests on the high mountains of southern Alberta, and is a common tree in Montana and Idaho. Westward, on the Cascades and Blue Mountains of Washington and Oregon, it is smaller in size, and is usually scattered amongst other trees; and on account of its slightly different habit, was distinguished by Lemmon as *P. columbiana*[2]; but both Sargent and Britton are unable to separate this even as a distinct variety. It is common on the Yellowstone plateau in Wyoming, and forms extensive forests in Colorado, Utah, eastern Nevada, and the San Francisco peaks of northern Arizona, reaching its most southerly point on the summit of the Mogollon Mountains in New Mexico.

(A. H.)

This tree is a conspicuous feature in the alpine forest of Alberta, where I often camped under its shelter in 1895. It grows from the foothills of the Rocky Mountains up to nearly timber line, which is here about 7000 feet. It seems to be a very slow-growing tree, for Wilcox[3] counted 400 rings on an old stump near Lake Louise, which was less than 3 ft. in diameter.

[1] A good photograph of a forest of this tree at Laggan is reproduced by Mayr, *Fremdländ. Wald- u. Parkbäume*, 325, fig. 103 (1906).

[2] The tree in Montana ascribed to this species by M. E. Jones, *Montana Botany Notes*, 10 (1910), is *P. albertiana*.

[3] *The Rockies of Canada*, 62 (1900).

Mr. F. R. S. Balfour, who saw it in the same region, writes to me as follows :—
" This alpine spruce covers immense tracts in the Rockies at high altitudes. The finest I have seen are near Lake Louise, where it grows to a height of 140 ft., mixed with *Abies lasiocarpa*. Indeed the two are singularly alike and difficult to distinguish, except for the red drooping cones of the Picea, and the small black erect ones of the Abies. The barks of these trees are very similar, of a grayish red broken into large loose scales. The leaves when crushed have a rather unpleasant smell. Wherever I have seen this tree, it has an arrowy appearance from the shortness of its branches. This is doubtless due to the weight of frozen snow which covers them in winter, and prevents lateral growth. The tree often fruits profusely when quite young and small, the leader then becoming bent with the weight of cones surrounding it. These are about two inches long and of a warm crimson when fully grown. The branches are produced in very regular whorls; and when young the bark is smooth and silvery before it becomes scaly."

This spruce was first distinguished by Dr. C. C. Parry, who found it in 1862 on Pike's Peak in Colorado, and sent seeds in the following year to the Botanic Garden of Harvard University, where the tree was first cultivated. It is said to have been introduced[1] into England in 1864; but the tree seems quite uncommon in this country. We have found no trees which are probably so old as this, except one at Highnam, which was bearing cones in 1909 and measured 38 ft. by 3 ft. 2 in. The bark of this separates into regular small scales. A thriving tree at Hatfield, Herts, which was planted in 1893, measured 21 ft. by 1 ft. in 1908. There are smaller trees in the same county at Bayfordbury, High Leigh, and Brickendon Grange.

(H. J. E.)

PICEA PUNGENS, Colorado Spruce

Picea pungens, Engelmann, in *Gard. Chron.* xi. 334 (1879) and xvii. 145 (1882); Masters, in *Gard. Chron.* xx. 725, fig. 130 (1883), and x. 547, figs. 73, 74 (1891); Kent, Veitch's *Man. Conif.* 448 (1900); Britton and Shafer, *N. Amer. Trees*, 60 (1908); Clinton-Baker, *Illust. Conif.* ii. 46 (1909).

Picea Parryana, Sargent, *Silva N. Amer.* xii. 47, t. 600 (1898), and *Trees N. Amer.* 44 (1905).

Picea Menziesii, Engelmann, in *Trans. St. Louis Acad.* ii. 214 (1863) (not Carrière); Sargent, in *Bot. Gaz.* xliv. 227 (1907).

Abies Menziesii, Engelmann, in *Amer. Journ. Science*, xxxiv. 330 (1862) (not Lindley), and in *Gard. Chron.* vii. 790 (1877); André, in *Gard. Chron.* vii. 562 (1877).

Abies Menziesii Parryana, André, in *Illust. Hort.* xxiii. 198 (1876), and xxiv. 53, 119 (1877).

A tree, attaining in America 150 ft. in height and 9 ft. in girth, usually considerably smaller. Bark reddish grey, fissuring on young stems into small oblong plates, on old trunks deeply divided into broad rounded scaly ridges. Young branchlets, stout, rigid, glaucous at first, gradually becoming orange brown. Buds $\frac{1}{4}$ to $\frac{5}{8}$ in. long, ovoid, rounded at the apex, with the tips of the upper scales rounded

[1] Veitch, *Man. Conif.* 69 (1881).

and scarious and usually reflexed ; terminal buds girt with a ring of keeled acuminate scales. Leaves on lateral branches in an imperfect radial arrangement, more crowded on the upper than on the lower side of the branchlet, all spreading forwards as well as outwards ; $\frac{3}{4}$ to $1\frac{1}{4}$ in. long, stout, rigid, incurved, tapering towards the hard sharp-pointed apex ; varying greatly in colour on different trees, bright green, bluish or silvery white ; quadrangular in section, with four to seven stomatic lines on each side.

Cones sessile, usually persistent till the second winter, about 2 to 4 in. long, 1 to $1\frac{1}{2}$ in. in diameter, cylindrical but slightly narrowed at both ends ; green tinged with red when growing, pale shining brown when mature : scales numerous, like those of *P. Engelmanni*, thin, tough and flexible, rhomboidal, narrowing towards the truncate denticulate apex, longer than broad, about $\frac{1}{2}$ in. wide : bract about $\frac{1}{6}$ in. long, with a denticulate ovate lamina. Seed blackish, $\frac{1}{8}$ in. long ; seed with wing $\frac{2}{5}$ in. long ; wing broadest near the truncate lacerate apex.

This species is readily distinguished by its radially arranged, rigid, sharp-pointed leaves, by its glabrous branchlets, and by the loose reflexed tips of the bud-scales.

Varieties

Seedlings differ very much in the colour of the foliage, which varies from almost a pure green to a silvery white. Trees with very blue glaucous foliage are distinguished as var. *glauca*, Regel,[1] and are much more ornamental than the green form, var. *viridis*, Regel,[1] which is less common in cultivation. Var. *argentea*, Waterer,[2] is a form with silvery foliage, which has longer and more slender needles than usual ; and on this account it is often erroneously known in gardens as *Picea Engelmanni glauca*,[3] from which species it is readily distinguished by its glabrous branchlets and peculiar buds with reflexed scales.

Var. *Kosteriana*[4] (var. *glauca pendula*[5]). This is a form with very pendulous branches and fine bluish foliage, which originated in Messrs. Koster's nursery at Boskoop in Holland.

Another variety, said to be vigorous in growth, and characterised by shining leaves, silvery white in colour and broader and longer than in the type, originated in the nursery of Herr Weise at Kamenz in Saxony, who sent it out as var. *König Albert von Sachsen*.[6]

Beissner describes two prostrate forms, var. *prostrata*[7] and var. *tabuliformis*.[8] According to Rehder,[9] a dwarf compact form originated about 1890 in the Arnold Arboretum, U.S.A.

[1] *Russ. Dendr.* i. 37 (1883).   [2] *Ex* Masters, in *Journ. Roy. Hort. Soc.* xiv. 223 (1892).
[3] This is *Abies Engelmanni glauca*, Veitch, *Man. Conif.* 69 (1881). Cf. Kent, Veitch's *Man. Conif.* 432 (1900).
[4] Masters, in *Kew Hand-List Conif.* 85 (1903).
[5] Beissner, *Nadelholzkunde*, 348 (1901). Bean, in *Kew. Bull*, 1908, p. 390, calls it var. *Kosteri pendula.*
[6] Ledien, in *Gartenflora*, xl. 69, fig. 22 (1891).   [7] *Mitt. Deut. Dend. Ges.* 1906, p. 141.
[8] *Ibid.* 1909, p. 268.   [9] In Bailey, *Cycl. Am. Hort.* 1334 (1901).

Distribution and Cultivation

P. pungens is very restricted in its area of distribution, and is nowhere abundant, growing usually as isolated trees or in small groves on the banks and terraces of streams at 6500 to 10,000 ft. elevation. It is met with in Colorado, eastern Utah, New Mexico,[1] and in Wyoming, extending in the latter state as far north as the Wind River mountains.

This species was discovered[2] in 1862 on Pike's Peak in Colorado by Dr. C. C. Parry, who sent seeds in the following year to the Botanic Garden of Harvard University. One of the earliest plants[3] raised had attained in 1883 16 ft. in height in Prof. Sargent's garden at Brookline, Massachusetts. Waterer[4] in 1877 took cuttings from this tree, which were propagated in his nursery at Knap Hill; and, doubtless, from these are derived most of the older specimens growing in England. Seeds[3] were again collected in Colorado by Roezl, from which some of the trees on the Continent may have originated.

It has been largely planted on account of its beautiful foliage; but has nowhere, so far as we know, attained large dimensions, and seems only suitable as an ornament in gardens. It has lately been attacked by a fungus.[5]

There are good specimens, about 30 ft. high, of both the green and glaucous varieties at Highnam, which were bearing cones in March 1910. A glaucous tree at Aldenham also bore cones in 1909. Mr. A. B. Jackson in 1911 reported two good trees, 35 and 33 ft. high, at Yattendon Court, Berks.

According to Schwappach,[6] this tree may prove valuable for planting in northern Germany, where it not only supports the cold of winter without injury, but is hardier[7] against late frosts than any other conifer. It grows well on moist soils, and thrives even on wet soils, which are unsuitable for *P. excelsa* or even for *P. sitchensis*; and on account of the sharp-pointed needles, it is not attacked by deer or squirrels. It has been tried at nineteen stations, the total area of the experimental plots being fifteen acres. It grows slowly at first, averaging in the fifth year 12 to 20 in. high, and in the tenth year $2\frac{1}{2}$ ft. to 4 ft. 8 in.

According to Sargent, it has been much planted in the United States as an ornamental tree, on account of its handsome pyramidal habit; but trees with bluish foliage lose in a few years much of their colour, and the older trees, 30 to 40 ft. in height, are losing their lower branches, so that their pyramidal habit is spoiled.

(A. H.)

[1] Britton and Shafer, *N. Amer. Trees*, 60 (1908), give New Mexico as a habitat for this species; but I have seen no specimens.

[2] According to Sargent in *Gard. Monthly*, quoted in *Woods and Forests*, 1885, p. 53, trees of this species were transplanted from Colorado to Iowa in 1860, and proved very hardy.

[3] *Gard. Chron.* xx. 725, fig. 130 (1883).   [4] *Gard. Chron.* vii. 48 and 562 (1877).

[5] Borthwick, in *Trans. Bot. Soc. Edin.* xxiii. 232 (1906), and *Notes R. Bot. Gard. Edin.* 1909, p. 260, plate 50, states that this fungus, which he names *Cucurbitaria piceae*, attacks the buds, and produces large black conical swellings, in which numerous fructifications of the fungus occur. The bud is either immediately destroyed, or it may produce a twisted cankered shoot which frequently dies off at an early period.

[6] *Anbauversuche mit Fremdländ. Holzarten*, 49 (1901).

[7] This is also the case in my experimental ground at Colesborne, where late frosts have severely injured all the other spruces.—H. J. E.

PICEA SPINULOSA, Sikkim Spruce

Picea spinulosa, Henry, in *Gard. Chron.* xxxix. 219 (1906); Beissner, in *Mitt. Deut. Dend. Ges.* 1906, p. 83; Clinton-Baker, *Illust. Conif.* ii. 50 (1909).

Picea morindoides, Rehder, in Sargent, *Trees and Shrubs*, i. 95, t. 48 (1903); Henry, in *Gard. Chron.* xxxix. 132, 219 (1906), and in *Trees of Great Britain*, i. 77 (1906); Masters, in *Gard. Chron.* xxxix. 218, fig. 84, and 274, fig. 113 (1906), and xli. 388 (1907); Brandis, *Indian Trees*, 720, 721 (1906); Stapf, in *Bot. Mag.* t. 8169 (1907).

Picea Alcockiana, Carrière, var. *morindoides*, Mottet, *Conif. et Taxac.* 273 (1902).

Abies spinulosa, Griffith, *Journals*, 259 (1847), and *Itin. Notes*, 145 (1848).

Abies Morinda, Lindley, in *Gard. Chron.* 1855, p. 334 (not *Picea Morinda*, Link).

Abies Smithiana, Hooker, *Himalayan Journals*, ii. 32 (1854) (not Lindley).

Pinus spinulosa, Griffith, *Icon. Pl. Asiat.* t. 363 (1854).

A tree, attaining in the eastern Himalayas over 200 ft. in height. Bark rough and scaling off in small quadrangular plates. Young branchlets slender, glabrous, yellowish grey. Buds, about $\frac{1}{4}$ in. long, ovoid, obtuse at the apex, brown, scarcely resinous, with glabrous obtuse scales. Leaves, in an imperfect radial arrangement, covering in closely imbricated ranks the upper side of the branchlets, those on the lateral sides directed outwards and forwards, those on the under side pointing downwards and forwards; $\frac{3}{4}$ to $1\frac{1}{4}$ in. long, $\frac{1}{20}$ in. broad, slender, acute at the apex, which is tipped with a sharp point; flattened, but keeled on both surfaces, so that the section is rhomboid-elliptic; ventral surface green without stomatic lines and directed towards the light; dorsal surface with two stomatic bands, each of 4 to 6 lines; resin canals, two (occasionally absent), dorsal, near the edges and close to the hypoderm.

Staminate flowers pink, $\frac{3}{4}$ in. long; connective with an orbicular denticulate appendix. Cones, about $2\frac{1}{2}$ in. to 3 in. long on cultivated trees, up to 4 in. long on wild trees, 1 to $1\frac{1}{4}$ in. in diameter, cylindric, obtuse at the apex, green with a purple border to the scales when growing, shining brown when mature: scales thin and flexible, suborbicular with a cuneate base, about $\frac{1}{2}$ to $\frac{5}{8}$ in. wide, bevelled in the upper margin, which is rounded,[1] entire, undulate, or slightly denticulate: bract ovate, acute, $\frac{1}{6}$ in. long. Seed, $\frac{1}{5}$ in. long, greyish brown; seed with wing $\frac{1}{2}$ in. long; wing broadest above the middle, rounded and denticulate at the apex.

This species is distinguishable from the other flat-leaved spruces with glabrous branchlets,—by the leaves somewhat radially arranged, distinctly keeled on both surfaces, slender, and ending in a sharp point. The leaves of *P. sitchensis*, which are similar in appearance, are arranged on the lateral branches as in the common spruce.

This species appears to be the only spruce occurring in the eastern Himalayas, where it has been found in Sikkim and Bhutan, at 8000 to 10,000 ft. altitude. There are no specimens of *P. Smithiana* from this region in the Calcutta and Kew herbaria.[2]

---

[1] In some wild specimens the scales are truncate in the upper margin.

[2] The specimens in the Calcutta Herbarium, which were sent on loan to Kew in 1910, comprise the following :—

"Sikkim, chief forest tree in Rinchingung; King, 1875."

"Sikkim, Lachen; King's collector in 1885."

"Sikkim, Lachung; Gammie, 1892." Referred to as *P. Morinda* in *Rec. Bot. Survey India*, i. No. ii. pp. 11, 19.

"Chumbi; Gamble, 1880."

"Chumbi; King's collector, 1884."

There are no specimens from Bhutan, either in the Kew or Calcutta herbarium.

It was discovered by Griffith, who found it growing on the slopes of the Rodoola Pass and in the Tung-chiew valley in northern Bhutan (about 91½° long. and almost 150 miles east of the Sikkim frontier). He described it as a tree usually of moderate size, but occasionally attaining 80 ft. in height, and growing in groups between 8500 and 10,000 ft. elevation. It resembled at a distance a larch in habit, with the lower branches deflexed and the upper branches spreading. He named it *Abies spinulosa*, and there is no doubt possible as to its identity. He refers to the pulvini from which the leaves arise, showing that it was a spruce and not a silver fir. His statement[1] that "the lower surface of the leaf is glaucous, but that probably this was the true upper surface turned downwards" is characteristic of the section *Omorica*, to which *P. spinulosa* belongs.

Sir Joseph Hooker found this spruce again in 1849 in the Lachen valley, at 9000 ft. altitude, in Sikkim; and Elwes saw it in the same place on 1st October 1870; but neither Hooker nor Elwes noticed it elsewhere in Sikkim. Hooker identified it with Griffith's *Abies spinulosa*; but, unfortunately, afterwards combined it with *Picea Smithiana*.[2]

It was again found at 9000 to 10,000 ft. altitude, in 1887, in the drier climate of the Chumbi valley (on the north-eastern frontier of Sikkim) by a native in the employment of the late Sir J. Ware Edgar, then Deputy Commissioner of Darjeeling. It has lately been rediscovered in this locality by Mr. E. H. C. Walsh, who accompanied the military expedition to Lhasa. There are also specimens in the Kew Herbarium collected at Yatung (27° 51′ N. lat., 88° 35′ E. long.) by Mr. H. E. Hobson.

Sir George King, who considered the Chumbi valley specimens to be a new species, which, however, he left undescribed, sent seeds in 1877 or 1878 to various botanical establishments in Europe; and a tree probably raised from this seed is now growing in the arboretum of M. Allard, at Angers in France. It was found here by Mr. Rehder, who described it as a new species of unknown origin, under the name *P. morindoides*. It bore cones which I gathered in 1906, and was then about 20 ft. high. M. Allard obtained it about 1891 from Van Houtte.

(A. H.)

Though I saw this tree in the Lachen valley of Sikkim during my first journey to the Tibetan frontier in 1870, I did not take special notice of it at the time, and certainly saw no such wonderful trees as are described by Mr. J. Claude White in his recently published book.[3] These grow in the Sebu valley, a tributary of the Lachung, and must be among the tallest trees in India. Mr. White says: "One fallen giant, a spruce that I measured, was 220 ft. from the roots to where it had broken off short, and there it measured 6 ft. in girth. What had become of the top I do not know, but it was a magnificent specimen."

---

[1] Griffith's MS. description at Kew, which is mutilated in the printed *Itin. Notes*, reads: "Ramulis fere omnibus deflexis, verrucis e quibus folia oriuntur exasperatis; folia undique patentia, linearea, mucrone spinulosa terminale, pungentia; pagina inferior glauca, superior an resupinata. Conis terminalibus, pendentibus, oblongis, fere cylindraceis, castaneo brunneis; squamis latiformis, obtusissimis, laevibus."

[2] *Flora Brit. India*, v. 653 (1888) under *P. Morinda*. Hooker gives on a drawing at Kew the height of the tree as 80 ft., and the locality, Lachen, 8000 to 10,000 ft. altitude.   [3] *Sikkim and Bhutan*, 79 (1909).

To confirm this, I may say that I have since heard from Mr. A. D. Hickley, who visited and re-measured the same tree. He informs me that it grew at a place called Chu-par-rab-dong on the left bank of the river just opposite Yakchi, at an elevation of about 10,000 ft. Five lengths of the fallen stem were missing, having been probably used by the natives; but without counting these he made the length 207 ft., and the girth as follows:—at the base, 12 ft. 7 in.; at 50 ft., 13 ft.; at 100 ft., 12 ft.; at 139 ft. (the place where the tree was first broken in its fall), 9 ft. 4½ in. The thinnest piece, without the bark, was 5 ft. 2 in. in girth. In the same valley not far away Mr. Hickley measured another spruce, also with the top broken, 197 ft. long; and there were many like it still standing. A silver fir (*Abies Webbiana*) in this valley girthed 15½ ft. at four feet from the ground; and a larch (*Larix Griffithii*) growing between two spruces, of which he sends me a photograph, was well over 125 ft., and very much taller than any previously recorded. The photograph, unfortunately, is not good enough for reproduction in this work, and I have vainly endeavoured to obtain a better one.

This species was first detected by Henry in cultivation in this country at Castlewellan, where there are three or four trees, which were raised by the late Earl Annesley from seed sent by the Calcutta Botanic Garden in 1890. The largest tree[1] was 22 ft. high and 18 in. in girth in 1906, and produced cones for the first time in 1907. It is a handsome tree (Plate 347), remarkably distinct in appearance, and perfectly hardy. Specimens have been sent from Castlewellan to the Botanic Gardens at Kew, Glasnevin, and Cambridge.

There are three trees at Leonardslee, Horsham, which produced cones in 1906 and 1907. Sir Edmund G. Loder informs me that these were planted by a former owner, and that their origin is unknown. The largest in 1910 measured 31 ft. 9 in. by 2 ft. 5 in. From the seed of these trees seedlings have been raised each year since 1908. Mr. H. Clinton-Baker found a tree about 25 ft. high at Menabilly, in August 1908, which was bearing cones. Seedlings were raised from their seed at Bayfordbury. In Mr. J. M. Robb's garden at Chiltley Place, Liphook, Hants, there is a tree with two stems about 26 ft. high, which was bearing young and old cones in June 1912. This is supposed to have been planted about twenty-five years ago; but has been crowded by other trees. A smaller specimen with a single stem, about twenty years old, is more thriving. I found another tree at Melbury, about 15 ft. high, which was bearing cones in September 1909. It is said to have been planted about ten years ago.   (H. J. E.)

---

[1] Figured in *Gard. Chron.* xxxix. 274, fig. 113 (1906).

JUNIPERUS

Juniperus, Linnæus, *Gen. Pl.* 311 (1737); Endlicher, *Syn. Conif.* 7 (1847); Parlatore, in De Candolle, *Prod.* xvi. 2, p. 475 (1868); Bentham et Hooker, *Gen. Pl.* iii. 427 (1880); Masters, in *Journ. Linn. Soc. (Bot.)* xxx. 12 (1893); Hickel, in *Bull. Soc. Dend. France*, 1911, p. 31.

Sabina, Haller, in Ruppius, *Fl. Jen.* 336 (1745); Garcke, *Fl. Deutschl.* 387 (1849); Antoine, *Cup. Gatt.* 35 (1857).

Thuiæcarpus,[1] Trautvetter, *Pl. Imag. Flor. Ross.* 11, t. 6 (1844).

Arceuthos, Antoine and Kotschy, in *Oestr. Bot. Wochenbl.* 1854, p. 249.

EVERGREEN shrubs or trees, belonging to the division Cupressineæ of the order Coniferæ. Bark usually thin, and scaling in longitudinal strips. Leaves on young plants always spreading and acicular; on adult plants acicular, or appressed and scale-like, different in the sections of the genus, where they are described.

Flowers monœcious or diœcious. Staminate flowers composed of numerous stamens on a central axis, with ovate or peltate scale-like connectives, each bearing two to six globose pollen-sacs. Pistillate flowers, surrounded at the base by minute scale-like bracts, which persist unchanged under the fruit; composed of three to eight opposite or ternate pointed scales, bearing either at their base or alternate with them one to two ovules. Fruit a succulent berry-like indehiscent strobile, composed of three to eight fleshy scales united together, covered by a membranous epidermis; ripening in the first, second, or rarely in the third year. Seeds, variable in number (one to twelve) and in shape; usually free, but in one species coalesced. Cotyledons two, or four to six.

The genus comprises about thirty-five species distributed over the northern hemisphere from the Arctic Circle to Mexico and the West Indies, Azores and Canary Islands, Northern Africa, Abyssinia, and the mountains of East Tropical Africa,[2] Himalayas, China, and Formosa.

The genus is divided into three sections :—

I. Leaves always acicular, spreading in whorls of threes, jointed at the base. Buds distinct, and with scale-like leaves. Flowers axillary, diœcious.

§ 1. OXYCEDRUS, Endlicher, *Syn. Conif.* 9 (1847).

Leaves always spreading, never appressed, linear, rigid, usually sharp-pointed, convex and green beneath, whitened above with one or two stomatic bands, entire in margin, without glands, not decurrent on the branchlets, which are glabrous and triangular in section.

Flowers solitary in the axils of the leaves. Staminate flowers, surrounded at

---

[1] Founded on a cultivated specimen of *J. communis*, L., var. *oblonga*, with abnormal fruit.
[2] Here extending south of the equator into the southern hemisphere.

the base by a few scales; stamens in ternate whorls, with oval connectives. Pistillate branchlets, composed of five to eleven ternate whorls of scales, with usually only the apical whorl fertile, each of its scales alternating with one of the three ovules, the tips of which protrude at the apex of the branchlet. Fruit, composed of three or six scales, marked at the apex by three radiating lines or furrows. Seeds, normally three, free and radially disposed in the centre of the berry, with depressions for large resin-glands.

A. *Leaves with one white stomatic band above.*

1. *Juniperus communis*, Linnæus. Europe, North Africa, Extratropical Asia, Canada, and the United States. See p. 1400.

Leaves, $\frac{2}{5}$ to $\frac{3}{5}$ in. long, slightly concave above, no trace of a green midrib being present except in rare cases near the base. In var. *nana*, leaves shorter, $\frac{1}{6}$ to $\frac{1}{3}$ in. long.

2. *Juniperus rigida*, Siebold and Zuccarini. Japan, Korea, Manchuria. See p. 1408.

Leaves $\frac{1}{2}$ to $\frac{3}{4}$ in. long, very slender, deeply concave above with the margins inflexed, forming a narrow median groove.

B. *Leaves with two white stomatic bands above.*
* *Shrubs, with leaves spreading horizontally outwards; branchlets usually not pendulous.*

3. *Juniperus Oxycedrus*, Linnæus. Mediterranean Region, Caucasus, Persia, Madeira. See p. 1409.

Leaves $\frac{1}{2}$ to $\frac{3}{4}$ in. long, gradually tapering from the middle to the sharp-pointed acuminate apex; upper surface with a conspicuous midrib, about half the width of the white bands, which are equal in width to the marginal green bands.

4. *Juniperus macrocarpa*, Sibthorp and Smith. Mediterranean Region. See p. 1412.

Leaves $\frac{3}{4}$ to 1 in. long, tapering from the base to the sharp-pointed acuminate apex; upper surface with a conspicuous midrib, less than half the width of the white bands, which are broader than the marginal green bands.

5. *Juniperus brevifolia*, Antoine. Azores. See p. 1413.

Leaves, oval-linear, very short, $\frac{1}{4}$ to $\frac{1}{3}$ in. long, with a rounded or acute and not acuminate apex; upper surface with a conspicuous midrib, and white bands broader than the green bands.

** *Trees, with leaves directed forwards towards the apices of the pendulous branchlets.*

6. *Juniperus Cedrus*, Webb and Berthelot. Canary Islands. See p. 1414.

Leaves, $\frac{1}{2}$ to $\frac{3}{4}$ in. long, with a conspicuous midrib above, about half the width of the white bands, which are equal in width to the marginal green bands.

7. *Juniperus formosana*, Hayata. China, Formosa. See p. 1415.

Leaves, $\frac{1}{2}$ to $\frac{3}{4}$ in. long; upper surface with a midrib much narrower than the white bands, which are broader than the marginal green bands.

§ 2. CARYOCEDRUS, Endlicher, *Syn. Conif.* 8 (147).

Leaves as in § *Oxycedrus*, but decurrent on the branchlets, which show between the whorls three raised pulvini, separated by grooves.

Staminate flowers, three to six in a head, on a scaly stalk arising in the axil of a leaf; stamens nine to twelve in each flower. Fruit, much larger than in the other sections, composed of six or nine ternate scales. Seeds consolidated into a thick globose three-celled bony mass. This section includes only one species.

8. *Juniperus drupacea*, Labillardière. Greece, Asia Minor, Syria. See p. 1417.

Leaves $\frac{1}{2}$ to $\frac{7}{8}$ in. long, widely spreading, very rigid, sharp-pointed; upper surface with a broad green midrib, deeply furrowed near the base, and two white bands, which are much broader than the marginal green bands.

II. Leaves (*a*) all scale-like and appressed, or (*b*) all acicular, or (*c*) often mixed; never jointed at the base, always adnate to the branchlet. No distinct leaf-buds. Flowers monœcious or diœcious, terminal on short axillary branchlets.

§ 3. SABINA, Spach, in *Ann. Sci. Nat.* xvi. 291 (1841).

Branchlets in most species like those of a Cupressus, terete or tetragonal in section, densely clothed with closely appressed imbricated scale-like leaves, which are usually in four ranks in decussately opposite pairs, or rarely ternate in six ranks, oval or triangular, adnate in the basal half, more or less free at the apex, glandular or marked with a depression on the back, entire or denticulate in margin.

In young plants of all the species, on occasional branches of adult trees of most species, and on all the branches of two species, acicular foliage occurs :— leaves linear-lanceolate, sharp-pointed, nearly appressed or more or less spreading, in whorls of threes or in opposite pairs, whitened on the ventral surface.

Staminate flowers with usually the stamens in opposite pairs, rarely ternate. Pistillate flowers, with usually opposite or rarely ternate scales, each of which bears one or two ovules. Fruit, with usually opposite or rarely ternate scales. Seeds, one to twelve, generally ovoid, with a broad base marked by a large hilum.

A. *Leaves on adult plants always acicular.*

9. *Juniperus recurva*, Buchanan-Hamilton. Himalayas. See p. 1419.

A tree with curved and pendulous branchlets. Leaves in threes, loosely appressed, $\frac{1}{8}$ to $\frac{1}{4}$ in. long, sharp-pointed, greyish green on the dorsal surface, which is channelled in the middle line near the base.

10. *Juniperus squamata*, Buchanan-Hamilton. Afghanistan, Himalayas, China, Japan, Formosa. See p. 1420.

A prostrate shrub. Leaves in threes, appressed or spreading, broader and shorter than in *J. recurva*, green on the dorsal surface, which is channelled from the base to near the apex.

J. procumbens, Siebold, is similar, but has larger leaves, which like the branchlets are glaucous in tint. See p. 1422.

B. *Leaves on adult plants scale-like, except on occasional branches.*
* *Leaves denticulate[1] in margin.*
(a) *Acicular leaves usually in whorls of threes.*

11. *Juniperus Wallichiana*, J. D. Hooker.   Himalayas.   See p. 1423.

Ultimate branchlets, tetragonal, $\frac{1}{25}$ in. in diameter; leaves closely appressed, narrowly ovate, acute, with a conspicuous linear furrow on the back from base to apex.   Acicular foliage usually present, similarly furrowed on the back. Fruit blue, ovoid, with one very large seed.

12. *Juniperus phœnicea*, Linnæus.   Mediterranean Region, Crimea, Canary and Madeira Islands.   See p. 1424.

Ultimate branchlets, terete, $\frac{1}{25}$ in. in diameter; leaves closely appressed, ovate-rhombic, blunt, with an inconspicuous oval depression on the back. Acicular foliage rare on adult trees.   Fruit, yellow or reddish brown, with fibrous yellow flesh, and three to nine seeds.

13. *Juniperus flaccida*, Schlechtendal.   Texas, Mexico.   See p. 1426.

A tree with long pendulous branches.   Leaves ovate-lanceolate, slightly spreading, $\frac{1}{12}$ in. long, sharp-pointed, with a linear sunken gland, exuding resin. Acicular foliage usually present, similarly glandular, mostly ternate, rarely in pairs.   Fruit reddish brown, minutely tuberculate, with six to twelve seeds.

14. *Juniperus pachyphlæa*, Torrey.   Arizona, Texas, New Mexico, Mexico. See p. 1429.

Tree with thick bark, divided into small square scaly plates, unique in the genus. Ultimate branchlets, tetragonal, $\frac{1}{25}$ in. in diameter; leaves appressed, ovate-rhombic, with a depressed oval gland, often exuding resin.   Acicular leaves often present.   Fruit reddish brown, tuberculate on the surface, with four seeds.

(b) *Acicular leaves in opposite pairs.*

15. *Juniperus thurifera*, Linnæus.   France, Spain, Portugal, Sardinia, Morocco, Algeria.   See p. 1427.

Ultimate branchlets, $\frac{1}{20}$ in. in diameter; leaves ovate, appressed but free at their acute or acuminate tips, with a conspicuous glandular depression on the back.   Acicular foliage often present.   Fruit blue, with two to four seeds.

** *Leaves entire in margin.*
(a) *Acicular leaves usually in whorls of threes.*

16. *Juniperus chinensis*, Linnæus.   China, Japan.   See p. 1430.

Ultimate branchlets, $\frac{1}{25}$ in. in diameter, marked with white crosses, due to the pale margins of the leaves, which are appressed, rhombic, obtuse, with an inconspicuous dorsal gland.   Acicular foliage usually present, either ternate or in pairs.   Fruit brown, covered with a white mealy bloom, subglobose, but widest and depressed at the summit, with usually two or three seeds.

17. *Juniperus bermudiana*, Linnæus.   Bermuda.   See p. 1434.

Ultimate branchlets, tetragonal, $\frac{1}{20}$ in. in diameter; leaves closely appressed, ovate, obtuse at the incurved apex, with a conspicuous dorsal furrow.   Acicular leaves usually present, furrowed on the back.   Fruit dark blue, covered with a glaucous bloom, with two or three seeds.

[1] The denticulations on the leaves are very minute, and can only be seen with a considerable magnifying power.

(b) *Acicular leaves in opposite pairs.*

18. *Juniperus virginiana*, Linnæus.   North America.   See p. 1435.

A tree.   Ultimate branchlets very slender, $\frac{1}{30}$ in. in diameter; leaves appressed, ovate, acute or acuminate, with often a small oval depression on the back. Acicular foliage usually present.   Fruit bluish, very small, $\frac{1}{4}$ in. in diameter, with one or two seeds.

19. *Juniperus Sabina*, Linnæus.   Europe, Caucasus, North America.   See p. 1443.

A shrub, distinguishable by the strong disagreeable odour of the foliage, when bruised.   Ultimate branchlets very slender, $\frac{1}{30}$ in. in diameter; leaves appressed, ovate, acute or acuminate, with a conspicuous depressed dorsal resin-gland.   Acicular foliage often present.   Fruit bluish, very small, $\frac{1}{5}$ in. in diameter, with usually two seeds.

20. *Juniperus excelsa*, Bieberstein.   Balkan States, Crimea, Syria, Asia Minor, Caucasus.   See p. 1446.

A tree.   Ultimate branchlets very slender, $\frac{1}{30}$ in. in diameter; leaves appressed, ovate-rhombic, acute or obtuse, marked on the back with a depressed gland. Acicular foliage rarely present.   Fruit dark purplish brown, $\frac{1}{3}$ to $\frac{1}{2}$ in. in diameter, with about six seeds.

The following species, of which I have seen no living adult specimens in cultivation, may be briefly mentioned, on account of their economic importance.

I. *Juniperus procera*, Hochstetter, *ex* Endlicher, *Syn. Conif.* 26 (1847).

A tree, widely spread throughout Abyssinia, Somaliland, and the highlands of equatorial Africa.   In botanical characters it closely resembles *J. excelsa.* Hutchins[1] gives a good account of this species in British East Africa, where it is called cedar and is a timber tree of great value, occasionally attaining an enormous size.   He figures a very old tree dividing into two stems, the largest of which is 110 ft. high and 12 ft. in diameter.   The wood is now imported into Hamburg in considerable quantity (about 2500 tons in 1911 from German East Africa), and is used for the same purposes as *J. virginiana*, selling in London at about 4s. per cubic foot.   *J. procera* is occasionally seen in the juvenile stage in conservatories in botanic gardens; but is not hardy in the open air in Britain. Koch[2] mentions a reputed tree of this species, growing in 1873, in Simon-Louis's nursery at Metz, which we cannot now identify.

II. *Juniperus occidentalis*, W. J. Hooker, *Fl. Bor. Amer.* ii. 166 (1839).

A tree, attaining 60 ft. in height, and 9 ft. in girth, readily distinguishable by the stout ultimate branchlets, covered with closely appressed scale-like leaves, arranged in six ranks, in whorls of threes, denticulate in margin, and conspicuously glandular on the back.   Fruit subglobose, $\frac{1}{3}$ in. long, bluish with a glaucous bloom, with resinous juicy flesh, containing two or three seeds.

This species occurs on mountain slopes and high prairies in western North America from Washington and Idaho to the Sierras of California.

I have seen no specimens of this species in the living state in England; and

---

[1] *Report of the Forests of British East Africa*, 20, 145 (1909), and *Scot. Geog. Mag.* 1909, p. 351.
[2] *Dendrologie*, ii. pt. ii. 132 (1873).

it is doubtful if it has ever been introduced, as Carrière states[1] that all the reputed plants of *J. occidentalis* which he saw were very doubtful; and Kent[2] does not appear to have recognised this species.   (A. H.)

JUNIPERUS COMMUNIS, Common Juniper

Juniperus communis, Linnæus, *Sp. Pl.* 1040 (1753); Loudon, *Arb. et Frut. Brit.* iv. 2489 (1838); Parlatore, in De Candolle, *Prod.* xvi. 2, p. 479 (1868); Bentley and Trimen, *Medicinal Plants*, iv. t. 255 (1880); Boissier, *Flora Orientalis*, v. 707 (1881); Willkomm, *Forstliche Flora*, 261 (1887); J. D. Hooker, *Fl. Brit. India*, v. 646 (1888); Köppen, *Geog. Verbreit. Holzgewächse Russlands*, ii. 396 (1889); Jack, in *Bot. Gaz.* xviii. 369, pl. 33 (1893); Sargent, *Silva N. Amer.* x. 75, t. 516 (1896), and *Trees N. Amer.* 86 (1905); Mathieu, *Flore Forestière*, 514 (1897); Ascherson and Graebner, *Syn. Mitteleurop. Flora*, i. 243 (1898); Franchet, in *Journ. de Bot.* 1899, p. 264; Kent, Veitch's *Man. Conif.* 170 (1900); Brandis, *Indian Trees*, 694 (1906); Kirchner and Schröter, *Lebengesch. Blütenpfl. Mitteleuropas*, i. 287 (1906).

A shrub or low tree, occasionally, however, attaining 40 ft. or more in height. Bark reddish brown, at first smooth, ultimately peeling in thin papery shreds. Young branchlets slender, triquetrous, with three projecting narrow ridges between the whorls of leaves. Buds about $\frac{1}{8}$ in. long, with a few loose ovate acuminate green scales. Leaves all acicular, persistent for three years, sessile, spreading, variable in length, averaging $\frac{2}{5}$ to $\frac{3}{5}$ in. long, linear-subulate, gradually tapering from near the jointed and swollen base to the slender spine-like apex; upper surface concave, with usually a single continuous broad white longitudinal stomatic band, no green midrib being present, except occasionally near the base, in which case the stomatic band is divided into two parts for a short distance; lower surface bluntly keeled, with usually a slight furrow in the middle line; resin-canal solitary, situated in the substance of the leaf below the central fibro-vascular bundle.

Flowers diœcious, rarely monœcious, formed in autumn in the axils of the lower whorls of leaves on the current year's shoot, opening in the following spring. Staminate flowers solitary, cylindrical, $\frac{1}{8}$ in. long, yellow; stamens in five or six whorls, three in each whorl, with ovate acute connectives, each of which bears three or four pollen-sacs. Pistillate flowers solitary, green, $\frac{1}{12}$ in. long; scales in six or seven whorls, three in each whorl; the upper three scales minute at the time of flowering, and alternating with three fleshy tubular ovules; the lower scales larger, ovate-acuminate, empty.

Fruit[3] ripening in the second or third year,[4] small and green in the first year; when mature, bluish or almost black, covered with a slight bloom, globose or slightly longer than broad, $\frac{1}{3}$ in. in diameter, on a short scaly stalk; smooth, marked at the summit by three very short radiating lines, below which are three shallow depressions overhung by three minute mucros, indicating the three scales of which

---

[1] *Conif.* 40 (1867).   [2] Veitch's *Man. Conif.* 178 (1900).

[3] Occasionally the three scales, of which the fruit is composed, do not unite at the summit of the fruit, but gape, showing the seeds inside. This abnormality, var. *thiocarpos*, Ascherson and Graebner, *op. cit.* 245, was described as a distinct genus, *Thuiæcarpus juniperinus*, by Trautvetter, *Imag. Pl. Ross.* 11, t. 6 (1844).

Another abnormality, var. *coronata*, Sanio, in *Deut. Bot. Monatsschrf.* i. 51 (1883), is occasionally met with, when the points of the scales unite together and form a projection at the summit of the fruit.

[4] Cf. Jack, in *Bot. Gaz.* xviii. 369-375, plate 33 (1893), who states that in America this species does not ripen its fruit till the autumn of the third year after blossoming.

the fruit is composed. Seeds usually three, rarely two, immersed in a soft resinous mealy sweet pulp; light-brown, elongated-ovoid, triquetrous, narrowed at the summit, which is compressed into a thin transverse ridge; broad at the base, above which on the sides are a few large depressions for resin-glands. Seedling, with two cotyledons, the primary leaves and those of the second year being arranged in whorls of fours.

VARIETIES

The common juniper in the wild state displays a considerable amount of variation in the length and breadth of the leaves, and in the size and shape of the fruit; and numerous varieties based on these characters have been named by Ascherson and Graebner and other botanists, most of which need not be even mentioned here. The most remarkable varieties are the following :—

1. Var. *nana*, Loudon, *op. cit.* 2489 (1838); Ascherson and Graebner, *op. cit.* 246 (1898); Kirchner and Schröter, *op. cit.* 303 (1906).

Var. γ, Linnæus, *Sp. Pl.* 1040 (1753).
Var. *saxatilis*, Pallas, *Fl. Ross.* ii. 12, t. 54 (1788).
Var. *montana*, Solander, in Aiton, *Hort. Kew.* iii. 414 (1789).
Var. *alpina*, Gaudin, *Fl. Helv.* vi. 301 (1830).
Var. *depressa*, Pursh, *Fl. Amer. Sept.* ii. 646 (1814).
Var. *sibirica*, Rydberg, in *Contrib. U.S. Nat. Herb.* iii. 533 (1896).
Juniperus sibirica, Burgsdorf, *Anleit.* ii. No. 272 (1787), and ii. 127 (1790).
Juniperus nana, Willdenow, *Berl. Baumz.* 159 (1796), and *Sp. Pl.* iv. 854 (1805).
Juniperus alpina, J. E. Gray, *Nat. Arr. Brit. Pl.* ii. 226 (1821).
Juniperus depressa, Rafinesque, *Medic. Pl.* ii. 13 (1830).
Juniperus canadensis, Loddiges, *Cat.* 1836, p. 47; Loudon, *op. cit.* 2490 (1838).
Juniperus pygmæa, Koch, in *Linnæa*, xxii. 302 (1849).

A low prostrate spreading shrub, seldom more than a foot in height. Branchlets stouter than in the type. Leaves shorter, less spreading, $\frac{1}{6}$ to $\frac{1}{3}$ in. long, $\frac{1}{24}$ to $\frac{1}{12}$ in. broad, linear-subulate, gradually tapering to the spine-like apex; upper surface very concave, with a continuous white stomatic band; lower surface convex, bluntly keeled. Fruit blue, glaucous, globose, $\frac{1}{5}$ in. in diameter; seeds one, two, or three, smaller than in the type.

This, which is the alpine and arctic variety[1] of *J. communis*, is considered by many botanists to be a distinct species; but Kirchner and Schröter adduce good reasons for supposing it to be only a form, due to climatic conditions. Connecting links[2] between the type and var. *nana* are found both in the Alps and in Lapland; and experimental sowings of the latter at Berlin and Zurich gave seedlings which resembled the common juniper in all respects. Similarly plants of common juniper from Fontainebleau, which were cultivated by Bonnier[3] on Mont Blanc at 6800 feet, assumed the habit of var. *nana* in three years.

[1] *J. dealbata*, Douglas, *ex* Gordon, in *Gard. Chron.* 1842, p. 652, is *J. communis*, var. *nana*. Cf. p. 1436, note 2.
[2] These have been distinguished as var. *intermedia*, Sanio, in *Deut. Bot. Monatsschrf.* i. 51 (1883); and as *J. intermedia*, Schur, in *Verh. Siebenb. Naturw.* v. ii. 169 (1850).
[3] *Assoc. Franc. Avanc. Sci., Compt. Rend.* 1892, pt. ii. 521.

2. Var. *hemisphærica*, Parlatore, *Fl. Ital.* iv. 83 (1867).

Juniperus hemisphærica, Presl, *Delic. Pragens.* 142 (1822).

Resembling var. *nana* in habit, but with longer leaves and larger fruits. It is said by Parlatore to occur on Mount Etna, where it grows[1] on sterile soil between 5000 and 7000 ft. elevation ; in the mountains of Calabria and Greece ; and on the Djurdjura range in Algeria.

Var. *echiniformis*,[2] which was introduced into gardens by Rinz of Frankfort, is usually identified with var. *hemisphærica*, but this is extremely doubtful. This cultivated variety, which is commonly known as the "hedgehog juniper," forms a globose bush, 1 to 2 ft. high, with densely crowded branches, branchlets, and leaves.

3. Var. *suecica*, Aiton, *Hort. Kew.* v. 414 (1813).

Var. *β*, Linnæus, *Sp. Pl.* 1070 (1753).
Var. *fastigiata*, Parlatore, in De Candolle, *Prod.* xvi. 2, p. 479 (1868).
Juniperus suecica, Miller, *Gard. Dict.* ed. 8, No. 2 (1758).
Juniperus hibernica, Loddiges, *ex* Loudon, *op. cit.* 2489 (1838).

Fastigiate in habit, with ascending branches, short leaves, and oblong fruit. This is occasionally found wild in the forests of Scandinavia and of east Prussia, where it often is a tree, 30 ft. to 40 ft. in height, resembling in appearance the fastigiate Mediterranean cypress. So far as I can learn, it has not been noticed[3] in the British Isles ; but many wild Junipers have a narrow columnar habit.

In gardens the fastigiate juniper is often seen under the name of Irish juniper, perhaps so called because it has the habit of the Irish yew. It is a neat shrub until it becomes thin and shabby in foliage, and often attains 10 to 15 ft. in height. The best specimen which we have seen is one at Westonbirt, about 25 ft. high in 1910. Another at Abercairney was about 20 ft. high in 1911.

4. Var. *compressa*, Carrière, *Conif.* 22 (1855).

Fastigiate, with short branches, and densely crowded very short leaves, forming a small bush, rarely exceeding 3 ft. in height. This is said to be tender, and is possibly of southern origin, as it was identified by Koch[4] with *J. hispanica*,[5] which was introduced into cultivation by Booth of Hamburg.

5. Var. *oblonga*, Loudon, *Arb. et Frut. Brit.* iv. 2489 (1838).

Var. *caucasica*, Endlicher, *Syn. Conif.* 16 (1847).
Juniperus oblonga, Bieberstein, *Fl. Taur. Caucas.* ii. 426 (1808) and iii. 634 (1819).

Leaves $\frac{3}{4}$ in. long. Fruit small, $\frac{1}{6}$ in. in diameter, oblong ; seeds solitary or two. This variety has been found growing wild in the Caucasus, and scarcely deserves mention, as individuals with the same characters are not uncommon elsewhere. According to Loudon, it was introduced from Paris into the Horticultural Society's Garden at Chiswick about 1826, and formed a bush 4 ft.

---

[1] Schouw, in *Ann. Sc. Nat.* iii. 243 (1845).
[2] *J. Oxycedrus echiniformis*, Knight, *Syn. Conif.* 11 (1850). Cf. also Koch, *Dendrologie*, ii. pt. ii. p. 115 (1873).
[3] Mr. R. A. Phillips has never seen the fastigiate form growing wild in Ireland, and knows of no records of its existence.
[4] *Dendrologie*, ii. pt. ii. p. 115 (1873).
[5] *J. hispanica*, Booth, *ex* Endlicher, *Syn. Conif.* 15 (1847).

high after being planted twelve years.[1] It has not been recognised by us as now in cultivation.

6. Var. *variegata aurea*, Carrière, *Conif.* 19 (1867).

Young shoots golden yellow, becoming green in the following year.

7. Var. *cracovia*, Knight, *Syn. Conif.* 11 (1850).

Juniperus cracovia, Loddiges, *ex* Loudon, *op. cit.* 2490 (1838).

This,[2] which was said to have been introduced from Poland, was described as a robust plant, with pendulous terminal branchlets. It appears to be very rare in England, the only specimen which we have seen being a tree at Bicton, about 25 ft. high, which has been much broken by the wind.

DISTRIBUTION

The common juniper is more widely distributed than any other tree or shrub in the northern hemisphere. In Europe it is common throughout the northern and central part of the continent, and occurs in the mountains of the countries bordering on the Mediterranean; while the alpine form is reported[3] to exist on the Djurdjura range in Algeria. Eastwards *J. communis* spreads through Siberia to Kamtschatka and the Kurile Isles, and through Turkestan to the Thianshan range in Mongolia; but it is not found in Japan,[4] where it is represented by *J. rigida*. It is also met with in Asia Minor, the Caucasus, Persia, Afghanistan, and the western Himalayas, as far east as Garhwal and Kumaon. In North America it extends from far north in Alaska and in Canada, southwards on the east to New Jersey, Pennsylvania, Michigan, and western Nebraska; in the Rocky Mountains, to New Mexico; and on the Pacific coast, to northern California; and, according to Sargent, only becomes a tree on the limestone hills of Illinois.

Kirchner and Schröter attempt to distinguish with great accuracy the limits of distribution of the common juniper and its var. *nana*; but the altitudes given show that the records of the two forms are much confused. In a general way, it may be stated that var. *nana* extends farther to the north in the plains of the holarctic region, and ascends higher in the mountains to the southward. It is said to occur at the following altitudes: in the Sierra Nevada, between 5400 and 9700 ft.; in the Alps, usually between 6000 and 8000 ft., attaining its highest point on Mt. Rosa at 11,900 feet; in the Caucasus, between 7600 and 9400 feet; in the western Himalayas, between 4300 and 12,000 feet.

J. communis, while widely distributed throughout Russia, is totally absent from the south, in the provinces where the " black earth " formation prevails; and is most common in the Baltic provinces, where in Courland, north-west Livland, Esthonia, and the islands of Oesel and Dagö, it covers large tracts of peaty sand. Kerner[5]

---

[1] Gordon, in *Gard. Chron.* 1842, p. 652, says that this variety is a robust spreading plant, 2 ft. or 3 ft. high, with leaves like the common juniper, but long and slender; fruit dark purple when ripe, and very small.

[2] Gordon, in *Gard. Chron.* 1842, p. 652, says that this variety is not different from *J. communis*.

[3] Chabert, in *Bull. Soc. Bot. France*, xxxvi. 30 (1889), who also states that the common form, 6 to 8 ft. in height, occurs, but is very rare, in Kabylia.

[4] Kawakami, in *Tokyo Bot. Mag.* 1900, p. 111, records it, however, for Rishiri Island, on the west coast of Yezo.

[5] *Pflanzenleben der Donauland* (1863).

long ago described a remarkable juniper forest in Hungary, between the Danube and the Thiess, north of the Bacser Canal, where on loose sand it forms extensive thickets of bushes, about 6 ft. in height. The juniper is also widely spread in certain regions as undergrowth in the pine forests, as in those of *P. sylvestris* in the Alps, of *P. austriaca* in Austria, Bosnia, and Servia, and of *P. Peuke* in Macedonia. It is very common on all kinds of heath land in northern Germany; and is elsewhere found on sunny rocky mountain slopes, as in the Jura, the Alps, etc.   (A. H.)

The juniper in Norway attains a very much larger size than it ever does in this country, many remarkable trees being mentioned by Schübeler, *Virid. Norv.* i. pp. 357-369. The cypress-like form, var. *suecica*, is not uncommon in a wild state; and he gives excellent illustrations of it, and mentions a tree found by Prof. Sexe in the Hardanger Fjord which was 40 ft. high. He figures a tree leaning very much on one side, which has the habit of a stunted spruce, growing near Vossevangen, which was 38 ft. high. Even as far north as Saltdalen (lat. $67\frac{1}{2}°$) it attains very large dimensions. Forstmeister Niewjaar told me that he had sent one from there to the Paris Exhibition in 1878 of great size, and knew of one still living which was eleven metres high and about one metre in girth. Schübeler figures a tree with a short thick stem 7 ft. 1 in. in girth, and a spread of branches 26 ft. in diameter, at Hohl, near Christiania, and another with a beautiful straight clean trunk dividing into three stems.

Such trees as those above mentioned are of great age; some of which the rings were counted were as follows :—

(1) 130 years old; longest diameter, 17 in.; shortest diameter, $14\frac{1}{2}$ in. at the base.
(2) 114 years old; longest diameter, $14\frac{1}{2}$ in.; shortest diameter, $11\frac{1}{2}$ in. at the top of the log.
(3) 150 years old; 33 ft. long; 4 ft. 3 in. in girth at the base, 3 ft. 3 in. at 5 ft., 2 ft. 8 in. at 10 ft., 2 ft. 1 in. at 15 ft., 1 ft. 6 in. at the top, where there were 47 annual rings; the total height of this tree was 37 to 38 ft.
(4) 216 years old; 14 in. in diameter.
(5) 300 years old; $12\frac{1}{2}$ in. in diameter.

Schübeler, *Tillæg til Virid. Norv.* 102 (1891) states that Lensmand Lund of Stryn parish, in North Bergenhus district (61° 55′ lat.) sent him a sketch of a juniper supposed to be the tallest and thickest in Norway. It grows at an elevation of 1200 to 1500 ft., and is 45 ft. high, with the trunk 6 ft. 3 in. in girth, and the crown $9\frac{1}{2}$ ft. in diameter. It is supposed to be 195 years old.

In the Swedish forestry journal, *Skogsvårdsfor. Tidskrift*, 1911, p. 132, fig. 1, a photograph is reproduced of a remarkable fastigiate juniper, growing in the parish of Tyrserum, south of Linköping in Sweden, which is 37 feet high.

Wittmack,[1] quoting from the *Tägl. Rundschau*, says that a juniper, one of the oldest trees in the world, grew in Kokenberg parish in Livland. It was so large that two men were unable to span it near the base. The stem when cut down was deposited in the Riga Museum, and is said to have shown 2000 annual rings.

---

[1] *Gartenflora*, xxxvi. 139 (1887). Willkomm, *Forstliche Flora*, 263 (1887) quotes the same account—except that the name of the parish is given as Ermas—from *Oesterr. Forstzeitung*, 1885, p. 137.

According to Koernicke and Roth[1] the juniper is the characteristic conifer of the Eifel Mountains at an elevation of 1000 to 2000 ft. Here it grows on all kinds of soil, but best on the heaths of grauwacke, where it often forms large groups and attains about 25 ft. in height. The plates show how variable is the form of the tree here, some being round bushes, and some narrow pyramids. Prof. T. Schube, of Breslau,[2] figures a tree in a forest meadow at Kuchelberg, about 23 ft. high, and 5 ft. 5 in. in girth at a foot above the ground, which he thinks is the finest wild juniper in Silesia.

CULTIVATION

Though it has now been to a great extent displaced by modern introductions, the juniper was a great favourite as a garden shrub in former times, and according to Evelyn[3] may be formed into most beautiful and useful hedges. He says : " The discreet loosening of the Earth about the Roots also, makes it strangely to prevent your Expectations, by suddenly spreading into a bush fit for a thousand pretty Employments. My Brother having cut out of one only tree, an Arbour capable for three to sit in ; it was at my last measuring seven foot square, and eleven in height, and would certainly have been of much greater altitude and farther spreading, were it not continually kept shorn. But what is most considerable, is the little time since it was planted, being yet hardly ten years, and then it was brought out of the Common a slender bush of about 2 ft. high." He adds : " I have raised them abundantly of their seeds (neither watering nor dunging the soil) which in two months will peep." My experience in raising the seeds of juniper, which agrees with that of Boutcher, and also with Loudon's statement, is that the seeds lie for one year before germinating, even when freshly sown. They keep for some years in the berry without losing their vitality.

REMARKABLE TREES

Though the common juniper seldom in Scandinavia and never in Great Britain attains the size of a timber tree, yet it is so striking a feature in the vegetation of some English hillsides that it cannot be passed by in silence. One of the best instances of its growth is seen on a dry oolite hillside at Hilcot, about two miles from Colesborne, on my own property. (Plate 348.) How old these trees are and how they originated, it is impossible to say ; and though the juniper is indigenous in the Cotswold hills, this is the only hillside I know of in the county where it is abundant, and as the trees have not perceptibly increased in size during the last fifty years, they must be of great age. The largest are 20 to 25 ft. high, but none have developed a single stem thicker than 6 to 9 in. in diameter, and though they produce berries freely, natural reproduction is entirely prevented by rabbits. In young plantations, however, protected by wire-netting, not far away, there are seedlings in the grass, which have grown from seeds dropped by birds. On many of the chalk-hills of the Chilterns, the South Downs, and in Surrey and Hants, scattered

---

[1] In Karsten and Schenck, *Vegetationsbilder*, v. tt. 5, 6 a and b (1908).
[2] *Mitt. Deut. Dend. Ges.* 1910, p. 47.   [3] *Silva*, 136 (1679).

bushes of juniper mixed with yew, holly, and thorn appear, but rarely exceed 8 to 10 ft. in height, and natural reproduction is in my own experience seldom found. Mr. Dykes informs us that many junipers grow on Shackleford Heath, and on the south side of the Hog's Back near Puttenham.

In a valley called Juniper Valley, about three miles east from Godalming, formerly the property of Col. H. H. Godwin-Austen, the juniper attains a greater size than I have seen it anywhere else in a state of nature in Great Britain. Col. Godwin-Austen tells me that a map in his possession shows the land to have been a sheep walk in 1733. The soil is a deep sand on the lower greensand formation, at an elevation of 350 ft., and is overgrown with rank ling and bracken. The trees are mostly branched near the ground, and attain 15 to 30 ft. in height, some having a fastigiate habit like the Irish juniper. They are mostly damaged and broken by heavy snow; and I could find no young seedlings.

H. Speight, *Romantic Richmondshire*, 239 (1897), states that " Down to the beginning of the last century there were many hundred acres of juniper and briar in the townships of Reeth, Helaugh, and Muker. The chips at one time were extensively used for fumigating, and during seasons of plague and sickness no house was found without them. The berries were used as a spice. The plant grows best on open elevated limestone country, and flourished amazingly in upper Swaledale. There are acres of it about Harkeside, above Marden Castle, and elsewhere. In Wensleydale it occurs but sparingly."

At Merton, Norfolk, there is a remarkable plantation of juniper which covers 14½ acres, and consists of numerous shrubs very variable in appearance, and averaging about 20 ft. high. The smallest are about 7 ft. high, while the majority range from 15 to 20 ft.; the tallest being 25 ft. This was planted about 1845. In a neighbouring plantation there is a fine specimen 35 ft. high.

The largest recorded juniper in England is one which grew at Farnham Castle, Surrey, and was said by Loudon to be of the Swedish variety, and 40 ft. high, but on a recent visit to Farnham I could find no trace of this tree. Loudon[1] figures a tree growing in a birch wood near Farningham which was 20 ft. high and 4 ft. in girth in 1838; but of this also I can find no trace.

At Langley Park, Norfolk, I measured a very fine bush about 20 ft. high, and 43 paces round the branches; and at Westonbirt there is the finest specimen I know of the fastigiate form, measuring 26 ft. high in 1909.

In Scotland the juniper is less common in a wild state than formerly, but in certain districts is still abundant. Mr. T. Cathie, forester at Aberuchill, Comrie, Perthshire, informs me that a tree no less than 40 ft. high was blown down near there in 1904, but that the tallest now living were only 17 ft. high.

The finest I have seen wild in Scotland are in the forest of Guisachan, near the house of the forester Donald Kennedy, who told me that thirty years ago some of the thickets of juniper were so dense that he could not pass through them. Now, however, many are broken down by the snow, and injured or killed by the stags cleaning their horns on them; and the tallest that I saw, on a grassy mound,

---

[1] *Arb. et Frut. Brit.* iv. 2492 (1838).

were not over 20 ft. high. A specimen log in the Forestry Museum, Cambridge, sent by the Earl of Portsmouth, shows a section about 7 in. in diameter at five feet from the ground, displaying 120 annual rings. Owing to decay, the rings at the base could not be counted. In Glen Urquhart there are large thickets of scrubby junipers, which in Gaelic are called *asten*, but I saw no tall ones in this locality. In Perthshire it is local and, according to Buchanan White,[1] commonest between Dunkeld and Ballinluig. It ascends to 2600 ft. in Breadalbane.

On Ford, a farm occupied by Mr. Wallace, near Capenoch, Dumfriesshire, there is a hillside facing south, on the Silurian formation, covered with junipers over an area of thirty to forty acres, which, as I was told by Mr. Hugh Gladstone who showed them to me, have been in their present condition as long as any one can remember. Prof. R. Wallace tells me that an old man, who died last year at a great age, said that he could remember them when they were quite small; and Mr. Paterson of Craigdarroch says that no class of stock would eat juniper, though Herdwick sheep are said to do so when they can get no other food. At Capenoch, though the ground is grazed by sheep and cattle, there are many young seedlings coming up among the heather. The largest bushes here were 15 to 18 ft., but the majority were 6 to 10 ft. high, and some had the fastigiate habit of the so-called Irish juniper. Except a few ash, holly, and rowan, I noticed no other trees but juniper on this hillside. Col. Kennedy of Milton Park Lodge, Dalry, Kirkcudbrightshire, says that juniper is very scarce in that district, and that in Inverness-shire, where it is common near Kingussie, grouse are very fond of taking shelter in it in hot weather.

The juniper is common as a wild plant in the west of Ireland, being recorded by Praeger[2] for Donegal, Sligo, Mayo, Galway, Tipperary, and Kerry, where it is confined to the mountains and lake shores. Mr. R. A. Phillips writes to us as follows:—" I doubt if var. *nana*[3] is really a variety or only a prostrate state of *J. communis*. In east Galway, on the limestone at Gort, and on the shores of Loch Derg near Portumna, where the species is abundant, it is an upright bushy plant when growing in fairly deep soil or in sheltered hollows; but close by on exposed rocks or bare ground it is perfectly prostrate. I have always failed in these localities to distinguish any difference in the leaves or fruits of the two forms. On the non-calcareous mountains of west Galway, Cork, and Kerry, principally Old Red Sandstone, the juniper is usually prostrate, and here its leaves are broader and more imbricate, with sometimes oval fruit, than the plant of the limestone; but even this form is in some sheltered lowland spots an upright bush. So far as I have seen, the juniper will not live in shade, and never forms undergrowth. Near Gort, the prostrate form covers large areas; and near Portumna, the upright form occurs in small groves, but never in such quantity as to deserve the name of a wood. The largest specimens which I have seen were about 12 to 15 ft. high, girthing near the ground 18 to 24 in." The finest that I have seen in Ireland is in the grounds of

[1] Buchanan White, *Flora of Perthshire*, 282 (1898). [2] In *Proc. Roy. Irish Acad.* vii. 288 (1901).
[3] Mr. Phillips is inclined to think that in Ireland, typical *J. communis*, as regards leaves and fruit, is a *calcicole*; whereas var. *nana* is a *calcifuge*.

Woodstock, Kilkenny, where in 1909 I measured a tree with four stems each measuring 3 to 4 ft. in girth and about 30 ft. high.

TIMBER

The wood of the juniper is whitish brown, hard, and takes a very good polish, but owing to its being usually too small for any but local uses is not often seen. It is, however, so strong and durable that the stems are commonly used for railway and other fences in Norway and Sweden, and are exported to Denmark from Sweden for the same purpose.

The fence of the Dyrhave or Royal Deerpark, near Copenhagen, is made from straight juniper poles about 3 to 4 in. in diameter, fixed on oak posts and rails. When I saw this fence in 1887, I was informed by the late Mr. O. Benson of Copenhagen that it had been erected about 100 years. I was so much struck by its appearance and durability that in 1904 I imported, through the kind assistance of Mr. E. Nilson of the Swedish Royal Forest Service, 1000 juniper poles, 2 metres long, at a cost of £15 free on board, and have put them up to fence off a part of my own deerpark for planting. The fence is made with stout oak posts, 4 yards apart, and the poles are kept in place by three strands of strained galvanised wire without crossbars, strong wire netting being fixed on the lower half to keep rabbits out and deer from getting their heads between the poles. It forms a strong, cheap, and probably a very durable deer fence.

Though the berries of the juniper are not now valued in medicine as much as they were in Evelyn's time,[1] yet they are still used for flavouring gin,[2] which owes its diuretic quality to them. An essential oil is also distilled from them. The berries are largely collected in the south-east of France and in Hungary for the wholesale druggists.[3]   (H. J. E.)

JUNIPERUS RIGIDA

Juniperus rigida, Siebold and Zuccarini, *Fl. Jap.* ii. 56, t. 125 (1844); Franchet and Savatier,
    Enum. Pl. Jap. i. 471 (1875); Masters, in *Journ. Linn. Soc. (Bot.)* xviii. 496 (1881), and
    xxvi. 543 (1902); Kent, Veitch's *Man. Conif.* 188 (1900); Shirasawa, *Icon. Ess. Forest. Jap.*
    i. t. 12, figs. 1-13 (1899); Komarov, *Fl. Mansh.* i. 207 (1901).

A small tree, attaining in Japan 20 to 30 ft. in height, often a low spreading bush. Bark thin and scaly. Young branchlets triquetrous, with three projecting ridges, becoming terete and scaly in the fourth year. Leaves all acicular, persistent three or four years, spreading, in whorls of threes, linear-subulate, $\frac{1}{2}$ to $\frac{3}{4}$ in. long, about $\frac{1}{25}$ in. broad, tapering from the middle to the very sharp cartilaginous apex, swollen

---

[1] Evelyn, *Silva*, 130 (1670) says :—" The berries afford (besides a tolerable pepper) one of the most universal remedies in the world to our crazy forester. The berry swallowed only, instantly appeaseth the worst collique, and in decoction most sovereign against an inveterate cough. They are of rare effect being steeped in beer. The water is a most singular specifique against the gravel in the reins."

[2] Bentley and Trimen, *Medicinal Plants*, t. 255 (1880), give a good account of the medicinal uses of juniper, and state that the gin ordinarily distilled in England is flavoured with oil of turpentine, whereas gin, made in Holland, is slightly flavoured with juniper berries, two pounds of berries being used to 100 gallons of gin.

[3] Flückiger and Hanbury, *Pharmacographia*, 626 (1879).

and jointed at the base; upper surface deeply concave with the margins inflexed, the narrow median groove whitened with a stomatic band scarcely so wide as the green margins on each side of it; lower surface green, prominently keeled, without glands. Buds ovoid, minute, with sharp-pointed scales, which persist brown and withered at the ends of the branchlets of the second year; lateral buds about $\frac{1}{25}$ in. long.

Flowers diœcious. Fruit ripening in the second year, globose, $\frac{1}{3}$ in. long, with six bracts in two whorls at the base, on a very short stalk, scarcely $\frac{1}{12}$ in. long; smooth, purplish brown, composed of three scales, separated at the apex by three radiating lines, each scale glaucous and with a minute mucro near the top. Seeds usually three, rarely two, in each berry, triquetrous, with three sharply angled sides, converging to a narrow apex, and marked near the base with three or four deep pits containing resin.

This species is a native of Japan,[1] where, according to Sargent,[2] it is usually a small tree or spreading bush. It is common on the barren land near Gifu, and is generally distributed at low elevations in central Japan, but usually is only found growing on dry sterile gravelly soil. Elwes saw it on dry hillsides in Kiushu, where it did not exceed 15 to 20 feet in height, and had usually a fastigiate habit. It is much cultivated by the Japanese, especially in temple gardens. *J. rigida* also occurs in Korea and in the provinces of South Ussuri and Kirin in Manchuria, where it often grows isolated in rocky situations; but it does not seem to occur in China.[1] It was introduced[3] into England in 1861 by J. Gould Veitch, who found it in the Hakone Mountains. Maximowicz[4] sent living plants in 1864 to the botanic garden at St. Petersburg, of a form, which he called var. *filiformis*; but this, if distinct and not a mere juvenile state, is unknown in England.

This species appears to be rather rare in gardens; but there are specimens at Kew, Tortworth, Casewick, Chipping Campden, Highnam, and other places in England, and at Hamwood and Glasnevin in Ireland. It nowhere attains more than 20 ft. in height, but it often produces fruit abundantly. (A. H.)

JUNIPERUS OXYCEDRUS

Juniperus Oxycedrus, Linnæus, *Sp. Pl.* 1038 (1753); Loudon, *Arb. et Frut. Brit.* iv. 2494 (1838); Parlatore, in De Candolle, *Prod.* xvi. 2, p. 477 (1868); Boissier, *Fl. Orient.* v. 707 (1881); Mathieu, *Fl. Forest.* 516 (1897); Ascherson and Graebner, *Syn. Mitteleurop. Fl.* i. 247 (1898); Kent, Veitch's *Man. Conif.* 179 (1900); Rikli, in Kirchner, Loew, and Schröter, *Lebenges. Blütenpfl. Mitteleurop.* i. 309 (1906).
Juniperus tenella, Antoine, *Cupress. Gattung.* 20, tt. 27, 29 (1857).
Juniperus Marschalliana, Steven, in *Bull. Soc. Nat. Mosc.* xxix. 244 (1856), and xxx. 397 (1857).
Juniperus heterocarpa, Timbal-Lagrave, *ex* Loret et Barrandon, *Fl. Montpellier*, ii. 610 (1876).
Juniperus Biebersteiniana, Koch, *Dendrologie*, ii. pt. ii. 112 (1873).

---

[1] The plant, collected by David in the Ourato Mountains in Mongolia, and referred to *J. rigida* by Franchet, *Pl. David.* i. 295 (1884), is considered by M. Hickel to be *J. communis*. Wilson's No. 370, collected in Hupeh, and referred to *J. rigida* by Masters, in *Journ. Bot.* xli. 268 (1903) is *J. formosana*. A plant, collected by Giraldi in Shensi, and referred to *J. rigida* by Beissner, in *Mitt. Deut. Dend. Ges.* 1897, p. 216, is also probably *J. formosana*.
[2] *Forest Flora of Japan*, 78 (1894).   [3] *Hortus Veitchii*, 340 (1906).
[4] Bretschneider, *Hist. Europ. Bot. Disc. China*, i. 610 (1898).

A shrub or small tree, occasionally attaining 30 ft. in height and 10 to 12 ft. in girth. Branchlets angled, becoming pale brown and scaly in the third year. Leaves all acicular, in alternate whorls of threes, spreading, linear, $\frac{1}{2}$ to $\frac{3}{4}$ in. long, $\frac{1}{24}$ to $\frac{1}{16}$ in. broad, gradually tapering from the middle to the acuminate cartilaginous point, swollen and jointed at the slightly narrowed sessile base, entire in margin; upper surface with a narrow elevated green midrib, on each side of which is a white stomatic furrow, bounded externally by a narrow green band; lower surface green, convex, with a projecting sharp keel.

Flowers dioecious. Staminate flowers solitary in the axil of a leaf, two or three in each whorl, sessile, ovoid, about $\frac{1}{5}$ in. long. Fruit ripening in the second year, solitary in the axils, on short stalks about $\frac{1}{16}$ in. long, globose, $\frac{3}{8}$ to $\frac{1}{2}$ in. in diameter; shining reddish brown when ripe, with or without a partial glaucous bloom; composed of three or six scales, each indicated by a minute mucro, the apex of the fruit having three radiating lines. Seeds normally three, radially disposed in the centre of the fleshy resinous pulp, reddish brown, triquetrous-oblong, with two sharp lateral edges; the upper edge usually thin, rounded, broad and emarginate, occasionally narrow and pointed; with two resin-glands at the base which occasionally mark the surface with depressions.

Both *J. Oxycedrus* and *J. communis* are frequently attacked [1] by the parasitic *Arceuthobium Oxycedri*, Bieberstein, a plant allied to the mistletoe, which is common in the Mediterranean region.

This species is very variable in the wild state, differences occurring in the size and shape of the fruit and of the leaves. It usually forms a compact shrub, with ascending or spreading branches and branchlets; but large trees occur on the Riviera, which bear long pendulous branches, and are similar in habit to *J. Cedrus*.

VARIETIES

The following varieties have been described:—

1. Var. *viridis*, Pospichal, *Fl. Oest. Küstenl.* i. 30 (1897). Fruit dull green when ripe. Has only been seen on the north bank of the Canal di Leme, near Rovigno in Istria.

2. Var. *umbilicata*, Rikli, *op. cit.* 315 (1906).

Juniperus macrocarpa, Tenore, *Syll. Fl. Neap.* 483 (1831) (not Sibthorp and Smith).
Juniperus umbilicata, Grenier et Godron, *Fl. de France*, iii. 158 (1855).

Fruit when young, glaucous and covered with a bluish bloom; when ripe, larger than in the type, chestnut brown, very variable in shape, often umbilicate at the base. This is considered by Rikli to be a variety of *J. Oxycedrus*; but by Ascherson and Graebner to be a variety of *J. macrocarpa*. It occurs in Italy, and is perhaps a hybrid between the two species.

---

[1] Prof. Ed. Henry in *Bull. Soc. Bot. France*, xlvii. 260 (1900) mentions an instance in which a branch of *J. communis*, attacked by this parasite, developed scale-like leaves, similar to those normally borne by *J. phœnicea*. Cf. also *ibid.* 155 (1900), where an instance is recorded of *J. phœnicea* being also attacked by *Arceuthobium Oxycedri*.

3. Var. *brachyphylla*,[1] Loret, in Billot, *Annot. Fl. France et Allemagne*, 282 (1855).

Juniperus heterocarpa, Timbal, *ex* Loret et Barrandon, *Fl. Montpellier*, ii. 610 (1876).

Leaves short, almost obtuse at the apex. Fruit large, pinkish, glaucous. Found on limestone rocks at Saint Béat in Haute Garonne.

4. Var. *maderensis*, Menezes, in *Bull. Acad. Internat. Géog. Bot.* xvii. Nos. 227-228, p. xii (1908).

Leaves very slender, $\frac{1}{3}$ to $\frac{2}{5}$ in. long, $\frac{1}{30}$ in. broad, rounded or acute and not mucronate at the apex. Fruit similar to the type in shape and size, brown with a glaucous tinge.

This is very rare, growing wild in Madeira at elevations of over 1500 feet, on the Serra do Faial, and on rocky slopes near Curral das Freiras. Menezes says that it is cultivated in the villages of Monte Camacho and Santo Antonio da Serra. Dr. Michael Grabham, from whom I obtained specimens,[2] says that it is an elegant small tree with pendulous branches, 20 to 30 ft. in height, and with a stem 15 in. in diameter. It is very distinct in appearance from *J. Cedrus*; and appears to be closely allied to a form of *J. Oxycedrus*, occurring in Portugal, of which I received a fruiting branch from Mr. H. Clinton-Baker. The latter has short and rather broad leaves.

DISTRIBUTION

This species[3] is widely distributed throughout the Mediterranean region, extending eastwards through Syria, western Asia Minor, the south coast of the Crimea, and the Caucasus to Armenia and the Elburz Mountains in northern Persia. It is very common in the shrub-covered waste called *maquis*, which is characteristic of much of the region where it is prevalent; and appears to be indifferent to the chemical nature of the soil, as it grows on limestone, sand, and other formations; but it thrives in warm arid soils, where larger trees do not succeed.

In Algeria and Tunis, it forms undergrowth in the forests of Aleppo pine and cedar, ascending to 5000 ft. on Teniet-el-Hââd. It is widely spread throughout Spain, extending to Traz os Montes in Portugal, and ascends in the woods of *Pinus sylvestris* on the Sierra Nevada to 6600 feet. It occurs throughout Italy, Sicily, and Sardinia, and is not uncommon in the mountains of Corsica. On the eastern coast of the Adriatic, it is the commonest shrub in Istria and Dalmatia, extending inland to Herzegovina, where in the Dinaric Alps it grows to 3000 ft. elevation, and ascends the Narenta valley to Stolac. Its most northerly point is in Servia, in the western branch of the Morava valley. On the Rhodope Mountains it ascends to 3600 feet; and southwards it is common in Albania, Macedonia, Greece, and the Ægean archipelago. (A. H.)

In France it is common in the departments bordering on the Mediterranean, as

---

[1] This variety is referred to *J. macrocarpa*, by Loret, in *Bull. Soc. Bot. France*, vi. 446 (1859).

[2] Through Dr. Herbert Watney.

[3] The distribution of the species and of *J. macrocarpa* is difficult to separate, both occurring in the same regions, but apparently occupying different altitudes. See under *J. macrocarpa*, p. 1412.

around Montpellier and on the Riviera ; but it has been found inland in Aveyron and Ardèche.   It is usually seen as a shrub, but is capable of becoming a tree in favourable situations, one recorded by Mathieu near Corbières being 11 ft. in girth. When I was at Montpellier in 1909, M. Marc Bazille showed me the largest tree known to exist in France.   This grows on the Ferme de la Rouvière, in the commune of Salincelles (Gard), twenty miles east of Montpellier.   It stands on dry rocky soil covered with scrub of *Quercus Ilex* and *Q. coccifera*, and though a very old tree, is in a good healthy condition.   It measured about 35 ft. high, with a trunk 8 ft. long, and 13 ft. 4 in. in girth, dividing into six or eight large limbs which spread over a space 58 paces round.   According to M. Pardé,[1] this species will not endure the severe winters of Paris, but it is cultivated at Les Barres.

It was cultivated as early as 1739 in the Physic Garden at Chelsea ; but we have seen no specimens of considerable size in England.   The largest which we have seen is a shrubby tree at Glasnevin in Ireland, evidently of great age and about 18 ft. high.   (H. J. E.)

JUNIPERUS MACROCARPA

Juniperus macrocarpa, Sibthorp and Smith, *Fl. Græc. Prod.* ii. 263 (1813) ; Loudon, *Arb. et Frut. Brit.* iv. 2494 (1838) ; Boissier, *Fl. Orient.* v. 706 (1884) ; Kent, Veitch's *Man. Conif.* 181 (1900) ; Parlatore, in De Candolle, *Prod.* xvi. 2, p. 476 (1868).

Juniperus Oxycedrus, Linnæus, sub-species *macrocarpa*, Ascherson and Graebner, *Syn. Fl. Mitteleurop.* i. 248 (1898) ; Rikli, in Kirchner, Loew, and Schröter, *Lebenges. Blütenpfl. Mitteleurop.* i. 315 (1906).

Juniperus Biasoletti, Link, in *Atti V. Riun. Sc. Ital. Napoli*, 878 (1845).

Juniperus attica, Orphanides, in Heldreich, *Nutzpfl. Griechen.* 13 (1862).

Juniperus Lobelii, Gussone, *Syn. Fl. Sicul.* ii. 635 (1844).

Juniperus sphærocarpa, Antoine, *Cupress. Gattung.* 11, t. 10 (1857).

This species is closely allied to *J. Oxycedrus*, with which it has been united as a sub-species by Rikli and by Ascherson and Graebner.   It differs in the longer and broader leaves, $\frac{3}{4}$ to 1 in. long, about $\frac{1}{12}$ in. broad, which gradually taper from the base to the acuminate sharp-pointed cartilaginous apex.   Fruit larger than in *J. Oxycedrus*, glaucous blue, turning purplish brown after ripening, about $\frac{1}{2}$ in. broad, and $\frac{5}{8}$ in. long, on a short stalk less than $\frac{1}{8}$ in. long ; seeds similar to, but larger than in *J. Oxycedrus*.   The fruits are either globose ; or more commonly pyriform in shape, gradually tapering to the base, constituting var. *ellipsoidea*, Neilrich, *Veget. Croat.* 52 (1868), and identical with *J. Lobelii*, Gussone.

This species has a similar distribution to that of *J. Oxycedrus*, extending throughout the Mediterranean region from Spain to Syria, and also occurring in Bulgaria.   It does not appear to be a native of southern France.   It grows on low hills, and in sandy tracts close to the sea-coast ; whereas, as a rule, *J. Oxycedrus* occupies more inland and higher elevated regions.[2]

[1] *Arb. Nat. des Barres*, 49 (1906).   It is commonly known in France as *genévrier cade* or *cadier*, and yields an oil, called *huile de cade*, which is much used in veterinary practice.   Cf. Legre, in *Bull. Soc. Bot. France*, xlviii. 129 (1901).

[2] This distribution of the two species is confirmed for Greece by Halacsy, *Consp. Fl. Græca*, iii. 455 (1904), and for Algeria, by *Rev. Hort. de l'Algérie*, iv. 176 (1900).

J. macrocarpa was introduced about 1838 by Strangways from Italy ; but does not appear to have succeeded in our climate. Rikli says that even on the coast of Istria it is often much injured in severe winters. (A. H.)

JUNIPERUS BREVIFOLIA

Juniperus brevifolia, Antoine, *Cupress. Gattung.* 16, tt. 20-22 (1857) ; Parlatore, in De Candolle
Prod. xvi. 2, p. 478 (1868) ; H. C. Watson, in *London Journ. Bot.* ii. 7, 9, 398, 401, 408
(1843) and iii. 606 (1844), and in Godman, *Nat. Hist. Azores,* 224 (1870) ; Trelease, in
Missouri Bot. Gard. 8th Ann. Rep. 169 (1897) ; Kent, Veitch's *Man. Conif.* 180 (1900).
Juniperus Oxycedrus, Linnæus, var. *brevifolia*, Hochstetter, in Seubert, *Fl. Azorica,* 26 (1844).
Juniperus rufescens, Link, var. *brevifolia*, Endlicher, *Syn. Conif.* 11 (1847).

A shrub or small tree in the Azores, with a stem often 3 to 4 ft. in girth. Branchlets numerous, short, densely clothed with foliage. Leaves all acicular, in alternate whorls of threes, very short and broad, $\frac{1}{4}$ to $\frac{1}{3}$ in. long, about $\frac{1}{12}$ in. wide, oval-linear, jointed and swollen at the slightly narrowed base, widest about the middle, whence they taper to a rounded or acute (non-acuminate) apex ; upper surface with a narrow green midrib not extending to the apex, on each side of which is a broad white stomatic furrow, bounded by an external green band ; lower surface green, with a prominent midrib ; margin entire.

Flowers diœcious. Fruit sub-globose, $\frac{1}{3}$ in. in diameter, on scaly stalks about $\frac{1}{16}$ in. long, dark reddish brown when mature ; scales three, separated at the apex by three radiating prominent lines, and each marked by a minute mucro. Seeds three, embedded in a scanty pulp, ovoid, triquetrous, broadest at the base, gradually tapering to an acute apex, light brown ; outer surface convex with two or three longitudinal furrows, not separated from the inner surface by a winged thin margin, as in *J. Oxycedrus.*

This species is remarkably distinct in the very short glaucous leaves ; and its seeds differ from those of the allied species.

This species is limited to the Azores, where it occurs on the islands of Corvo, Flores, Fayal, San Miguel, and Pico, ascending to 5000 ft., and rarely descending below 1000 ft. It is locally known as *cedro,* and is usually a compact shrub or small tree, becoming a prostrate bush with interlacing branches on exposed hill summits. Formerly it appears to have been a tree of considerable size, as large logs have been found deeply buried under the secondary volcanic debris in the Grotto do Enferno of the large crater known as Caldeira des Sette Cidades. A slab of this was presented to the Kew Museum by Dr. Goeze, which was reported to have been excavated from a depth of 100 metres.[1]

So far as we know this beautiful species has not been introduced[2] into cultivation in England. (A. H.)

---

[1] Cf. Goeze in *Gard. Chron.* 1867, p. 929, and Masters in *Journ. R. Hort. Soc.* xvii. 3 (1894), who identified this wood with *Cupressus lusitanica.* The slab at Kew is undoubtedly the wood of this species of juniper. Cf. our Vol. V. 1179, note 1.

[2] Gordon, *Pinetum,* 131 (1880), says it is tender in England ; but he gives no particulars as to its introduction.

JUNIPERUS CEDRUS

Juniperus Cedrus, Webb and Berthelot, *Hist. Nat. Isles Canar.* iii., *Phytog. Canar.* ii. 277, t. 217 (1840); Antoine, *Cupress. Gattung.* 14, tt. 16-19 (1857); Parlatore, in De Candolle, *Prod.* xvi. 2, p. 478 (1868); Kent, Veitch's *Man. Conif.* 180 (1900).
Juniperus Oxycedrus grandifolius, Buch, *Phys. Beschr. Can. Inseln*, 109, 159 (1825).
Juniperus canariensis, Knight, *Syn. Conif.* 13 (1850).
Juniperus Webbii, Carrière, *Conif.* 13 (1855).

A tree, attaining a large size in the Canary Islands, with wide-spreading branches and long pendulous branchlets. This is an insular form of *J. Oxycedrus*, differing mainly from the Mediterranean tree in habit and not in technical characters. Leaves directed towards the apex of the branchlet, and not widely spreading, resembling in this respect *J. formosana*, thinner in texture than in *J. Oxycedrus*, and becoming acute or rounded at the apex in adult trees; glaucous and not bright green on the lower surface, the glaucous tint being present on the narrow midrib and the borders external to the white stomatic bands of the upper surface. Fruit similar to that of *J. Oxycedrus*, with two or three seeds, which are often acute and not broad and emarginate at the apex, the resin-pits being usually more developed than in the Mediterranean species.

Copious specimens received from Dr. Perez show that this tree is scarcely separable as a distinct species from *J. Oxycedrus*; and trees of the latter, with pendulous branchlets, which occur on the Riviera, are very similar to, if not identical with *J. Cedrus*.

J. Cedrus is a native of the Canary Islands, where it has been nearly exterminated by the inhabitants, who value its timber highly. It still exists on Teneriffe, Grand Canary, and Palma, but is extinct on Gomera.[1] Dr. Perez writes that it was common on Teneriffe at the end of the eighteenth century,[2] as Humboldt mentions it as occurring all the way up from Orotava to the *cañadas*, growing with *Pinus canariensis*; but only a few stunted specimens now survive, which grow on inaccessible rocks about the cañadas, from 7000 to 9000 ft. altitude. Schenck,[3] who gives the latest published account, quotes Fritsch,[4] who mentions a noble juniper, which formerly grew on the south-west side of the Peak at 8000 ft. elevation. This tree was 18 ft. in girth and nearly 100 ft. high.

It also occurs in the crater of Tirijana on Grand Canary, but seems to be most abundant now on Palma,[5] where it grows on the inaccessible inner walls of the crater, and outside it to the north-east on isolated rocks at 7000 ft. elevation. Dr. Burchard [6]

---

[1] Dr. Christ, in Engler, *Bot. Jahrb.* vi. 487, 500 (1885).

[2] Buch, *Phys. Beschr. Canar. Inseln*, 109 (1825), mentions a few trees which were then growing at 9000 ft. altitude amidst the lava at the foot of the last cone of the Peak.

[3] *Wiss. Ergeb. Deut. Exped.* 'Valdivia,' ii. pt. 1, p. 375, figs. 63, 64 (1907).

[4] *Reisebilder*, 6 (1879).

[5] Bornmüller, in Engler, *Bot. Jahrb.* xxxiii. 398 (1904), states that he found trees also on the south side of the crater, in the Barranco de las Angustias at 1300 and 2700 ft. elevation; and at Cumbrecita at 4700 ft.

[6] In *Mitt. Deut. Dend. Ges.* 1911, p. 296, fig.

says that these trees have trunks almost completely bare of bark, and over a metre in diameter; and reproduces the photograph of a fine old female tree, which he took in June 1910. Schenck[1] also figures a very old tree, with a short bole, a few snaggy branches, and very little foliage. (A. H.)

It is doubtful if this plant was introduced until recently, as it was not mentioned by Loudon[2] in 1838, and was included by Knight and Perry in 1850 amongst the kinds of juniper of which little was known. Kent says that it is not hardy in England; and the only plant now living at Kew is one in the Temperate House, about 7 ft. high, raised from seed sent by Sir Daniel Morris in 1893. Sir John Ross-of-Bladensburg, K.C.B., however, informs me that a plant in his garden at Rostrevor survived without protection the winter of 1909-1910, which was exceptionally severe in the north of Ireland. As it grew well during the following summer, he looks upon it as hardy; in December 1911 it was 6 ft. high with a good leader, and slightly pendulous branches. A small plant at Glasnevin bore last winter 12° of frost without injury.

J. Cedrus is readily propagated by cuttings at Kew; but there seems to be a great difficulty in raising it from seed.[3] Beissner,[4] however, has raised young plants from seed which I saw in his collection at Bonn in 1908. (H. J. E.)

JUNIPERUS FORMOSANA

Juniperus formosana, Hayata, in *Journ. Coll. Sci. Tokyo*, xxv. art. 19, p. 209, pl. 38 (1908).
Juniperus oblonga pendula,[5] Knight and Perry, *Syn. Conif.* 11 (1850); Carrière, *Conif.* 20 (1867).
Juniperus taxifolia, Masters, in *Journ. Roy. Hort. Soc.* xiv. 215 (1892), and *Journ. Linn. Soc. (Bot.)* xxvi. 543 (1902) (in part); and in *Journ. Bot.* xli. 268 (1903) (not Hooker and Arnott); Kent, Veitch's *Man. Conif.* 191 (1900).

A tree, attaining in China 40 ft. in height. Branchlets triquetrous, with three narrow ridges, yellowish green in the first year, reddish brown in the second year. Leaves all acicular, spreading, in whorls of threes, linear-subulate, about $\frac{1}{2}$ in. long and $\frac{1}{20}$ to $\frac{1}{12}$ in. broad, jointed and swollen at the base, ending in a sharp spine-like

---

[1] *Beit. Kennt. Veget. Canar. Inseln*, fig. 63, in *Wiss. Ergeb. Deut. Exped. 'Valdivia,' 1898-1899* (1907). This figure is a reproduction of a photograph taken by Prof. Simony. A figure of a similar tree, also growing on Palma, is given by Webb and Berthelot, *Hist. Nat. Isles Canar.*, Atlas, t. 8 (1838).   [2] Cf. note 5 below.

[3] Dr. Geo. V. Perez of Orotava, Teneriffe, wrote in *Gard. Chron.* xl. 14 (1906), and xli. 134 (1907), that none of the seed which he sent to Kew and elsewhere germinated. He finds that seeds, soaked in water at 70° Fahr. for 15 to 30 days, germinate freely in about six weeks. Probably germination occurs, under ordinary conditions, in the second year after sowing, or in the wild state after the seeds have been eaten and voided by birds. Correvon, in *Gard. Chron.* xlii. 209 (1907), reports that seeds, which had been soaked in a weak solution of acetic acid, germinated well at Geneva.

Dr. Perez informs me that Mr. Lister has raised, from seed sent from Orotava, eight plants in the Government nursery at Pretoria, which are now 4 ft. high.—(A. H.)   [4] *Mitt. Deut. Dend. Ges.* 1906, p. 91.

[5] Knight and Perry's plant is undescribed, but is stated to have come from China and Japan, and is evidently the Chinese species here described, as is confirmed by Carrière's description. It is apparently not the same plant as *J. communis oblonga pendula*, Loudon, *Arb. et Frut. Brit.* iv. 2489, fig. 2345 (1838), applied to a shrub at Kew, then 5 ft. high, with fastigiate branches and pendulous branchlets. Gordon, in *Gard. Chron.* 1842, p. 652, describes the latter as: "Trained to a single stem, if left to nature, it will not rise more than three feet, but will spread over a large space of ground. It is quite hardy and a native of the Caucasus." Webb, *Phyt. Canar.* ii. 277 (1840), identified this plant with *J. Cedrus*, which is, however, not hardy at Kew. It is impossible now to identify Loudon's plant, but in all probability it was a pendulous variety of *J. communis*.

point; lower surface convex, keeled; upper surface concave, with two broad white stomatic bands, separated by a very narrow green or glaucous midrib, extending from the base to near the apex, where the white bands coalesce.

Fruit globose, ⅓ in. in diameter, ripening in the second year, shining dark reddish brown when ripe; smooth on the surface, with three deep radial furrows at the summit, in the centre of each of which is a dark line, showing the separation of the three scales of which the fruit is composed; outer edge of each furrow with a thin mucro, overhanging a slight depression; base of the fruit hollowed out at the insertion of the short scaly stalk. Seeds three, elongated-ovate, triquetrous, mucronate at the narrow thin apex, with several circular resin-pits at the base, above which on the outer surface are three or four larger elongated oval pits.

In cultivated specimens in Europe the branches are ascending, but the branchlets are very pendulous, giving a weeping appearance to the tree. In China it assumes various habits, but is often very pendulous, and occasionally shrubby.

The Chinese species now described, as pointed out by Hayata, is distinct from *J. taxifolia*,[1] Hooker and Arnott, with which it has been confused. The latter, so far as we can learn, has never been introduced into cultivation, and is confined to Bonin Isle, where it was discovered by Capt. Beechey in 1827, and to the Liu Kiu Islands. I cannot find any particulars of the size and habit of this species.

J. formosana is widely spread throughout the mountains of China, and is also commonly cultivated in temple grounds, being known as the *Tz'e Poh*, or "prickly cypress." It is represented in the Kew herbarium by numerous wild specimens from the provinces of Szechwan, Hupeh, Chekiang, and Fokien; and was collected in Shensi by Père Giraldi. It has lately been found on Mt. Morrison and the adjacent ranges in Formosa, between 8000 and 13,000 ft. altitude.[2]

This species, which is usually known in cultivation as *J. oblonga pendula*, is stated by Kent to have been introduced from China by Fortune in 1856; but this is incorrect as regards the date. It was for sale in Knight and Perry's nursery in 1850, and was probably one of the plants sent home by Fortune in 1844. Knight and Perry describe it as a very elegant drooping shrub from China and Japan, which they supposed to be possibly identical with *J. rigida*; but the latter was not introduced till 1861.

J. formosana is now rare in collections, the best specimen that we have seen being a tree at Bicton, 30 ft. by 2 ft. A vigorous tree at Eastnor Castle is about 28 ft. high. A smaller specimen at Bayfordbury is reputed to have been planted

[1] *Juniperus taxifolia*, Hooker and Arnott, *Bot. Cap. Beechey's Voyage*, 271 (1841); Siebold and Zuccarini, in *Abh. Akad. Wiss. München*, iv. 3, p. 233 (1846); Miquel, *Prol. Fl. Japonica*, 331 (1867); Parlatore, in De Candolle, *Prod.* xvi. 2, p. 481 (in part); Masters, in *Journ. Linn. Soc. (Bot.)* xviii. 496 (1882); Hayata, in *Journ. Coll. Sci. Tokyo*, xxv. art. 19, p. 210, fig. 6 (1908).
Apparently a shrub. Leaves thicker than those of *J. formosana*, rounded or acute at the apex, and not ending in a spine-like point, about ⅝ in. long; upper surface with two white stomatic bands, separated by an elevated green midrib from base to apex. Fruit globose, ¼ in. in diameter, yellowish, rugose on the surface, with three prominent radial ridges at the apex, overhanging three furrows, external to each of which is a mucro. Seeds three, similar in shape and resin-pits to those of *J. formosana*. This species is closely allied to *J. formosana*, mainly differing in the blunt and not spine-tipped leaves.
[2] Specimens kindly sent me from Formosa by Mr. T. Kawakami are identical with specimens which I collected in central China in the mountains of Hupeh. Elwes gathered it in February 1912, at about 8000 ft. on the ridges above Arisan; but saw it only in a bushy form.

in 1845. Another at Glasnevin, which bore fruit in 1911, was obtained some years ago from Messrs. Veitch, who have, however, no longer the plant for sale.

It has been much confused with *J. rigida*; and all the reputed trees[1] of *J. oblonga pendula* on the continent are referable to this Japanese species.   (A. H.)

JUNIPERUS DRUPACEA

Juniperus drupacea, Labillardière, *Icon. Pl. Syr.* ii. 14, t. 8 (1791); Loudon, *Arb. et Frut. Brit.* iv.
2494 (1838); Lindley, in *Gard. Chron.* 1854, pp. 387, 455, fig.; Webster, in *Gard.
Chron.* xix. 519, fig. 80 (1896); Kent, Veitch's *Man. Conif.* 173 (1900).
Arceuthos drupacea, Antoine and Kotschy, in *Oester. Bot. Wochblatt.* iv. 249 (1854).

A diœcious tree, attaining about 60 feet in height, usually in cultivation columnar in habit, but in the wild state broadly pyramidal.[2] Young branchlets triangular, with three prominent linear ridges and three grooves, due to the decurrent bases of the leaves. Older branchlets, from which the leaves have fallen, terete and smooth, with a brown scaly bark. Buds, one terminal and usually two or three in the axils of the leaves on the branchlet of the first year, about $\frac{1}{8}$ inch long, surrounded by minute sharp-pointed lanceolate scales, the outermost of which persist on the apex of the second and third year's branchlets.

Leaves, all acicular, spreading in whorls of threes, about $\frac{1}{2}$ to $\frac{7}{8}$ in. long, $\frac{1}{10}$ to $\frac{1}{5}$ in. broad, jointed at the base, decurrent on the branchlet to the next whorl, linear-lanceolate, widest near the base, gradually tapering to the apex, which ends in a sharp cartilaginous point; lower surface green, convex, with a linear prominent keel; upper surface concave, with a broad green midrib deeply furrowed near the base, and two white stomatic bands, not extending to the margin, which is entire.

Staminate flowers, five or six in a head, on a short scaly stalk, arising in the axil of a leaf on the second year's branchlet; stamens nine to twelve in each flower.

Fruit ripening in the second year, larger than in any other species, $\frac{3}{4}$ to 1 in. in diameter, on a short scaly stalk, ovoid or nearly spherical, brown or bluish with a glaucous bloom, usually composed of nine fleshy scales, in whorls of threes, united together, ovate, thickened and often mucronate at the apex; enclosing a large globose hard bony stone,[3] with three small cells, each containing a minute oblong kernel, one of which is often aborted.

This remarkable species is a native of the mountains of Asia Minor, Syria, and Greece. It is found throughout the Taurus range in Asia Minor, several localities being mentioned by Boissier,[4] between Karaman and Ermenek, in the Bulgardagh, and in the Akkerdagh, close to Marasch. It grows at elevations of 1600 to 5600 feet, and either forms small pure woods or is scattered amidst the forests of cedar and

---

[1] Specimens sent by Späth from Berlin, and by Pardé from Nancy and Les Barres.

[2] Walter Siehe, in *Gartenflora*, xlvi. 207 (1897), states that this tree in the Cilician Taurus never assumes the narrow columnar form which is so common in cultivation. Young trees in the wild state are pyramidal in habit; whilst older trees, which are mainly females preserved by the peasants on account of their fruit, have a rounded head of foliage.

[3] Antoine and Kotschy, *loc. cit.* consider the bony stone to arise from the union of the testa of the seeds with the inner part of the three upper scales.   [4] *Flora Orientalis*, v. 706 (1881).

Abies cilicica.   Siehe mentions[1] enormous trees between Namrun and Güllek in Cilicia, which are over 3 ft. in diameter.   It also occurs in Syria, on Mount Cassio, and on the Lebanon and Anti-Lebanon ranges.   In Greece it appears to be confined to Mount Malevo, in the southern part of the Morea, where, according to Halacsy,[2] it forms a small wood at 3700 to 4000 feet altitude.   According to Boissier,[3] the reported occurrence of this species in Crete is erroneous.

It is known to the Turks as *Andys* or *Habhel,* and is called *Duffran* by the natives of Syria, who collect and eat the fruits, which have a pleasant though resinous flavour.

This species is said by Loudon[4] to have been introduced in 1820, but he acknowledges that he had only seen young plants, and these were probably incorrectly named, as Lindley[5] in 1854 speaks of *J. drupacea* as a new plant.   It is generally believed to have been for the first time introduced into western Europe in that year by Kotschy, who collected it in Asia Minor in 1853.

So far as we know it has never produced fruit in England, where all the trees in cultivation are supposed to be males.   It is normally diœcious, but M. Allard[5] states that a tree in his arboretum at Angers, which bore staminate flowers for a long time, ultimately produced fruit, and afterwards remained monœcious.   A female plant at Angers has, however, never produced staminate flowers.   M. Mottet[6] states that a small tree at Verrières, only 6 ft. high, has produced fruit, though no other tree of the same species is near it, and he supposes that it must have been fertilised by the pollen of another species.   (A. H.)

This species usually forms a narrow column and is one of the most beautiful of the junipers, yet is rarely seen in collections.   It is perfectly hardy, and, judging from the way it grows at Colesborne, thrives in a limestone soil.   The best specimen which I have seen is at Eastnor Castle, where there is a well-shaped tree, 31 ft. by 3½ ft. in 1908.   This produced staminate flowers in May 1899, of which there are specimens in the British Museum.   At Scorrier, in Cornwall, a fine specimen was 36 ft. by 1 ft. 9 in. in 1911.   There are two good trees at Kew about 30 ft. high. Henry saw at Holkham in 1911 two very narrow columnar trees, about 40 ft. high by 3 ft. in girth.   At Brickendon Grange, Hertford, a fine specimen measured 36 ft. high in 1912.   Smaller trees occur at Highnam, Tortworth, Chiltley Place near Liphook, Young's nursery at Milford near Godalming, and other places.   In Ireland the best specimen that we have seen, a tree about 30 ft. high, is growing at Woodstock, Kilkenny.

In France it attains a larger size, and at Angers and Montpellier has produced fruit.   A tree in M. Allard's arboretum at Angers, nearly 40 ft. high, is figured in *Bull. Soc. Dend. France,* 1908, p. 109.   Another[7] at Antibes, about 30 ft. high, has a leaning stem and wide-spreading branches, being very different in habit from the usual form of this species in cultivation.   (H. J. E.)

---

[1] In *Mitt. Deut. Dend. Ges.* 1911, p. 305.

[2] *Comp. Fl. Græc.* iii. 455 (1904).   There is a specimen in the Cambridge Herbarium collected by Orphanides on Mount Malevo.   [3] *Flora Orientalis,* v. 706 (1881).

[4] *Encycl. Trees,* 1084 (1842).   [5] *Gard. Chron.* 185₁, p. 455.

[6] Cf. Mottet, in *Rev. Hort.* 1904, p. 356, figs. 147, 148, where a tree is figured in the park of Baron Mallet at Château des Côtes, near Versailles, which Elwes found to be 36 ft. high in 1909.

[7] Erroneously labelled *J. Oxycedrus,* var. *macrocarpa.*

JUNIPERUS RECURVA

Juniperus recurva, Buchanan-Hamilton, *ex* Don, *Prod. Fl. Nepal.* 55 (1825); Loudon, *Arb. et Frut. Brit.* iv. 2504 (1838); Masters, in *Gard. Chron.* xix. 468 and 574, fig. 69 (1883), and in *Journ. Linn. Soc. (Bot.)* xxvi. 542 (1902); J. D. Hooker, *Fl. Brit. India*, v. 647 (1888); Kent, Veitch's *Man. Conif.* 185 (1900); Gamble, *Indian Timbers*, 698 (1902); Brandis, *Indian Trees*, 694 (1906).

Sabina recurva, Antoine, *Cupress. Gatt.* 67, tt. 88, 90, 91 (1857).

A tree, attaining in the Himalayas 30 ft. or more in height. Bark brown, thin, peeling off in long fibrous strips. Branches curved, more or less pendulous. Young branchlets marked between the whorls by the decurrent bases of the leaves, which are separated by three grooves. Older branchlets from which the leaves have fallen smooth, with a light reddish-brown scaly epidermis.

Leaves all acicular, densely imbricated in whorls of threes, their basal part being decurrent on the branchlet, loosely appressed, directed forwards, scarcely spreading; linear-lanceolate, $\frac{1}{8}$ to $\frac{1}{4}$ in. long, $\frac{1}{50}$ to $\frac{1}{25}$ in. broad, ending in a sharp cartilaginous point; outer surface convex, channelled longitudinally from the insertion to about the middle, greyish green; inner surface concave, whitened throughout. The leaves persist for several years, becoming brown in the third and fourth years, the mixture of green and brown leaves giving the foliage a peculiar appearance.

Flowers monœcious.[1] Staminate flowers, terminal or solitary in the axils of one or two of the leaves on the ultimate branchlets, oblong, about $\frac{1}{6}$ in. long, with twelve to sixteen stamens. Fruit axillary, ripening in the second year, subsessile, bracteate at the base, dark purplish brown, ovoid, about $\frac{3}{8}$ in. long, composed of three or six united scales, each bearing a triangular spreading mucro, prominent near the umbilicate apex of the fruit. Seed solitary, occupying the greater part of the berry, ovoid, with a thin narrowed apex, marked on the surface with two or three large depressions for resin-glands.

This species is a native of the eastern Himalayas, occurring in Sikkim and Bhutan, between 9000 and 12,000 feet. It is represented in the north-western Himalayas, China, and Japan by the closely allied species or variety, *J. squamata*.

(A. H.)

In the very moist climate of the interior of Sikkim, where it is common in the Lachen and Lachoong valleys, from about 10,000 to above 13,000 feet, Sir Joseph Hooker, whose sketch[2] of it has been reproduced in Veitch's *Coniferæ*, fig. 58, gives 30 ft. as its height; but if my recollection is correct, I saw much larger trees above Lachoong; and G. A. Gammie, in his account of a botanical tour in Sikkim,[3] says that in the Sebu valley he saw large trees at 11,000 feet; and at 13,000 feet in the same valley it was the only arborescent vegetation.

---

[1] This species appears to be always monœcious. Kent, Veitch's *Man. Conif.* 187 (1900), states that a tree at Fota is a male. We have specimens of this bearing both staminate and pistillate flowers on different branchlets of the same branch.

[2] *Himalayan Journals*, ii. 45, fig. (1854).

[3] Published as a Government paper, No. 41 B.S.I., dated Calcutta, 26th July 1893, reprinted in *Kew Bulletin*, October-November 1893, p. 311.

According to Gamble its growth is slow, about twenty-two rings per inch of radius for the Sikkim tree; and the wood is very good, equal to the best pencil cedar, but is not used except to burn as incense in the Buddhist temples.

This is the most ornamental of the junipers on account of its graceful drooping habit; and though introduced[1] in 1830, and hardy enough to grow well and ripen its berries in Scotland, it is not common in cultivation, and is seldom found in nurseries.[2] Its success in cultivation seems to depend principally on sufficient moisture in summer, all the best specimens that I have seen being in districts where the rainfall is heavy.

The largest I know in England is at Bicton, where there are two trees about 40 and 35 ft. high by 3 ft. 4 in. in girth, which bear abundance of berries. At Hafodunos in Denbighshire I saw in 1911 a very fine tree with three stems from the ground, almost equal in height, 40 ft. to 41 ft., and each about 2 ft. in girth. At Bodorgan in Anglesea there is a good-sized tree. Even in the drier climate of Gloucestershire there is a thriving tree at Highgrove, near Tetbury, the seat of Arthur Mitchell, Esq., from the berries of which I have raised plants; and another occurs at Highnam. There are also good specimens at Pencarrow and Menabilly in Cornwall, Mamhead in Devon, Bayfordbury and High Canons, Herts, Rotherfield Park, Hants, and Holkham, Norfolk.

In Scotland there is a small tree at Murthly, from which I have raised seedlings; and a thriving shrub at Drumtochty Castle. Another at Dalkeith was 14 ft. high in 1907.

In Ireland there is a large bushy tree[3] with nine main stems, 40 ft. high and thirty-seven paces round at Castlewellan (Plate 349). At Salterbridge, Co. Waterford, the seat of Major Chearnley, I saw in 1910 another of the same type and almost as large; and at Fota I measured a tree 38 ft. high in 1910. (H. J. E.)

JUNIPERUS SQUAMATA

Juniperus squamata, Buchanan-Hamilton, in Lambert, *Genus Pinus*, ii. 17 (1824); Don, *Prod. Fl. Nepal.* 55 (1825); Loudon, *Arb. et Frut. Brit.* iv. 2504 (1838); Endlicher, *Syn. Conif.* 18 (1847); Koch, *Dendrologie*, ii. pt. ii. 121 (1873).

Juniperus religiosa,[4] Royle, *Illust. Him. Plants*, i. 351 (1839) (name only).

Juniperus densa,[5] Gordon, *Pinet. Suppl.* 32 (1862).

Juniperus recurva, Don, var. *squamata*, Parlatore, in De Candolle, *Prod.* xvi. 2, p. 482 (1868); Brandis, *Forest Flora N.W. India*, 536 (1874), and *Indian Trees*, 694 (1906); Hooker, *Fl. Brit. India*, v. 647 (1888); Masters, in *Journ. Linn. Soc. (Bot.)* xxvi. 543 (1902).

Juniperus morrisonicola, Hayata, in *Gard. Chron.* xliii. 194 (1908), in *Journ. Coll. Sci. Tokyo*, xxv. art. 19, p. 211, fig. 7 (1908), and xxx. art. i. p. 307 (1911), and in *Journ. Linn. Soc. (Bot.)* xxxviii. 298 (1908).

Sabina squamata, Antoine, *Cupress. Gatt.* tt. 89, 90 (1860).

A shrub, with long decumbent stems, running over and under the surface of the ground, from which arise numerous short erect branches. Young branchlets green,

---

[1] According to Loudon, *Trees and Shrubs*, 1089 (1842). Seeds were subsequently sent home to Kew by Hooker in 1850.

[2] It is known in some nurseries as *J. repanda*, Hort. *ex* Carrière, *Conif.* 27 (1867).

[3] Figured by Earl Annesley, *Beautiful and Rare Trees*, 54 (1903).

[4] This is identified with *J. squamata* by Hooker, *Fl. Brit. India*, v. 647 (1888).

[5] Gordon's account is confused, as he states that the berries are three-seeded; otherwise his description applies to *J. squamata*.

with three grooves separating the decurrent pulvini of the leaves. Leaves all acicular, densely imbricated in whorls of threes, appressed or slightly spreading, decurrent on the branchlets, broader and shorter than in *J. recurva*, the free part $\frac{1}{6}$ in. long and $\frac{1}{24}$ in. wide, curved, gradually tapering to an acute apex, which is tipped with a sharp cartilaginous point; ventral surface concave, whitened, usually with a faint or obsolete midrib; dorsal surface convex, green, with a median furrow extending from the base to near the apex. Older branchlets stout, reddish brown, covered with persistent reddish brown acicular leaves.

Fruit ellipsoid, reddish brown at first, turning black when ripe in the second year, smaller and of a different shape from that of *J. recurva*, about $\frac{1}{4}$ to $\frac{1}{3}$ in. long, somewhat less in diameter, composed of three or six scales, each with a triangular mucro, umbilicate at the apex. Seed solitary, ovoid, broadest above the base, and tapering to an apiculate apex, nearly filling the cavity of the fruit, with about four ridges running from base to apex, and three or four depressions below the middle for resin-glands.

J. squamata, differs mainly from *J. recurva* in habit and in having stouter broader needles; but it is readily distinguishable, and has a much wider distribution. It varies considerably in the colour of the leaves, and appears occasionally to become an erect instead of a prostrate shrub.

It occurs in Afghanistan, the Himalayas, and the mountains of China and Formosa. It grows at a high elevation in the Himalayas, being most common in the north-west; but is also found in Sikkim, where Gammie states that it attains 15,000 ft. altitude. Brandis describes it as a gregarious shrub, often covering large areas, either pure or mixed with *J. communis*, with decumbent stems, at times six inches in diameter, running over the ground and giving off numerous short branches, which make it very difficult to traverse such thickets.

It is also found at high elevations in China, in the provinces of Hupeh, Szechwan, and Yunnan. In Hupeh, where I saw it in 1888, it is a shrub about a foot high, usually growing on rocky ground, and spreading over the surface to a radius of six feet or more. It resembles in habit the dwarf form of *J. communis*, but is readily distinguished by its broader shorter leaves and one-seeded berries. *J. squamata* grows in Formosa on Mt. Morrison, near the summit at 13,200 ft. altitude.

J. squamata was introduced[1] into England about 1836, and is occasionally cultivated in rockeries, being known occasionally as *J. pseudosabina*,[2] *J. densa*, etc.

In its typical form, it has leaves of a pure green tint, which are occasionally nearly as long as those of *J. procumbens*, Siebold. There are specimens at Kew, Bicton, Bayfordbury, and Glasnevin.

The following is probably a variety of *J. squamata*; but in the absence of fruit I hesitate to assign it to that species :—

---

[1] Gordon, in Loudon, *Gard. Mag.* xvi. 10 (1840), states that it was raised in the Chiswick Garden from Indian seed sent three or four years previous to 1840.

[2] *J. pseudosabina*, Fischer and Meyer, is a Turkestan shrub. Cf. p. 1423, note 1.

Juniperus procumbens, Siebold, in *Ann. Soc. Hort. Pays-Bas*, 1844, p. 31, and in Siebold and Zuccarini, *Fl. Jap.* ii. 59, t. 127, fig. iii. (1870) (not Sargent[1]).

Juniperus chinensis, var. *procumbens*, Endlicher, *Syn. Conif.* 21 (1847).

A prostrate shrub similar to *J. squamata* in habit, but differing in the branchlets being glaucous-white on the edges of the pulvini : leaves longer, their free part ⅓ in. long, gradually tapering to an acuminate spine-like apex ; upper surface concave and covered except along the margins with a white stomatic band, divided except near the apex by an elevated and usually green midrib ; lower surface convex, bluish, spotted with white, and with a median furrow which is variable in length. Fruit not seen.

This beautiful shrub was first described by Siebold,[2] who stated that it was wild in the mountains of Japan, and was cultivated in gardens and temple woods at Nagasaki. It has been collected since only by Faurie,[3] who found it at high elevations in Hondo.

Siebold considered it to be perhaps *J. nipponica*, Maximowicz,[4] a species with which it has no affinity ; and subsequent botanists confused it with *J. chinensis*,[5] a totally different species. It resembles *J. squamata* very closely, differing only in the glaucous tint of the leaves and branchlets ; but in the absence of fruit cannot be safely united with that species.

J. procumbens is said[6] to have been introduced, by living plants, into the Botanic Garden at St. Petersburg in 1864 ; but does not appear to have been known in England until of late years.[7] It is now imported largely from Japan, and was a striking feature in the exhibit of Japanese plants at the Anglo-Japanese Exhibition of 1909. It is the most ornamental of the creeping junipers, and is occasionally sold under the erroneous name of *J. litoralis*,[8] a totally distinct species. So far as I know it has not yet produced fruit in England. (A. H.)

[1] *J. procumbens*, Sargent, *Forest Flora of Japan*, 78 (1894), and in *Garden and Forest*, x. 421 (1897), is a variety of *J. chinensis*, described on p. 1432.

[2] One of Siebold's original specimens, a branch without fruit, of *J. procumbens* is in the Kew herbarium, where it was sent from the Leyden Museum.

[3] Masters, in *Bull. Herb. Boissier*, vi. 274 (1898), refers two specimens collected by Faurie, "No. 47, summit of Sennintoge, and No. 3409, summit of Ckokkai," to *J. recurva*, var. *squamata*. No. 3409 is in the Kew herbarium, and is identical with *J. procumbens*, Siebold.

[4] *J. nipponica*, Maximowicz, in *Mél. Biol.* vi. 374 (1867), is a remarkably distinct species, of which little is known, except the original specimen described by Maximowicz. This species has not been introduced into Europe.

[5] Siebold's plant has been much confused with *J. chinensis*, var. *japonica*, which is also cultivated in Japan. Gordon's specimen of *J. japonica procumbens* in the Kew herbarium is *J. chinensis*.

[6] Bretschneider, *Hist. Europ. Bot. Disc. China*, 610 (1898).

[7] The plant at Kew was introduced from Japan in 1893.

[8] *J. litoralis*, Maximowicz, in *Mél. Biol.* vi. 375 (1867), is a sea-shore plant, which grows abundantly on the shore of Hakodate Bay in Yezo, and near Honjo on the west side of Hondo, where it was found by J. Veitch in 1892. It also grows in Kiusiu and the Liu Kiu Islands. It has three-seeded berries, and has some affinity with *J. rigida*. It has never been introduced into England so far as we are aware. Bretschneider, *op. cit.* 610, referring to it as *J. conferta*, Parlatore, says that it was introduced into St. Petersburg in 1864, along with *J. procumbens*, Siebold.

JUNIPERUS WALLICHIANA

Juniperus Wallichiana, J. D. Hooker, *ex* Parlatore, in De Candolle, *Prod.* xvi. 2, p. 482 (1868);
    Brandis, *Forest Flora N.-W. India*, 537 (1874), and *Indian Trees*, 695 (1906).
Juniperus pseudosabina, J. D. Hooker, *Fl. Brit. India*, v. 646 (1888) (not Fischer and Meyer[1]);
    Kent, Veitch's *Man. Conif.* 184 (1900); Gamble, *Indian Timbers*, 698 (1907).

A tree, attaining in the Himalayas 60 ft. in height. Leaves dimorphic. Adult foliage with tetragonal ultimate branchlets, about $\frac{1}{25}$ in. in diameter, densely covered with scale-like leaves, which are arranged in four ranks in decussately opposite pairs, closely appressed, narrowly ovate, about $\frac{1}{16}$ in. long, tapering to an acute apex, bright green with a whitish margin, marked on the back with a linear glandular furrow extending from the base to near the apex. Leaves on the main axes, larger, up to $\frac{1}{4}$ in. long, tipped with acuminate points. Juvenile foliage, often preponderant on adult trees; leaves acicular, in threes, decurrent, densely clothing the branchlet in successive whorls, slightly spreading, about $\frac{1}{6}$ in. long, sharply mucronate, whitened on the inner (upper) surface, usually marked on the back with a longitudinal furrow.

Flowers dioecious. Fruit, ripening in the second year, on the ends of short curved branchlets, ovoid, $\frac{2}{5}$ in. long, $\frac{1}{3}$ in. broad near the base, dark purplish brown, becoming quite blue when ripe; smooth on the surface except for the minute mucros which indicate the three to five component scales; depressed at the summit with a minute transverse rhomboidal apiculate umbo. Seed, one in each fruit, large for the genus, $\frac{1}{4}$ in. long, ovoid, compressed, with a narrow thin pointed apex, and two or three depressions for resin-glands about the middle of each surface.

This species is a native of the Himalayas from the Indus to Bhutan, occurring between 9000 and 15,000 ft. elevation. In the western part of its range, it is a large gregarious shrub; but in Sikkim, it becomes a large tree, sometimes 60 ft. in height, with a stout trunk and dark branches and foliage. An illustration of it is given by Hooker,[2] who calls it the "Black Juniper." Mr. J. Claude White[3] saw a large "weeping cypress," at Chalimaphe in Bhutan, which was 50 ft. round the trunk at five feet from the ground; and this remarkable tree in all probability was *J. Wallichiana.*

J. Wallichiana was introduced in 1849, when Sir J. D. Hooker sent seeds from India to Kew.[4] It is very rare in cultivation, the only specimens which we have seen being one in the Juniper collection at Kew, about 20 ft. high; and another of the same size at Leonardslee, which bore fruit in 1911. (A. H.)

---

[1] *J. pseudosabina,* Fischer and Meyer, in *Index Sem. Hort. Petrop.* 65 (1841), and *Plant. Schrenk.* ii. 13 (1842), differs in appearance from the Himalayan tree, the scale-like leaves being less acute, and the fruits smaller and often globose. It was described from specimens gathered in the Altai and the Tarbagatai mountains in Turkestan. It appears to be a low shrub, like *J. Sabina* in habit, and has not apparently been introduced.

[2] *Him. Journ.* ii. 55, fig. (1854).

[3] *Sikkim and Bhutan,* 131 (1909). No specimens of this enormous tree appear to have been collected.

[4] See Kew archives, "*List of Seeds received from Dr. Hooker during his Travels in India,*" where "No. 78, 1849 (No. 152), *Juniperus,* large tree," is evidently *J. Wallichiana.*

JUNIPERUS PHŒNICEA

Juniperus phœnicea, Linnæus, *Sp. Pl.* 1040 (1753); Loudon, *Arb. et Frut. Brit.* iv. 2501 (1838);
    Parlatore, in De Candolle, *Prod.* xvi. 2, p. 486 (1868); Vallot, in *Journ. de Bot.* ii. 329 (1883);
    Mathieu, *Flore Forestière*, 517 (1897); Ascherson and Graebner, *Syn. Mitteleurop. Fl.* i. 250
    (1898); De Coincy, in *Bull. Soc. Bot. France*, xlv. 432 (1898); Kent, Veitch's *Man. Conif.*
    182 (1900); Kirchner and Schröter, *Lebenges. Blütenpfl. Mitteleuropas*, i. 316 (1906); Albert
    and Jahandiez, *Pl. Vasc. du Var*, 451 (1908).

Juniperus Lycia, Linnæus, *Sp. Pl.* 1039 (1753); Loudon, *Arb. et Frut. Brit.* iv. 2502 (1838).

Sabina phœnicea and *Lycia*, Antoine, *Cupress. Gattung.* tt. 42, 44 (1860).

A shrub or tree, attaining usually about 20 ft. in height. Foliage dimorphic. Leaves on young plants, and very rarely on isolated branches of adult trees, acicular, spreading in whorls of threes, not jointed at the base, decurrent on the branchlet, about $\frac{1}{4}$ in. long, with two stomatic lines on both the upper and lower surfaces. On adult trees, branchlet systems two- to three-pinnate; ultimate branchlets terete, about $\frac{1}{25}$ in. in diameter; leaves scale-like, either in four ranks in opposite decussate pairs, or in six ranks in alternating whorls of threes, closely appressed, ovate-rhombic, about $\frac{1}{25}$ in. long, blunt at the apex, serrulate in margin, rounded on the back, which is often marked with a longitudinal or oval furrow.

Flowers usually monœcious, rarely diœcious.[1] Fruit very variable in size and shape, ripening in the second year, on short scaly stalks, shining yellow or reddish brown, with remarkably fibrous yellowish flesh; composed of six to eight scales, with no distinct lines of separation between them, each marked by a minute or obsolete mucro; in the typical form, globose or sub-globose, $\frac{1}{4}$ in. to $\frac{1}{2}$ in. in diameter. Seeds variable in number, three to nine, shining brown, separable with great difficulty from the adherent yellow flesh, triquetrous, furrowed longitudinally with two or three depressions for the closely adherent glands.

This species is remarkably variable in the size, shape, and colour of the fruits. Five or six varieties can be distinguished in specimens gathered by Jahandiez[2] near Hyères, which were sent to me by Lord Walsingham. The typical form of the species has globose berries; but these vary in size from $\frac{1}{4}$ in. to $\frac{1}{2}$ in. in diameter, and in colour from dark reddish brown to yellow or orange brown.

1. Var. *turbinata*, Parlatore, in *Fl. Ital.* iv. 91 (1867).

Juniperus turbinata, Gussone, *Fl. Sic. Syn.* ii. 634 (1844).

Juniperus oophora, Kunze, in *Flora*, xxix. 637 (1846).

Fruit ovoid or shortly conical. Seeds deeply furrowed from base to apex, and compressed at the summit into a sharp transverse edge. This variety is met with in

---

[1] This is De Coincy's opinion. Most authors say that it is usually diœcious and occasionally monœcious. Some of the trees we have seen in cultivation are certainly monœcious.

[2] Albert and Jahandiez, *Pl. Vasc. du Var*, 451 (1908), state that shrubs with large globose berries grow on the maritime sands, whilst those with small berries occur in rocky situations. De Coincy, in *Bull. Soc. Bot. France*, xlv. 432 (1898), refers the form with large globose berries to *J. Lycia*, Linnaeus, which may be named, if considered worth distinguishing, as *J. phœnicea*, var. *Lycia*, Loiseleur, *Nouv. Duham.* vi. 47, t. 17 (1812).

nearly all the localities where the typical form occurs, and, like it, is variable in the size, shape, and colour of the berries.

2. Var. *filicaulis*, Carrière, *Conif.* 51 (1855) and *Conif.* 52 (1867).

Juniperus myosuros, Sénéclauze, *Catalogue*, 1854, p. 35.

A shrub with elongated twisted branches and slender pendulous branchlets. Leaves scale-like, as in the type; but occasional branchlets bear acicular juvenile foliage. The parent plant, which was 3 ft. high in 1867, is said by Carrière to have originated from a seed of *J. phœnicea*, which was sown by Sénéclauze in his nursery at Bourg-Argental (Loire), sometime before 1854. Sénéclauze, however, in his *Catalogue*, 1867, p. 11, calls this plant *J. thurifera hybrida myosuros*; and its origin must be considered doubtful. We have seen no specimens.

J. phœnicea is widely spread throughout the Mediterranean region, occurring in Spain and Portugal, south-eastern France, Corsica, Sardinia, Italy, Sicily, Dalmatia, Greece, Rhodes, Cyprus, Crete, and the Crimea; but appears to be unknown in Asia Minor. It is also common in Algeria, Morocco, the Canary and Madeira Islands. It usually grows in arid situations on rocky hills, and often forms extensive and impenetrable thickets, as in La Camargue at the mouth of the Rhone. It ascends in the Riviera to 4500 feet. In Algeria[1] it is common on the coast, and on the southern slopes of the mountains of the interior, where it is often the only arborescent vegetation, ascending to 6000 feet.

In the Canary Isles, *J. phœnicea* was formerly one of the characteristic trees of the coast-region between 600 and 2000 ft. elevation; but has been much destroyed on account of its use for firewood. Dr. Burchard[2] states that it is still plentiful on the north side of Gomera and Hierro; but is nearly extinct on Grand Canary and Teneriffe, where only a few specimens remain in the south. On Gomera, it is usually seen as a globose bush on the cliffs, but becomes a tree when old. On the west point of Hierro, there are specimens supposed to be 1000 years old. Dr. Burchard[2] reproduces photographs of two of these remarkable trees, which have short stems, 4 to 5 ft. in diameter, with enormous crowns, spreading for an immense distance on one side of the trunk, as the result of the continuous influence for centuries of the north-east trade-wind.

Dr. Grabham tells us that in Madeira, *J. phœnicea* was formerly widely distributed from sea-level to the highest summits of the mountains, but is now nearly extinct. The wood of this species is still to be seen in enormous beams and slabs in old buildings, and its fragrant roots of great size are often found underground.

According to Aiton, *J. phœnicea* was first cultivated in Britain in 1683 by James Sutherland, curator of the Edinburgh Botanic Garden. It usually forms a pyramidal shrub or low tree, dense in habit; but is now rare in cultivation in this country. There are specimens at Highnam, Bicton, and Rostrevor. These bear small globose orange-coloured fruits.

It loves a warm climate, and is scarcely hardy in Germany; but a specimen,[3]

---

[1] Lefebvre, *Les Forêts de l'Algérie*, 431 (1900).   [2] In *Mitt. Deut. Dend. Ges.* 1911, pp. 286, 287.

[3] Kirchner and Schröter, *op. cit.* 316 (1906).

sheltered by other trees and 6 ft. high, at Tübingen, bore, without any injury but a slight browning of the leaves, a minimum temperature in winter of $-29\frac{1}{2}°$ Cent.

(A. H.)

JUNIPERUS FLACCIDA

Juniperus flaccida, Schlechtendal, in *Linnæa*, xii. 495 (1838); Sargent, *Silva N. Amer.* x. 83, t. 519 (1896), and *Trees N. Amer.* 89 (1905); Kent, Veitch's *Man. Conif.* 177 (1900).
Juniperus fœtida, var. *flaccida*, Spach, in *Ann. Sc. Nat.* xvi. 300 (1841).
Juniperus gracilis, Koch, *Berl. Allg. Gartenzeit.* 1858, p. 341 (not Endlicher).
Sabina flaccida, Antoine, *Cup. Gatt.* 37, tt. 49, 50 (1857).

A tree, attaining 30 to 40 ft. in height, with brown bark separating into thin scales; branches widely spreading, with long pendulous branchlets. Leaves dimorphic. Adult foliage: leaves in opposite decussate pairs, slightly spreading, ovate-lanceolate, about $\frac{1}{12}$ in. long, decurrent, ending in a sharp cartilaginous point, rounded on the back, which is marked with a linear sunken gland, often exuding resin. Juvenile foliage, usually on the ends of some branchlets of adult trees, acicular-subulate, spreading, usually in whorls of threes, rarely in pairs, decurrent, about $\frac{1}{4}$ in. long, gradually tapering from the base to the very sharp cartilaginous apex; upper surface concave, with inflexed margins, and with two narrow stomatic lines; lower surface marked near the base with a linear gland, often exuding resin. Similar spreading acicular leaves, in pairs or in threes, are borne on the main axes of the branchlet-systems, and like these turn reddish brown in the third and fourth year, and fall in succeeding years, leaving the branchlets smooth with a scaly bark.

Flowers monœcious. Fruit, ripening in the second year, on a short ($\frac{1}{16}$ in. long) scaly stalk, four- to six-bracteate at the base, sub-globose, about $\frac{1}{2}$ in. in diameter, reddish brown with a glaucous bloom, and marked on the surface with a few minute tubercles; composed of six to eight opposite scales, each indicated by a reflexed triangular mucro. Seeds, six to twelve, several often aborted, embedded in a resinous pulp; cotyledons two. (A. H.)

This species is a native of Texas and Mexico. It is limited in Texas to the Chisos Mountains, where it was discovered in 1888 by Dr. V. Harvard. It is common in north-eastern Mexico, at elevations of 6000 ft. to 8000 ft., on the hills to the east of the tablelands, ranging from Coahuila to Oaxaca, and extending eastward to about a hundred miles from the coast.

It was introduced[1] in 1838 from Mexico by Hartweg, but is probably too tender for our climate in most parts, as the only specimen which we know of in Britain is a fine tree at Bicton, which I found to be about 40 ft. by 3 ft. 10 in. in 1906. It grows in a sheltered hollow, and bears fruit regularly, which is smaller in size than in native specimens, and contains only imperfect seed.

Carrière states[2] that it is not hardy at Paris, but he mentions a tree at Angers

---

[1] Loudon, *Gard. Mag.* xv. 241 (1839), and xvi. 10 (1840). [2] *Conif.* 49 (1867).

10 in. in diameter. It is said[1] to be occasionally cultivated in the south of France and in Algeria; but we have seen no specimens. There is a small tree in the Botanic Garden at Genoa, and a larger one in the Botanic Garden at Naples, which was bearing ripe fruit in March 1910. (H. J. E.)

JUNIPERUS THURIFERA

Juniperus thurifera, Linnæus, *Sp. Pl.* 1039 (1753); Loudon, *Arb. et Frut. Brit.* iv. 2503 (1838); Parlatore, in De Candolle, *Prod.* xvi. 2, p. 487 (1868); Laguna, *Fl. Forest. Hispan.*, i. 103 (1883); De Coincy, in *Bull. Soc. Bot. France*, xlv. 430 (1898); Kent, Veitch's *Man. Conif.* 191 (1900).

Juniperus hispanica, Miller, *Gard. Dict.* ed. 7, No. 13 (1757), and ed. 8, No. 13 (1768).

Juniperus sabinoides, Endlicher, *Syn. Conif.* 23 (1847) (in part).

Juniperus cinerea, Carrière, *Conif.* 35 (1867).

A tree, attaining in Spain a height of 30 ft. to 40 ft. Leaves dimorphic. Adult foliage, with flattened branchlet-systems, pinnately divided mostly in one plane. Young branchlets tetragonal, slender, $\frac{1}{20}$ in. in diameter; leaves in opposite pairs in four ranks, appressed but free at their acuminate apices, ovate, about $\frac{1}{16}$ in. long, adnate to the branchlet in their basal half, marked on the back with an oblong glandular depression, minutely denticulate in margin. Juvenile foliage often present on adult trees; leaves in opposite pairs in four ranks, spreading, acicular, decurrent, $\frac{1}{12}$ to $\frac{1}{4}$ in. long, whitened on the upper surface.

Flowers diœcious. Fruit on short scaly stalks, ripening in the second year, sub-globose, $\frac{1}{3}$ in. or a little more in diameter, dark blue with a slight glaucous bloom when mature; composed of six scales in opposite decussate pairs, two at the base, the upper four scales meeting at the apex, which is marked with their lines of separation; each scale with a minute mucro. Seeds, two, three, or four, immersed in a granular sweet fragrant flesh, ovate, triquetrous, $\frac{1}{5}$ in. long, shining brown, smooth, narrowed at the apex to a curved point, with two or three oblong resin-pits at the base.

1. Var. *gallica*, De Coincy, in *Bull. Soc. Bot. France*, xliv. 232 (1897), and xlv. 430 (1898); Holmes, *Pharmac. Soc. Museum (London) Report*, 1907, p. 26.

A small tree, apparently differing only from the type, in the one to three seeds, being striate and not smooth on the surface, more prominent at the apex, and less angular.

This variety, which is the French form of the species, was first noticed in 1830 by Mutel, who gave it the name of *J. Sabina*, var. *arborea*.[2]

It appears to be confined to the Dauphiné, where it occurs at a few stations in the valley of the Isère, in the immediate neighbourhood of Grenoble; and in the valley of the Durance, chiefly near Embrun. Near Grenoble it is found on the

---

[1] Sargent, *Silva N. Amer.* x. 83 (1896).

[2] Mathieu, *Flore Forestière*, 519 (1897) refers the trees at Saint Crépin to *J. Sabina*. Cf. also Vidal, in *Bull. Soc. Bot. France*, xliv. 51 (1897).

mountains of Comboire, Néron, and Saint Eynard.   In the valley of the Durance it is more abundant, and grows at Guillestre, Saint Clément, Saint Crépin, in the valley of Ubaye, near Gap, and at Remollon.   M. Ph. Guinier informs us that in all these localities it grows on dry limestone soil, and usually as isolated trees. At Saint Crépin, however, it forms a small wood above the village, at 3500 ft. to 4000 ft. altitude.   It is usually a small tree, 20 ft. to 25 ft. high, but in rare cases attains 40 ft. in height.   The trunk is short, irregular, and deeply furrowed; and is frequently 6 ft. to 10 ft. in girth—one tree at Saint Crépin being as much as 17 ft. in girth.   It attains a great age, a section in the Museum of the Forestry School at Nancy, 0.94 metres in girth, showing 175 annual rings, and another 1.48 metres in circumference showing 169 annual rings.

J. thurifera is a native of south-eastern France, Spain, Portugal, Sardinia,[1] Morocco, and Algeria.   It is common in the mountains of central and southern Spain, occasionally forming pure open woods, one of which in the Sierra de Albarracin is figured by Willkomm;[2] or growing mixed with other conifers, as in the fine forests of *Pinus Laricio* in the Serrania de Cuença.   Here,[2] on the Muela de S. Juan, near Tragacete, it attains 25 ft. to 35 ft. high, and 10 ft. to 13 ft. girth.[3]   Laguna, who gives many localities for this species in Spain, states that it never ascends to the high altitudes occupied by *J. Sabina*, nor descends to the hot and sandy plains, where *J. phœnicea* is often seen.   It inhabits the zone between 2500 and 3500 ft. altitude, where there are abrupt changes of temperature, which it supports well.   It is always met with on soils which are either pure limestone, or contain lime in considerable quantity.

In Algeria,[4] *J. thurifera* grows mainly in the cedar forests at high elevations, where it is usually a small tree, not exceeding 20 ft. in height.   It was collected in southern Morocco by Sir J. D. Hooker.   (A. H.)

Although this species was cultivated by Miller in 1752, it has never become common, and according to Kent only thrives in warm and sheltered situations. Loudon records a tree at Boyton 28 ft. high in 1837, and another at Croome, forty years planted, which was 30 ft. high in 1838, but we have not found these specimens now living.   The largest tree in England is one at Bicton, about 40 ft. high, 4 ft. 4 in. in girth, conical in shape, and bearing male flowers in April 1911.

There are two trees at Kew, about 30 ft. high, which were planted in 1870. Another at Bayfordbury, 30 ft. high, was planted in 1841.   Smaller trees exist at Highnam and Leonardslee.   I saw a tree bearing ripe fruit at Simon-Louis's nursery, Metz, which was about 40 ft. by 3 ft. in 1908.   (H. J. E.)

---

[1] Grisebach, *Veg. der Erde*, i. 572 (1872), states that it occurs in Sardinia.   Cf. Nyman, *Consp. Fl. Eur.* iii. 676 (1881).

[2] Willkomm, *Pflanzenverb. Iber. Halbinsel*, 160, 185, fig. 11 (1896).

[3] Dillwyn, *Hort. Collinson.* 27 (1843), quotes a letter written to Collinson in 1766 by Bowles, an engineer in Spain, who states that large trees, girthing 14 ft., with wide-spreading branches like a beech, grew in the mountains near the source of the Tagus.   Willkomm confirms this.   [4] Lefebvre, *Forêts de l'Algérie*, 431 (1900).

JUNIPERUS PACHYPHLÆA

Juniperus pachyphlæa, Torrey, *Pacific R. R. Rep.* iv. pt. v. 142 (1858) ; Sargent, *Silva N. Amer.* x. 85, t. 520 (1896), and *Trees N. Amer.* 90 (1905) ; Kent, Veitch's *Man. Conif.* 181 (1900) ; Britton and Shafer, *North American Trees*, 113 (1908).
Juniperus plochyderma, Parlatore, in De Candolle, *Prod.* xvi. 2, p. 492 (1868).
Sabina pachyphlæa and *plochyderma*, Antoine, *Cupress. Gatt.* 39, 40, t. 52 (1857).

A tree, attaining in America 60 ft. in height and 15 ft. in girth. Bark, different from that of all the other junipers, $\frac{3}{4}$ to 4 inches thick, deeply divided into small square scaly plates. Branchlets slender, angled, becoming light brown, terete, and scaly after the fall of the leaves. Leaves dimorphic : on vigorous branchlets, acicular, spreading, in threes and in opposite pairs, $\frac{1}{8}$ to $\frac{1}{4}$ in. long, tipped with slender elongated cartilaginous points ; upper surface concave and whitened, lower surface greyish green and keeled. The juvenile foliage gradually passes into the adult foliage ; ultimate branchlets tetragonal, $\frac{1}{25}$ in. in diameter, with scale-like leaves in opposite pairs, imbricated, closely appressed, about $\frac{1}{16}$ in. long, ovate-rhombic, rounded at the narrowed apex, minutely toothed in margin, convex on the back, which is marked with a depressed oval gland, often exuding resin ; leaves on the older branchlets tipped with a sharp point.

Flowers dioecious. Fruit ripening in the second year, sub-globose, nearly $\frac{1}{2}$ in. in diameter, sub-sessile, ebracteate, reddish brown covered with a glaucous bloom, tuberculate on the surface, with six to eight scales each marked by a slightly reflexed mucro. Seeds four, nearly filling up the cavity of the fruit, ovoid, angled, shining brown ; flesh scanty, fibrous, yellow.

Reputed juvenile forms of this species, vars. *conspicua, elegantissima*, and *ericoides*, differing in habit and with blue or whitish-blue foliage, have lately been obtained by Barbier [1] at Orleans ; and are now in cultivation at Kew [2] and Glasnevin.

This species grows on dry arid mountain slopes, at 4000 to 6000 feet elevation, from the Eagle and Limpio Mountains in south-western Texas, westward along the desert ranges of New Mexico and Arizona, and southwards into Mexico, where it occurs along the Sierra Madre to the state of Jalisco and over the mountains of northern Sonora and Chihuahua.

It was discovered in 1851 by Dr. S. W. Woodhouse in eastern New Mexico, and is considered by Sargent to be the most beautiful of all the west American Junipers, its thick checkered bark being unlike that of any other species.

It is uncertain when it was introduced into England. It is extremely rare, the only specimen which we have seen being a tree in Kew Gardens, about 20 ft. high, showing the peculiar bark, and producing on its stem several epicormic branches. This has not as yet produced flowers. (A. H.)

[1] *Mitt. Deut. Dend. Ges.* 1910, pp. 139 and 289. [2] *Kew Bulletin*, 1911, p. 101.

JUNIPERUS CHINENSIS

Juniperus chinensis, Linnæus, *Mantissa*, i. 127 (1767); Loudon, *Arb. et Frut. Brit.* iv. 2505 (1838);
Siebold et Zuccarini, *Fl. Jap.* 58, tt. 126, 127, (1844); Parlatore, in De Candolle, *Prod.* xvi. 2,
p. 487 (1868); Franchet et Savatier, *Enum. Pl. Jap.* i. 472 (1875); Masters, in *Journ. Linn.
Soc. (Bot.)* xviii. 497 (1881), and xxvi. 541 (1902), and in *Journ. Bot.* xli. 268 (1903);
Beissner, in *Mitt. Deut. Dend. Ges.* 1896, p. 69, and 1898, p. 32, and in *Bull. Soc. Bot. Ital.
Firenze*, 1898, p. 167; Kent, Veitch's *Man. Conif.* 169 (1900); Shirasawa, *Icon. Ess. Forest.
Japon*, i. text 29, t. 12, figs. 14-27 (1899); Diels, in Engler, *Bot. Jahrb.* xxix. 220 (1901).
Juniperus cernua and *dimorpha*, Roxburgh, *Fl. Ind.* iii. 839 (1832).
Sabina chinensis, Antoine, *Cupress. Gattung.* 54, t. 75 (1857).
Sabina Cabiancæ, Antoine, *Cupress. Gattung.* 41, t. 54 (1857).

A tree, attaining in China and Japan a height of 60 ft. Leaves of two kinds: on adult trees scale-like; ultimate branchlets $\frac{1}{25}$ in. in diameter, clothed with four ranks of leaves in opposite pairs, which are imbricated, closely appressed, narrowly rhombic, $\frac{1}{16}$ in. long, tapering to rather an obtuse apex, adnate to the stem, entire in margin; outer surface convex, green with a pale margin, and marked with a depressed oval or oblong gland; interiorly concave, with a raised narrow midrib, glaucous. On older branchlets the scale leaves are larger, about $\frac{1}{10}$ in. long, conspicuously glandular on the back, persistent four or five years. On young trees and on occasional branches of old trees, the juvenile foliage is linear-acicular, $\frac{1}{3}$ in. long, spreading, either in whorls of threes or in opposite pairs, tipped with a rigid spine-like point, adnate to the branchlets, swollen on the upper surface near the base, but not jointed; concave above, with a green midrib and two glaucous bands; green and convex beneath.

Flowers diœcious. Staminate flowers bright yellow, very numerous.[1] Fruit ripening in the second year, borne on the ends of short branchlets, which are covered with ordinary scale-leaves; brown covered with a thick white mealy bloom; variable in shape, commonly sub-globose, but widest and usually depressed at the summit, averaging $\frac{1}{3}$ in. in diameter, composed of four to eight scales. Seeds two or three, rarely four or five, immersed in a resinous pulp, shining deep chestnut brown, smooth, broadly ovoid, with a wide base, gradually tapering to a sharp thin-edged apiculate apex, compressed from before backwards, each surface convex, with a longitudinal groove near the thinner outer edges.

This species is readily distinguishable by the pale margins of the scale-like leaves, which mark the ultimate branchlets with a series of white crosses. In nearly all adult trees, acicular foliage with the leaves either ternate or in opposite pairs can be found on some of the branches.

VARIETIES

I. This species is very variable in habit in the wild state; and, as Beissner[2] points out, in the mountains of Shensi in China, both male and female trees exist,

---

[1] On certain trees at Kew, and in wild specimens of Shensi (*fide* Beissner) staminate flowers are borne on branchlets with acicular as well as with scale-like foliage; and this seems peculiar to *J. chinensis*.
[2] In *Mitt. Deut. Dend. Ges.* 1896, p. 69, and 1898, p. 32.

which bear exclusively acicular foliage; whilst others occur in both sexes with the leaves mostly scale-like. There are no grounds for supposing that the sexes are distinguished in nature by any peculiar habit; but in cultivation, owing to long-continued propagation by cuttings from trees of different habit, many female trees differ in appearance from that commonly met with in male trees. This is by no means universal, as there are two trees of the same habit, but of different sexes, in the Cambridge Botanic Garden. A common staminate form, with preponderating acicular foliage, and dense branches, forming a conical pyramid, was formerly distinguished as *J. struthiacea*, Knight, *Syn. Conif.* 12 (1850). A pistillate form, known at first as *J. flagelliformis*, Loudon, *Trees and Shrubs*, 1090 (1842), was introduced from Canton in 1839 by J. Russell Reeves, and was subsequently named *J. Reevesiana*, Knight, *Syn. Conif.* 12 (1850).

II. The following are either closely allied species or varieties of *J. chinensis* :—

1. *Juniperus sphærica*, Lindley, in Lindley and Paxton, *Flower Garden*, i. 58, fig. 35 (1850).

> *Juniperus chinensis*, Linnæus, var. *Smithii*, Gordon, *Pinetum*, 119 (1858) (not Loudon [1]).
> *Juniperus Fortunii*, Van Houtte, *ex* Gordon, *Pinetum*, 119 (1858).

A tree, 30 to 40 ft. high, discovered by Fortune[2] in the hills north-west of Ningpo and near Shanghai, where it is frequently planted around graves. The type specimen, preserved in the British Museum, does not differ from *J. chinensis* in the foliage, which is all scale-like, no acicular leaves being present; but is monœcious, and bears fruit, quite spherical in shape and larger than that of *J. chinensis*, $\frac{7}{10}$ in. in diameter, smooth, dark purple, scarcely glaucous, containing five seeds, which are larger than, but similar in shape and colour to those of *J. chinensis*.

This is kept separate from *J. chinensis*, but with some doubt, by Parlatore,[3] Kent,[4] and Masters;[5] and is probably only a variety of that species, differing mainly in the larger spherical fruit, not covered with a whitish bloom, and containing numerous seeds. The branch collected by Fortune is monœcious; but this is perhaps an abnormality.

A specimen (No. 6576) which I collected in Fang district in the province of Hupeh, with large spherical glaucous berries, resembles Fortune's plant, but is diœcious and with only three seeds in each fruit. Wilson found in the same province another specimen with smaller four-seeded fruits.[6]

Fortune sent seeds in 1850 to Standish and Noble, who probably raised *J. sphærica* in their nursery; but I have found no living specimens, either monœcious or with the large spherical berries of Fortune's plant. The trees now known in cultivation either as *J. sphærica* or *J. sphærica Sheppardi*, Veitch, *Man. Conif.* 290 (1881), usually prove to be female trees of *J. chinensis*, with a rather spreading habit.

---

[1] *J. chinensis Smithii*, Loudon, *Arb. et Frut. Brit.* iv. 2505 (1838), described (long before Fortune's discovery of *J. sphærica*) as monœcious with angular fruits, was supposed to be of Nepalese origin, and may have been *J. religiosa*.

[2] *Residence amongst the Chinese*, 63, 140 (1857).

[3] In De Candolle, *Prod.* xvi. 2, p. 488 (1868).

[4] Veitch's *Man. Conif.* 190 (1900).

[5] In *Journ. Linn. Soc.* (*Bot.*) xxvi. 543 (1902).

[6] The fruits of cultivated trees of *J. chinensis* are usually three-seeded; but occasionally four or five seeds are present, the fruits in this case being small, covered with whitish bloom, and depressed at the apex, and not in the least like the large spherical bluish fruits of *J. sphærica*.

2. Var. *Sargenti*, Henry (var. *nova*).

Juniperus procumbens, Sargent, *Forest Flora Japan*, 78 (1894), and in *Garden and Forest*, x. 421 (1897) (not Siebold).

A sea-shore plant,[1] forming dense mats, and sending out for long distances prostrate creeping stems, which bear foliage similar to that of *J. chinensis*: branchlets tetragonal, covered with minute scale-like appressed leaves, furrowed on the back; no acicular leaves being present on adult plants. Berries bluish, covered with a slight glaucous bloom; seeds three, like those of *J. chinensis*.

This is said by Sargent, who has kindly sent a dried specimen, to grow on the coasts of Korea and Japan, on low grassy bluffs freely exposed to the ocean gales. A few plants were raised in the Arnold Arboretum from seeds gathered in 1892 near the Aino village of Horobetsu on the coast of Yezo. I have seen no living specimens.

III. The following varieties are of horticultural origin :—

3. Var. *albo-variegata*, Veitch, *Man. Conif.* 288 (1881).

A compact shrub, differing from the type in many of the branchlets being creamy white at the tips. It usually bears adult scale-like foliage; but a form with acicular foliage is also in cultivation. It is said to have been first introduced from Japan by Fortune, and subsequently by J. Gould Veitch. It is known by several names, as var. *variegata*, Fortune; var. *argentea*, Gordon; and var. *argenteo-variegata*, Rehder.

4. Var. *aurea*, Young, *ex Gard. Chron.* 1872, pp. 8, 1193.

An upright form, with adult scale-like foliage, having the whole of the young growth suffused with a deep golden yellow, which gradually turns green in the summer. The colour is heightened by exposure to the sun. This originated in Maurice Young's nursery at Milford, Godalming, where the original plant was 12 ft. high in 1872, and when Elwes saw it in 1909 was still a small tree. The best specimens we know of this are at Burnham Park, the residence of Sir Harry J. Veitch, and are about 18 ft. high.

5. Var. *japonica aurea*, Masters, in *Journ. R. Hort. Soc.* xiv. 211 (1892).

Juniperus japonica aurea, Carrière, *Conif.* 32 (1867).

A straggling shrub, with long decumbent branches; branchlets tinged with golden yellow. This is said to have been first introduced by Fortune from Japan, and subsequently by J. Gould Veitch.

6. Var. *japonica aureo-variegata*, Masters, in *Journ. R. Hort. Soc.* xiv. 211 (1892).

Juniperus japonica variegata, Carrière, *Conif.* 31 (1867).

A dense dwarf shrub, with many of the branchlets of a deep golden yellow. Also of Japanese origin.

7. Var. *Pfitzeriana*, Späth, *Catalogue*, No. 104, p. 142 (1899).

A broad pyramidal shrub, with dense horizontal branches, and long and slightly pendulous branchlets, clothed with glaucous foliage. This originated in Späth's nursery at Berlin, where the original plant was 10 ft. high in 1901.

8. Var. *japonica*, Vilmorin, in *Hortus Vilmorin.* 58 (1906).

Juniperus japonica, Carrière, *Conif.* 31 (1855).

---

[1] *J. Thunbergii*, Hooker and Arnott, *Bot. Beechey's Voyage*, 271 (1841), gathered in the Liu Kiu Islands, is represented at Kew by a specimen with acicular leaves, which bears no fruit. It cannot be identified with certainty; but may be a form of *J. chinensis*.

A diffuse bushy plant, with mostly juvenile acicular foliage in threes; some of the terminal branchlets being covered with adult scale-like leaves, and occasionally bearing fruit.[1]

This is a juvenile form of *J. chinensis*, of which the Japanese make dwarf plants, that are frequently imported into Europe. It was erroneously identified by Carrière with *J. procumbens*,[2] Siebold; and this mistake has been copied by Kent and other writers. Sargent describes,[3] as *J. japonica*, a plant of compact habit, with many erect branches and acicular bluish-green needles, which is often cultivated in Japanese gardens, and is very hardy and distinct in appearance. It retains its peculiar compact juvenile habit for several years, but often becomes thin and ragged before it is 12 ft. high, and loses its value as an ornamental plant. Sargent adds that this is one of the most difficult conifers to transplant. Judging from the description, the variety alluded to by Sargent is *J. japonica pyramidalis*, Carrière, *Conif.* 32 (1867), but I have seen no specimen.   (A. H.)

DISTRIBUTION

J. chinensis is a native of China, Mongolia, and Japan. In China, it is frequently cultivated in temple grounds; but appears to be truly wild in the mountains of Hupeh, Shensi, and Szechwan, where it is usually found growing solitary on cliffs, but occasionally as underwood[4] in the forests. Père David[5] found it abundant on the Moni-ula range of the Ourato territory in south-western Mongolia.

This is a favourite tree in the parks and temples of Peking, where it attains a great age. The largest I saw were at the temple of Confucius, and were said to be over 700 years old. In a double avenue here, one tree on the left-hand side was about 40 ft. by 17 ft.; and another whose trunk was covered with burrs was 14 ft. in girth. At the Ming Tombs there are many very old junipers and fine specimens of arbor vitæ, together with numerous pine trees (*Pinus Bungeana* and *P. funebris*).

In Japan, it is also common in cultivation; but is recognised as a native tree by Japanese botanists[6] and foresters. Shirasawa states that it is wild in the mountains of the Shinano province in central Hondo, mixed with *Pinus densiflora* and *Quercus serrata*, and forming a tree 30 to 40 ft. in height, with straggling contorted branches and greyish green foliage. It occasionally attains a large size, as Sargent[7] mentions two venerable trees at the temple of Zenkogi in Nagano,[8] which are 70 to 80 ft. high with hollow trunks about 6 ft. in diameter. I never saw it wild; but I saw several handsome trees in the ancient temple of Tennoji at Osaka, with fine large round heads, the best with a trunk 10 ft. high and 10 ft. in girth, and with a spread of branches of about 14 yards. One had a very twisted and fluted stem. This species is known to the Japanese as *Bya Kushin*.

---

[1] This is *J. japonica*, Carrière, and what is cultivated under that name in Veitch's nursery at Coombe Wood.

[2] *J. procumbens*, Siebold, is a very distinct species. Cf. p. 1422.   [3] In *Garden and Forest*, x. 421 (1897).

[4] Diels, in Engler, *Bot. Jahrb.* xxix. 220 (1901).

[5] Franchet, *Pl. David.* i. 291 (1884), describes, as var. *pendula*, a form with elongated pendulous branches, found by Père David in Shensi.

[6] Matsumura, *Index Pl. Jap.* 10 (1905), gives as localities, Kunasiri in the Kuriles, Rebunsiri in Yezo, Hakoda in Nippon, and the Liu Kiu Islands.   [7] *Forest Flora of Japan*, 78 (1894).

[8] Shirasawa states "the temple of Kenchoji in the province of Sagami."

This tree was first described in 1767 by Linnæus, who states that it was then cultivated at Upsala. It appears, however, to have been first introduced into England by William Kerr, who sent plants from Canton to Kew in 1804. Next to the Virginian juniper, it is the species now most commonly cultivated in nurseries and private gardens. It is absolutely hardy everywhere, seems quite indifferent to soil, and in many places is a very ornamental shrub or small tree. It ripens seed, which, so far as my observations go, germinate the year after they are sown. The finest trees we have seen are :—At Arley Castle, several old specimens, the largest of which was 48 ft. by 5 ft. 3 in. in 1907; at Eastnor Castle, a well-shaped tree, 48 ft. by 3 ft. which was bearing fruit in 1908; at Hardwicke, near Bury St. Edmunds, a fine tree, 38 ft. by 3 ft. 10 in., with abundant ripe berries in 1905; at Redleaf in Kent, a tree about 35 ft. high in 1907; at Westonbirt, a tree 32 ft. high and growing fast, which in 1909 was covered with fruit. A tree growing at Rood Ashton, Wilts, about 25 ft. high, wide-spreading in habit, was figured in *Gardeners' Chronicle*, xlii. 163, fig. 63 (1907).   (H. J. E.)

JUNIPERUS BERMUDIANA

Juniperus bermudiana, Linnæus, *Sp. Pl.* 1039 (1753); Loudon, *Arb. et Frut. Brit.* iv. 2498 (1838); W. J. Hooker, in *London Journ. Bot.* ii. 141, t. 1 (1843); Endlicher, *Syn. Conif.* 29 (1847); Parlatore, in De Candolle, *Prod.* xvi. 2, p. 490 (1868) (in part); J. M. Jones, *Botany of Bermuda*, 272 (1873); Hemsley, in *Gard. Chron.* xix. 656, figs. 105, 106 (1883), in *Journ. Bot.* xxi. 259 (1883), and in *Voy. Challenger, Bot.* i. 81, t. 5 (1885); Sargent, in *Garden and Forest*, iv. 289, figs. 51, 52 (1891); Masters, in *Journ. Bot.* xxxvii. 1-11 (1899); Kent, Veitch's *Man. Conif.* 166 (1900).

Juniperus oppositifolia, Moench, *Meth.* 698 (1794).

Juniperus pyramidalis, Salisbury, *Prod.* 397 (1796).

A tree, attaining 50 ft. in height in the Bermudas, with dark red bark and spreading branches. Foliage of two kinds: on adult trees scale-like; ultimate branchlets tetragonal, about $\frac{1}{20}$ in. in diameter, densely covered with imbricated leaves, which are usually in four ranks, about $\frac{1}{12}$ in. long, ovate, obtuse at the narrow incurved apex, greyish green or glaucous on the back, which is usually marked with a longitudinal furrow, entire in margin; on older branchlets, in four ranks or ternate in six ranks, those on the main axes always ternate, up to $\frac{1}{7}$ in. long, and becoming acuminate at the apex. Juvenile foliage, occasionally present on some branches on old trees, in alternate whorls of threes, about $\frac{1}{3}$ in. long, acicular, slightly spreading; upper surface whitened with a raised midrib; lower surface greyish green, very convex, and marked with a longitudinal furrow.

Flowers diœcious.[1] Fruit ripening in the first year, sub-globose, about $\frac{1}{4}$ in. in diameter, dark brown, covered more or less with a bluish bloom, with six to eight scales, each marked by a depression with a minute mucro. Seeds, two to three, immersed in fleshy pulp, shining chestnut brown, ovoid, oblique at the broad base, tapering to the thin-edged apex, with two furrows on the outer surface.

[1] Stewardson Brown, in *Proc. Acad. Nat. Sc. Philadelphia*, lxi. 488 (1909), says that the tree is diœcious: "At the time of flowering in March and April the staminate trees are a golden colour, presenting a strong contrast with the rich blue-green of the pistillate tree." Most authorities say it is monœcious, but specimens with fruit in the British Museum bear no male flowers.

This species is confined[1] to the Bermuda Islands, where it is the only indigenous exogenous tree. It was formerly abundant[2] on the islands, thriving both on the dry limestone hills and in the brackish swamps. The trees grow to a large size in the salt-water marshes, and have much darker heartwood than those on the hills, but not nearly so durable.[3] Large trees are no longer common. Sargent[4] gives an illustration of one in the churchyard of Devonshire parish, which was about 50 ft. high and 15 ft. in girth, and states that only two larger trees were known to exist. Another illustration shows the habit of this species in the Devonshire marshes. The wood was formerly much used in shipbuilding and in making beautiful furniture. "Cedar" chests and cabinets over two hundred years old are preserved as heirlooms by the descendants of the old Bermuda families, who live in houses finished with this wood, which becomes with age a rich dark colour like mahogany.

The Bermuda juniper was cultivated[5] in England as early as 1684, but it is not hardy in the climate of London. Knight and Perry[6] state that it was hardy in their day in Devonshire, and that plants remained uninjured in the open air during the winter of 1849 in Oxfordshire. We have seen, however, no living specimens in England except a shrub at Bicton, about 2 ft. high, which is not thriving, and small plants which were received at Kew[7] in 1910. Reputed specimens of this species at Castlewellan,[8] as well as some plants that were formerly cultivated at Kew under the name *J. bermudiana*, turned out to be *Cupressus funebris*.

This species is cultivated in the south of France, Italy, and the Canary Isles. There is a good specimen, which bears fruit regularly, in Dr. Perez' garden at Orotava. Dr. Perez,[9] as the result of numerous experiments, finds that seeds of this species germinate speedily when immersed in boiling water for three, six, or ten seconds, and at the end of the time are plunged into water at 65° to cool. Longer exposure to boiling water destroys the embryos. (A. H.)

JUNIPERUS VIRGINIANA, Pencil Cedar

Juniperus virginiana, Linnæus, *Sp. Pl.* 1039 (1753); Loudon, *Arb. et Frut. Brit.* iv. 2495 (1838); Parlatore, in De Candolle, *Prod.* xvi. 2, p. 488 (1868); Sargent, *Silva N. Amer.* x. 93 (in part), t. 524 (1896), and *Trees N. Amer.* 94 (1905); Kent, Veitch's *Man. Conif.* 192 (1900); Mohr, *U.S. Forestry Bull.* No. 31 (1901); Pinchot, *U.S. Forestry Circ.* No. 73 (1907); White, *U.S. Forestry Circ.* No. 102 (1907); Clinton-Baker, *Illust. Conif.* ii. t. 74, fig. 4 (1909).

Juniperus caroliniana, Miller, *Dict.* ed. 8, No. 4 (1768).

Juniperus arborescens, Moench, *Meth.* 699 (1794).

Juniperus fragrans, Salisbury, *Prod.* 397 (1796).

Sabina virginiana, Antoine, *Cupress. Gattung.* 61, tt. 83, 84 (1857).

A tree, attaining in North America 100 ft. in height and 12 ft. in girth, often

---

[1] Kent states that pieces of its wood were found 50 ft. below low-water mark, during dredging operations undertaken for the construction of a dock.

[2] J. M. Jones, *Botany of Bermuda*, 272 (1873), states that the trees are becoming extinct, no longer growing in the salt-marshes; but Sargent's later account does not confirm this.

[3] A. Haycock, in *Gard. Chron.* xxv. 176 (1899). Capt. L. Clinton-Baker informs us that the best trees in 1911 averaged 40 to 50 ft. in height and 9 ft. in girth. [4] *Garden and Forest*, iv. 289, figs. 51, 52 (1891).

[5] For the early history of this species, see Hemsley's account, cited at the head of this article. [6] *Syn. Conif.* 11 (1850).

[7] A tree in the Temperate House at Kew, which was about thirty years old, was cut down in 1905.

[8] *List of Plants Hardy at Castlewellan*, 65 (1897). [9] In *Gard. Chron.* 1. 127 (1911).

with the trunk fluted at the base.   Bark, $\frac{1}{4}$ in. thick, reddish brown, shredding off in long strips.   Leaves of two kinds: on adult trees scale-like; ultimate branchlets very slender, $\frac{1}{30}$ in. in diameter, clothed with four ranks of leaves in opposite pairs, which are imbricated, appressed but free towards the apex, ovate, acute or acuminate with a short point, $\frac{1}{16}$ in. long, adnate, entire in margin; green and glabrous on the back, which is often marked with a small oval glandular depression.   On older branchlets the scale-leaves are broadly ovate, acute, larger, about $\frac{1}{12}$ in. long, and become brown and withered, ultimately disappearing in the fifth or sixth year. Acicular leaves often present on occasional branches of adult trees, spreading in pairs, $\frac{1}{8}$ to $\frac{1}{4}$ in. long, ending in a spine-tipped apex, adnate, swollen at the base, but not jointed; concave and glaucous above; green and convex beneath.   The acicular leaves on seedlings and young plants are arranged in alternate whorls of threes.

Flowers usually diœcious, rarely monœcious.   Fruit ripening in one year, borne at the ends of short branchlets, which are furnished with ordinary adult scale-leaves; sub-globose, but usually longer than broad, about $\frac{1}{4}$ in. long, bluish, covered with a glaucous bloom, composed of four or six scales.   Seeds, one or two, immersed in a resinous flesh, ovoid, broad at the base, tapering towards the apex, smooth, shining chestnut brown, with two indistinct longitudinal furrows, and with or without resin-pits at the base.   Seedling [1] with two ligulate cotyledons, which have no resin-canals; primary needles with one resin-canal immediately below the dorsal ridge.

ALLIED SPECIES AND VARIETIES

This species was formerly supposed to spread over the greater part of the North American continent; but the geographical forms of the west and south are now considered by American botanists to be two distinct species.

I. *Juniperus scopulorum*, Sargent, in *Garden and Forest*, x. 420, fig. 54 (1897), *Silva N. Amer.* xiv. 93, t. 739 (1902), and *Trees N. Amer.* 96 (1905).

Juniperus excelsa, Pursh, *Fl. Amer. Sept.* ii. 647 (1814) (not Bieberstein).

Juniperus dealbata, Loudon, in *Gard. Mag.* xvi. 639 (1840), and *Trees and Shrubs*, 1090 (1842) (not Douglas [2]).

(?) *Juniperus fragrans*, Knight and Perry, *Syn. Conif.* 13 (1850); Carrière, *Conif.* 57 (1855).

Juniperus bacciformis, Carrière, *Conif.* 56 (1855).

Juniperus Henryana, Brown (Campst.), in *Trans. Bot. Soc. Edin.* ix. 377 (1868), and in *Gard. Chron.* 1873, p. 8.

Juniperus virginiana, Linnæus, var. *scopulorum*, M. E. Jones, in *Bull. Univ. Montana, Biol. Ser.* No. 13, p. 12 (1910).

Sabina scopulorum, Rydberg, *Flora of Colorado*, 10 (1906).

A tree about 40 ft. high and 9 ft. in girth, often divided near the base into several stems.   Adult foliage like that of *J. virginiana*, but with a disagreeable pungent smell, and with stouter branchlets and leaves marked on the back by a conspicuous glandular pit.   Fruit ripening in the second year, globose, $\frac{1}{4}$ in. in diameter, bright blue covered with a glaucous bloom.   Seeds, two, triquetrous, reddish brown, prominently angled, and with one longitudinal groove.

This tree grows on dry rocky ridges and, except near the coast, usually at over

---

[1] Hill and De Fraine, in *Ann. Bot.* xlii. 696 (1908).

[2] Gordon, in *Gard. Chron.* 1842, p. 562, states that *J. dealbata*, Douglas, is the same as *J. nana*, Willdenow.   Cf. p. 1401, note 1.

5000 feet elevation, from the eastern foothills of the Rocky Mountains from Alberta to Texas, westward to the coast of British Columbia and Washington, and to eastern Oregon, Nevada, and northern Arizona. This species was introduced into England from north-western America in 1839, under the name *J. dealbata*, Loudon; and was erroneously identified with *J. occidentalis*, Hooker, a species that has apparently never been in cultivation in this country. Soon after its introduction, *J. scopulorum* seems to have been known as *J. fragrans*, a name which still exists in nursery catalogues. It is a rare tree; but there is a specimen at Kew, not very thriving and about 15 ft. high, which is labelled *J. occidentalis fragrans*. We obtained in 1911 a fruiting branch from a small plant, named *J. fragrans*, in Dicksons' nursery, Chester. It is cultivated in Germany,[1] at Darmstadt and at Tübingen, where it has borne a temperature of − 24° Cent.

II. *Juniperus barbadensis*, Linnæus, *Sp. Pl.* 1039 (1753); Loudon, *Arb. et Frut. Brit.* 2504 (1838); Mohr, *U.S. Forestry Bull.* No. 31, p. 37, plate ii. (1901); Sargent, *Silva N. Amer.* xiv. 89, t. 738 (1902), and *Trees N. Amer.* 95 (1905).

> *Juniperus virginiana*, Linnæus, var. *australis*, Endlicher, *Syn. Conif.* 28 (1847).
> *Juniperus virginiana barbadensis*, Gordon, *Pinetum*, 114 (1858) (in part).
> *Juniperus bermudiana*, Lunan, *Hort. Jamaic.* i. 84 (1814) (not Linnæus).
> *Sabina barbadensis*, Small, *Flora S.E. United States*, 33 (1903).

A tree, attaining 50 ft. in height and 6 ft. in girth; branches and branchlets pendulous. Adult foliage similar to that of *J. virginiana*, but branchlets more slender, and leaves smaller, acuminate, and conspicuously marked on the back by an oblong or linear oil-gland. Flowers diœcious. Fruit ripening in the first year, sub-globose, $\frac{1}{8}$ to $\frac{1}{6}$ in. in diameter, bluish with a glaucous bloom; seeds, one or two, ovoid, pointed, ridged.

This species occurs in inundated coastal river swamps from southern Georgia southward to the Indian River, Florida; and on the west coast of Florida from Charlotte Harbour to the Appalachicola River, often forming thickets under the shade of larger trees. It is often planted in the cities and towns near the coast from Florida to western Louisiana, and is now said to be naturalised on the Gulf Coast. This species also occurs in the West Indies, in San Domingo, the Bahamas, Antigua, St. Lucia,[2] and the mountains of Jamaica. It appears to be now extinct in Barbadoes.[2]

The "red cedar" which grows on the northern Bahama Islands is usually referred to this species, but it has lately been separated [3] as *Juniperus lucayana*, Britton, on account of its depressed globose and somewhat laterally flattened fruit. Little is left of it on account of its use formerly for construction and in more recent times for making pencils. The juniper of eastern Cuba is closely related to, if not identical with, that of the Bahamas.[3]

III. *Juniperus Bedfordiana*, Loudon, *Trees and Shrubs*, 1090 (1842).

> *Juniperus gracilis*, Endlicher, *Syn. Conif.* 31 (1847).
> *Juniperus virginiana*, Linnæus, var. *Bedfordiana*, Knight, *Syn. Conif.* 12 (1850); Parlatore, in De Candolle, *Prod.* xvi. 2, p. 489 (1868); Veitch, *Man. Conif.* 284 (1881); Kent, Veitch's *Man. Conif.* 193 (1900).
> *Juniperus virginiana*, Linnæus, var. *gracilis*, Sargent, in *Silva N. Amer.* x. 96, note 1 (1896).

---

[1] *Mitt. Deut. Dend. Ges.* 1906, p. 37, and 1908, p. 144.
[2] Cf. Stapf, in *Kew Bull.* 1911, p. 377.   [3] Britton and Shafer, *North American Trees*, 121 (1908).

A dense low tree, columnar in habit, with slender elongated pendulous branchlets; leaves bright green, acicular, like the juvenile foliage of *J. virginiana*. This handsome tree was first mentioned by Loudon, who merely states that it closely resembles *J. virginiana*. Its origin is unknown,[1] but it is usually considered to be identical with *J. barbadensis*, and the fact that it is rather tender in England supports this view. As it has only juvenile foliage, and apparently never bears fruit, the question cannot be decided.

IV. Many varieties of *J. virginiana*, which are always propagated by cuttings, have arisen in nurseries, no less than twenty-one varieties being enumerated in the Kew Hand-List. According to Loudon, it varies much when raised from seed, as at White Knights, where there were hundreds of trees in 1838, differing much in appearance. Some were low and spreading, others were tall and fastigiate, and some had pendulous branches. The foliage varied much in colour, being light green, dark green, or glaucous. The fruit also differed in size. The most important varieties are :—

1. Var. *pendula*, Knight, *Syn. Conif.* 12 (1850).

This exists in at least three forms :—

(*a*) Branches spreading, branchlets pendulous, leaves scale-like, bearing staminate flowers.

(*b*) Var. *pendula viridis*. Branches and branchlets pendulous ; leaves scale-like, bright green.

(*c*) Var. *Chamberlainii*, Knight, *loc. cit.* Branches spreading and reflexed ; branchlets pendulous ; leaves mostly acicular, of a greyish tint.

2. Var. *pyramidalis*, Carrière, *Conif.* 47 (1867).

Columnar in habit, with either glaucous or bright green foliage.

3. Var. *dumosa*, Carrière, *Conif.* 45 (1855).

A dense rounded shrub, with both acicular and scale-like leaves.

4. Var. *Schottii*, Gordon, *Pinetum*, 157 (1875).

A narrow pyramidal tree, dense in habit, with bright green scale-like leaves.

5. Var. *tripartita*, R. Smith, *ex* Gordon, *Pinetum*, 157 (1875).

A low spreading bush with dense branches, and bright green, usually acicular foliage.

6. Var. *glauca*, Knight, *Syn. Conif.* 12 (1850).

Differs from the type in its beautiful glaucous silvery foliage, which is most pronounced in spring.

7. Var. *Triomphe d'Angers*, Beissner, *Nadelholzkunde*, 127 (1891).

Young branchlets creamy white, contrasting well with the dark bluish-green of the older foliage.

8. There are several variegated varieties, as *alba variegata* and *aurea variegata*, Gordon, *Pinetum*, 157 (1875), which are usually poor in colour.

9. Var. *elegantissima*, Beissner, *Nadelholzkunde*, 128 (1891).

A pyramidal tree, with the tips of the young branchlets golden yellow.

[1] There appears to be no ground for Loudon's statement, *Trees and Shrubs*, 1118 (1842), that it is identical with *J. gossainthanea*, Loddiges, as the latter and *J. Bedfordiana* are kept distinct by Knight and Perry, *Syn. Conif.* 11, 12, 13 (1850). Endlicher states that *J. Bedfordiana* is a native of Mexico.

10. The dwarf forms are often like *J. Sabina*, and are hard to distinguish in the absence of fruits, except by the much stronger disagreeable odour of the bruised branchlets of the latter species.   (A. H.)

DISTRIBUTION

The distribution of this species, as now limited by Sargent,[1] is as follows :— From southern Nova Scotia and New Brunswick southward, often close to the sea-coast, to Georgia, southern Alabama and Mississippi, westward to the valley of the lower Ottawa river, eastern Dakota, eastern Nebraska, Kansas, Indian Territory, and eastern Texas ; not ascending the mountains of New England and New York, nor the high southern Alleghanies ; in middle Kentucky and Tennessee and northern Alabama and Mississippi, covering great areas of low rolling limestone hills with nearly pure forests of small bushy trees.

In New England it is very common in the south, rarer in Maine, New Hampshire, and Vermont ; but nowhere, so far as I saw, attains the size of old trees in England.   Dame and Brooks[2] give 25 to 40 ft. with a trunk diameter of 8 to 20 in. as the average size, and I saw none larger.   It grows here on principally dry, rocky, and exposed hills, but also sometimes in wet ground ; and on the abandoned cultivated fields which are so numerous in the hilly and poorer parts of Massachusetts is taking possession of the soil in many places.   At Boston I noticed that both in Prof. Sargent's own grounds and in the Arboretum, pencil cedar was coming up freely from seed ; and I have no doubt it will be planted largely in suitable localities farther south.   The rapidly increasing demand for its useful wood has cleared out the accessible timber already in many districts.

In Canada, it is a comparatively rare tree, and is confined[3] to the limestone districts in the St. Lawrence valley and along Lake Ontario to the Niagara peninsula, where considerable areas were covered with it in 1888.   All the timber of any value has now been cut here, as it has been in New England generally.

Mohr says that there is hardly any tree in the Eastern States which is so indifferent to soil and climate as the juniper.   It thrives in the valley of the St. Lawrence and in New England, often growing on barren hillsides where few trees succeed ; on the exposed arid regions of Kansas and Nebraska, in air and climate with great extremes of heat and cold ; on the limestone plateaux and hills of the south-western States, and on the deep soil of the coast of Georgia, but not ascending the mountains or descending to the alluvial river bottoms.   It attains its maximum development south of lat. 36°, where in Alabama it is sometimes as much as 100 ft. in height, but is much oftener 60 to 70 ft. high, and in the north rarely exceeds 40 to 50 ft. and is often a mere shrub.   As a rule it is scattered among other trees and forms a small proportion of the forest ; but in the so-called " Cedar Barrens " of Tennessee, it formerly formed an almost pure forest extending over large areas ;

---

[1] Sargent, in *Garden and Forest*, viii. 61, fig. 9 (1895), gives an excellent illustration of a mature tree near Wawa Station, Delaware County, Pennsylvania.

[2] *Trees of New England*, 27 (1902).

[3] Britton and Shafer, *North American Trees*, 117 (1908), state that it also occurs in poor and rocky soil in Nova Scotia and New Brunswick ; and there is a specimen in the Kew herbarium from Newfoundland.

and in the rocky hills of the Tennessee valley grows in mixture with ash, maple, and oak, and in the prairies of Alabama, with magnolia, lime, and hickory.

It grows best on a light loamy soil containing lime, and does not come to perfection on clay or sand. It reproduces itself freely from seed, which it bears every year, the berries being a favourite food of many birds, which scatter it widely. The seedlings endure shade, and spread over abandoned farms in New England and in the south; but Mohr says that its habit of reproducing itself from suckers seems to have weakened the vitality of the seed, and that under the best conditions only 15 to 25 per cent of the seed will germinate.

Excellent illustrations are given in Mohr's paper showing the botanical details and the structure of the wood, with a map giving the distribution of both this species and *J. barbadensis*.

CULTIVATION

Though described by Parkinson in 1640, and introduced to England before 1664 by Evelyn,[1] it has never become an abundant tree in England; and was much more generally planted a century ago than it is now. In most old places it may be found in a more or less damaged condition, for though a long-lived tree it is often broken by wind and snow; and it varies so much from seed that it is often mistaken for other species. As, however, it is very hardy, and will grow in almost any kind of dry and well-drained soil, but only to a large size where this is also deep and fertile, it should be planted more generally, and in some parts of the south of England might be tried for the sake of its very valuable timber. It is easily raised from seed,[2] and grows faster than the common juniper, but it does not produce seed freely in most places; and I am unable to say what part of America seed suitable for English planting is most likely to come from. Though in the northern United States and Canada it is usually a small and scrubby tree as compared with the large size it attains in the south, I can find no evidence that under cultivation this difference is reproduced. Prof. N. E. Hansen of the South Dakota Agricultural College, however, states, in a paper[3] on the "Breeding of Cold-Resistant Fruits," that "The red cedar was formerly brought in large quantities from Tennessee, which is well to the south; northern nurserymen have learned that they must cultivate only the northern form of the red cedar to avoid total failure."

It seems to require fairly close planting, as well as pruning, in order to get a clean trunk; but I do not know to what extent it is capable of bearing shade in this climate. Under favourable conditions in America the growth at first is very rapid, trees only twelve years old having attained in Alabama 25 ft. high and 1 ft. in diameter. Up to from seventy to one hundred and twenty years the increase continues good, but after that age the trees often begin to decay, though they may live for several centuries, and attain a diameter at the butt of 2 to 4 ft.

[1] Aiton, *Hort. Kew.* v. 414 (1813).

[2] Sargent, in *Garden and Forest*, viii. 61 (1895) says that the seeds should be gathered in autumn and then moistened and mixed with sand, and kept in a pit till the following autumn, or the second spring, when they will produce strong plants, 6 to 8 in. high at the end of their second season and ready for transplanting.

[3] *Report of Conference on Genetics* (*Roy. Hort. Soc.*), 1906, p. 402.

Sargent says[1] that it bears pruning well, and is suitable for formal gardening; but is not good as a hedge plant, as its branches die when they come in contact with those of a neighbouring tree. There is, however, a thick hedge of this species in Barbier's nursery at Orleans, which is said to grow at the rate of a foot per year.

Trees in the American forest are often attacked by two species of *Polyporus*, which cause white rot and red rot of the timber, spoiling it for commercial purposes. These diseases have been described by Schrenk.[2]

As a rule, this species, like other junipers, is seen in a bushy form, but on good soil it is capable of producing a clean trunk of considerable length, and this would no doubt be more often the case if planted closer and pruned when young. Mr. A. D. Webster assures me that a tree which grew on sandy loam at Esher, had a trunk with a clean and well-rounded stem free from branches, for 33 ft. in length, and when measured by him contained fully 51 cubic ft. of timber.

REMARKABLE TREES

Among the finest specimens I have seen in England the largest is an immense old tree at Pains Hill, close to the cedar figured on Plate 128. It measures 13 ft. 9 in. in girth, with a bole about 5 ft. high, dividing into several stems, more or less broken, but one attains 68 ft. in height. At Woolbeding, in Col. Lascelles' grounds, there is a fine tree 65 ft. by 6 ft. 8 in. in 1906. At Bagshot Park a tree with a clean trunk measured 64 ft. by 7 ft. in 1907. At Sherborne, Dorsetshire, a clean-stemmed tree on the ruins of the old castle, 60 ft. by 7 ft. 3 in., is long past its prime. At Coolhurst, Sussex, there is a symmetrical tree 62 ft. by 7 ft. 3 in., and another, 56 ft. by $4\frac{1}{2}$ ft. with a clean trunk of 25 ft. At Arley Castle,[3] a tree in 1910 measured 68 ft. by 4 ft. 10 in. It is supported by a wire, as it was blown over and pulled upright again several years ago. At Raglan Castle, Monmouthshire, inside the ruins there is a fine old tree 53 ft. by 4 ft. 10 in. in 1906. At Wimpole, Cambridgeshire, a very handsome tree with pendulous branchlets was 50 ft. high in 1908.

In Scotland, the tree does not seem to attain so large a size, the best I have seen being one at Moncrieffe, Perthshire, which[4] was 47 ft. by 6 ft. 10 in. in 1907. Another, at Murthly, in the Tayside walk, which in 1906 was bearing many berries, measured 40 ft. by $3\frac{1}{2}$ ft.

We have seen no large trees in Ireland; but there is a good specimen of var. *glauca* at Hamwood.

The most remarkable trees that I have seen in Europe are two in the grounds of the Trianon, at Versailles, one of which, not more than about 30 ft. high, has immense spreading branches, which cover an area 57 paces round. The other, close to the château, I could not measure, but estimated it to be 75 ft. high. At Colombez, near Metz, there is an avenue of about fifty old trees, 40 to 50 ft. high, with trunks 4 to 5 ft. in girth, growing in an exposed situation.

[1] *Garden and Forest*, x. 142 (1897).   [2] *U.S. Dept. Agric. Bull.* No. 21 (1900).

[3] Woodward, *Hortus Arleyensis*, No. 4 (1907).

[4] Mentioned by Hunter, *Woods, Forests, and Estates of Perthshire*, 136 (1883), as *Cupressus thyoides viridis*.

In some parts of Germany the tree grows very well, but does not seem to have any economic value, and in Dr. Mayr's [1] opinion is only likely to be useful in the south of Europe.

TIMBER

The timber of the common and of the southern species present no essential difference and are not distinguished in commerce, though at the present time the greater part of that imported to Europe comes from the Southern States, that from Florida produced by *J. barbadensis* being considered the best.

Its great resistance to decay makes it very valuable, and formerly, when commoner, it was largely used in America for fencing, telegraph poles, boat- and house-building. Now, however, trees large enough for such purposes can hardly be found in the north, and the principal use for which it serves is to make the casing of lead-pencils. Mohr states that for this purpose alone 500,000 cubic feet are annually used in the United States, and 75,000 more exported, most of this going to Germany.

At Greenville, Alabama, the logs are cut into pieces of five standard sizes, varying from $\frac{3}{4}$ to $2\frac{1}{4}$ in. thick, which are packed for shipment in square wooden cases. The waste and sawdust from the mills is made into fine shavings, used for protecting furs and woollen goods against moths, and into paper for underlaying carpets.

Cigar-boxes are also made from this wood in Germany; but most of the so-called cedar in which Havana cigars are packed is the wood of *Cedrela odorata*, a very different tree of the West Indies.

Though in former times the wood was commonly used in this country for the finer kinds of joiner's work and interior decoration, under the name of cedar, red cedar, or pencil cedar, yet it has, during the last generation, become so scarce and dear, that its use is almost confined to the making of pencils. The greater part of the logs imported at present are from Jamaica, Alabama, and Georgia; and are usually small and faulty. The few large and clean ones which arrive are worth from 6s. to 8s. per cubic foot, whilst small logs and billets are sold at 2s. to 6s. per cubic foot in London.

The heart-wood is of a pinkish brown, becoming darker with age, and fading on exposure to the sun, and the sap-wood is whitish. It is distinguished by its fragrance, which, however, is fainter than that of Lebanon cedar, Lawson cypress, or camphor wood; and not too strong to use for the panelling or ceilings of living rooms. No wood is better adapted for delicate mouldings or carvings, though it is too soft to be used in any positions exposed to friction or contact with furniture. In some cases the heart-wood of old trees is beautifully variegated and twisted, and such pieces would be of the highest value for cabinetmaking, if procurable; but I have hardly ever seen the waved or curly grain, which is so ornamental in some other conifers, as pitch pine, redwood, or American cypress.

One of the best examples of the use of this wood for ornamental work is in the

[1] *Fremdländ. Wald- u. Parkbäume*, 292 (1906). Cf. Schwappach, in *Zeitschr. Forst- u. Jagdwesen*, xliii. 602 (1911), and in *Mitt. Deut. Dend. Ges.* 1911, p. 11.

library of Lord Llangattock's house, The Hendre, near Monmouth, which was panelled and ceiled by Messrs. Norman and Burt, from the design of Sir Aston Webb. Overmantels of this wood have been taken out of old London houses, where they have been for probably two centuries, and have realised very high prices; and on account of its scent, it was a favourite wood in early Victorian times for lining wardrobes, or for matchboarding bathrooms in country mansions. The roof of the fine old church at Bitton, near Bath, is entirely lined with pencil cedar, which was executed under Canon Ellacombe's direction, with wood purchased from a ship wrecked in the Bristol Channel; and though the odour is not strong enough to be very noticeable, except in damp weather, the effect is very good.

Oil of cedar, for which there is a large demand in the United States, is distilled from sawdust and other refuse of the wood, at Cedar Keys in Florida.[1] The wood contains as much as 4 or 5 per cent of this oil, which is used as a taenifuge. The shoots of *J. virginiana* are sometimes used medicinally in the United States, as a substitute for the true savin, but contain considerably less essential oil.[2]

(H. J. E.)

JUNIPERUS SABINA, Savin

Juniperus Sabina, Linnæus, *Sp. Pl.* 1039 (1753); Loudon, *Arb. et Frut. Brit.* iv. 2499 (1838); Parlatore, in De Candolle, *Prod.* xvi. 2, p. 483 (1868); Bentley and Trimen, *Medicinal Plants*, iv. t. 254 (1880); Mathieu, *Flore Forestière*, 518 (1897); Ascherson and Graebner, *Syn. Mitteleurop. Flora*, i. 251 (1898); Kent, Veitch's *Man. Conif.* 189 (1900); Kirchner and Schröter, *Lebengesch. Blütenpfl. Mitteleuropas*, i. 320 (1906).
Sabina officinalis, Garcke, *Fl. Nord- u. Mitteldeutschl.* 387 (1858).

A shrub, attaining about 15 ft. in height, with foliage of a strong and disagreeable odour, and bitter to the taste. Leaves of two kinds; on adult shrubs scale-like; ultimate branchlets very slender, tetragonal, $\frac{1}{30}$ in. in diameter, clothed with 4 ranks of leaves in opposite pairs, which are imbricated, appressed, ovate, acute or blunt at the apex, about $\frac{1}{20}$ in. long, adnate in their lower half, entire in margin, rounded on the back, which usually bears an elliptic depressed resin-gland. On older branchlets, the leaves are more elongated, about $\frac{1}{8}$ in. long, acuminate, becoming brown and withered in the third and fourth years. On young plants, and on isolated branches of adult shrubs, the juvenile foliage is acicular, slightly spreading, in opposite pairs, about $\frac{1}{6}$ in. long, acuminate at the apex, adnate and not jointed at the base; upper surface concave, glaucous, and with a prominent midrib; lower surface, green, convex, marked with a longitudinal depressed gland.

Flowers monœcious or diœcious. Fruit ripening in the autumn of the first year or in the following spring, borne on the ends of short scaly recurved branchlets; irregularly globose or ovoid, about $\frac{1}{5}$ in. in diameter, brownish blue, covered with a glaucous bloom, composed of four to six scales, each marked with an obsolete

---

[1] *Garden and Forest*, ii. 301 (1889).   [2] Flückiger and Hanbury, *Pharmacographia*, 628 (1879).

mucro.[1] Seeds usually two, rarely one or three, immersed in a resinous flesh, ovoid-triquetrous, compressed, narrowed towards the apex, shining brown, with two or three longitudinal furrows, and verrucose on both surfaces towards the summit.

VARIETIES

The Savin, like most species of juniper, is variable in habit in the wild state in Europe, either occurring as a low prostrate shrub with the branches widely extended and lying on the ground, or as a tall upright dense pyramidal shrub, with horizontal or ascending branches.   A fastigiate form has also been observed, similar in appearance to the Swedish variety of the common juniper.

The leaves on the adult plant are usually small, scale-like, blunt at the apex, and closely appressed, acicular leaves being not very common in the typical form. The following varieties, differing in foliage, have been described:—

1. Var. *lusitanica*, Ascherson and Graebner, *Syn. Mitteleurop. Flora*, i. 253 (1898).

> *Juniperus lusitanica*, Miller, *Gard. Dict.* ed. 8, No. 11 (1768).
> *Juniperus sabinoides*, Grisebach, *Spicil. Fl. Rum.* ii. 352 (1848) (not Nees[2]).

An upright shrub, with scale-like leaves, which are sharply acuminate at the apex.   This variety is common in southern Europe.

2. Var. *tamariscifolia*, Solander, in Aiton, *Hort. Kew.* iii. 414 (1789).

A low spreading shrub, with mostly persistent juvenile acicular foliage; leaves on the ultimate branchlets in opposite pairs, slightly spreading, glandular on the dorsal surface, bright green in tint, about ⅛ in. long; on the older branchlets, occasionally in whorls of threes.   This variety, which has been known in cultivation for at least 200 years, is occasionally met with in the wild state in Europe.

3. Var. *variegata*, Carrière, *Conif.* 36 (1855).

A dwarf shrub, with adult scale-like foliage, the tips of some of the young branchlets being creamy white.   This is mentioned by Loudon, and was known over a century ago in gardens.   It is often planted in rockeries.

4. Var. *prostrata*, Loudon, *Arb. et Frut. Brit.* iv. 2499 (1838).

> *Juniperus prostrata*, Persoon, *Syn. Pl.* ii. 632 (1807); Kent, Veitch's *Man. Conif.* 183 (1900); Britton and Shafer, *N. Amer. Trees*, 120 (1908).
> *Juniperus Sabina*, var. *procumbens*, Pursh, *Fl. Amer. Sept.* 647 (1814); Jack, in *Bot. Gaz.* xviii. 372 (1893).
> *Juniperus repens*, Nuttall, *Gen. Amer.* ii. 245 (1818).

A depressed, usually procumbent shrub, seldom more than 3 ft. high.   Leaves usually of two kinds; the scale-like leaves appressed in four ranks, mucronate at the apex, with a well-marked depressed resin-gland, and similar to those of the type;

---

[1] The berries are in rare cases open and not coalesced at the summit, the tips of the seeds protruding.   This sport is known as var. *gymnosperma*, Schröter, *op. cit.* 333, fig. 176 (1906).

[2] *J. sabinoides*, Nees, in *Linnæa*, xix. 706 (1847), is better known as *J. mexicana*, Sprengel, *Syst.* iii. 909 (1826), and is a native of Texas and Mexico.

the acicular leaves on a few of the branchlets, in opposite pairs, slightly spreading, with a longitudinal depressed gland on the back, about $\frac{1}{6}$ in. long. Fruit, on recurved stalks, light blue and scarcely glaucous, $\frac{1}{4}$ in. in diameter, with one to four seeds.

This, which is the American form[1] of the Savin, is considered by some botanists to be a distinct species. It is distributed from southern Maine to the shores of Hudson Bay, and westward from Newfoundland and northern New England through New York along the shores of the Great Lakes and through northern Minnesota and south Dakota to the eastern slopes of the Rocky Mountains in Alberta, Montana, and Wyoming. The American Savin grows mainly on sandy soil, as on the seashore of the Atlantic coast and on inland dunes and barrens, and thus differs remarkably from the lime-loving savin of Europe.

The American Savin is said to have been first introduced by Loddiges, who called it *J. hudsonica* in his catalogue of 1836. The shrub cultivated as var. *prostrata* is low and prostrate, bearing only acicular very glaucous foliage; leaves in their free part $\frac{1}{6}$ in. long, spreading, the glaucous bloom appearing on the dorsal surface near the base; ventral surface whitened; branchlets of the second and third year bright reddish brown, with persistent needles of the same colour. I have not seen this cultivated shrub in fruit or bearing scale-like leaves; but it has the disagreeable odour of the savin, and in all probability is, as reputed, of American origin. Sargent states[2] that it is the hardiest and most beautiful of all the prostrate junipers that can be grown in New England gardens.

DISTRIBUTION

The Savin is widely distributed, occurring in central and southern Europe, the Caucasus, and North America. It occurs mainly in Europe in extensive thickets on dry rocky sunny mountain slopes; but is also met with as undergrowth in many pine forests, as those of *Pinus sylvestris* in the Sierra Nevada in Spain, and of *P. leucodermis* in Herzegovina. It grows usually on limestone; but is occasionally seen on other formations. In Europe it is most common in Spain and Portugal, and in the Balkan peninsula. It is widely spread throughout the whole Alpine mountain system, but is rare towards the north, though it is met with in the Bavarian Alps and in a few stations in Switzerland. Its distribution in Russia is remarkable, as it occurs in isolated spots throughout the great plain, reaching as far north as the Baltic coast; but is a mountain plant, as usual elsewhere, in the Crimea and in the southern part of the Ural range. It also occurs in the Caucasus and the mountains of northern Persia; but is not met with in Asia Minor, and is totally absent from northern Africa. Its occurrence in Siberia is attested by Russian botanists; but we have seen no specimens. (A. H.)

The Savin was early introduced into England, as it is mentioned in Turner's *Names of Herbes*, published in 1548.

---

[1] Rehder, in Bailey, *Cycl. Amer. Hort.* 850 (1900), says that it is sometimes called in America the Waukegan juniper.

[2] In *Garden and Forest*, x. 421 (1897). Sargent adds that the European savin, if it has ever been tried in gardens in Massachusetts, has probably not proved hardy.

The finest specimen in Britain is probably one growing in the garden of Stourton Court, Stourbridge, the residence of R. Matthews, Esq. This is well figured in *Journ. R. Hort. Soc.* xxxiii. 327 (1908), where it is stated to measure 6 ft. high and 57½ ft. in circumference. The soil is Old Red Sandstone.

The young green shoots of the Savin are used in medicine, and yield a volatile oil, which is officinal and possesses extremely active properties.[1]   (H. J. E.)

JUNIPERUS EXCELSA

Juniperus excelsa,[2] Bieberstein, *Beschr. Länd. Casp.* 204, App. No. 72 (1800), and *Fl. Taur. Cauc.* ii. 425 (1808); Parlatore, in De Candolle, *Prod.* xvi. 2, p. 484 (1868); Boissier, *Fl. Orient.* v. 708 (1881); Siehe, in *Gartenflora*, xlvi. 208, t. 26 (1897); Kent, Veitch's *Man. Conif.* 174 (1900).

Juniperus Sabina, Linnæus, var. *taurica*, Pallas, *Fl. Ross.* ii. 15 (1788).

Juniperus Sabina, Linnæus, var. *excelsa*, Georgi, *Besch. Russ. Reichs*, iii. 1358 (1802).

Juniperus fœtida, var. *excelsa*, Spach (excl. syn. Amer.), in *Ann. Sc. Nat.* xvi. 297 (1841).

Juniperus polycarpos and *isophyllos*, Koch, in *Linnæa*, xxii. 303, 304 (1849).

Juniperus Olivierii, Carrière, *Conif.* 57 (1855).

Juniperus ægæa, Grisebach, *Veg. der Erde*, 378, 572 (1872).

Sabina excelsa, *polycarpos*, and *isophyllos*, Antoine, *Cupress. Gatt.* 45, 47, 48 (1857).

A tree, occasionally attaining in Asia Minor a height of 70 to 100 ft. Leaves dimorphic in wild specimens; but juvenile foliage is rarely seen on cultivated adult trees of the typical form. Adult foliage; ultimate branchlets very slender, $\frac{1}{30}$ in. or less in diameter; leaves scale-like, closely appressed, in four ranks in opposite decussate pairs, ovate-rhombic, about $\frac{1}{24}$ in. long, acute or obtuse, marked in the middle on the back with a depressed oval or linear gland; leaves on older branchlets, in pairs or in threes, spreading, mucronate, glandular on the back. Juvenile foliage, when present[3]; leaves acicular, spreading, in opposite pairs, $\frac{1}{5}$ to $\frac{1}{4}$ in. long, concave above with two stomatic bands, marked on the lower surface with a linear gland at the base.

Flowers monœcious or diœcious. Fruit, ripening in the second year, on short scaly stalks, globose, $\frac{1}{3}$ to $\frac{1}{2}$ in. in diameter, smooth, dark purplish brown, covered with a bluish bloom when ripe, composed of four or six scales, each marked by a minute mucro. Seeds, about six in each fruit, oblong, more or less triquetrous, apiculate at the apex.

1. Var. *stricta*, Rollisson, *ex* Gordon, *Pinetum*, 144 (1875).

A tree, narrowly pyramidal in habit, with glaucous juvenile foliage; leaves acicular, slightly spreading, about $\frac{1}{8}$ in. long including the basal decurrent part, whitened with a stomatic band above, marked with a minute gland near the base on the lower surface.

This originated in Messrs. Rollisson's nursery at Tooting, and appears to be

---

[1] Cf. Fluckiger and Hanbury, *Pharmacographia*, 628 (1879).

[2] *J. excelsa*, Loudon, *Arb. et Frut. Brit.* iv. 2503 (1838), includes a mixture of junipers from Siberia, the Himalayas, and North America, and does not appear to refer to the true plant from Asia Minor.

[3] Described from a native specimen collected by Hausknecht.

perfectly hardy, retaining its characters in old age. There are three trees at Kew, 25 to 30 ft. high, which were obtained from Rollisson in 1868.

Var. *Perkinsii* and var. *venusta*, Gordon, *Pinetum*, 144 (1875), are similar to var. *stricta*, only differing in the more glaucous leaves. A specimen of var. *Perkinsii* at Kew is almost columnar in habit.

J. excelsa is a native of the Balkan States, Island of Thasos,[1] Crimea, Asia Minor, Syria, Armenia, and the Caucasus. In Thrace, Macedonia, and the Rhodope mountains of Rumelia it is, according to Adamovic,[2] a shrub rather than a tree, ascending occasionally to 4000 ft. It is recorded for one station in Bulgaria, near the village of Beli Iskar. Köppen[3] states that in the Crimea it is common on the coast side of the mountains, forming pure woods of considerable extent; but never attaining a large size, the tallest tree noted being about 30 ft.

J. excelsa appears to attain its greatest development in Asia Minor, where it forms extensive woods in the mountains, either pure or mixed with Lebanon Cedar and *Abies cilicica*. Siehe[4] gives an illustration of a tree in the Cilician Taurus, nearly 100 ft. high, and states that it assumes two forms, being either a tall narrow pyramidal tree, or a shorter tree with wide-spreading branches; occasionally with a trunk $4\frac{1}{2}$ ft. in diameter. There are very fine woods at Namrun, Efrenk, and Güllek in Cilicia, where the trees are tall, slender, and dense upon the ground. Siehe considers that the timber will prove of great value for railway sleepers. The Turkish name of the tree is *arytsch*.

The date of introduction is uncertain, as Loudon's account of *J. excelsa* does not relate to this species; but it was probably brought into England about 1836. It usually forms a narrow columnar or pyramidal tree. A tree at Arley Castle, said to have been planted in 1877, measured[5] 32 ft. by 3 ft. 1 in. in 1904. Another at High Canons, Herts, was 32 ft. by $2\frac{1}{2}$ ft. in 1908. There are trees of similar size in the botanic gardens at Kew and Cambridge. We have also seen good specimens at Hardwicke, Tortworth, Westonbirt, Highnam, and Eastnor. In Ireland there are trees at Glasnevin and Powerscourt.

J. excelsa has been much confused with the two following species, which are not apparently in cultivation. These are remarkably distinct from *J. excelsa* both in foliage and fruit.

I. *Juniperus macropoda*, Boissier, *Flora Orientalis*, v. 709 (1881).

A tree, occasionally attaining 70 ft. in height, but often shrub-like, which is a native of Persia, Afghanistan, and Baluchistan, where it forms extensive open forests, east of Quetta.[6]

This has much coarser foliage than *J. excelsa*, resembling that of *J. chinensis*. Fruit globose, $\frac{1}{3}$ in. in diameter, brownish purple, tinged with a glaucous bloom, each of the four to six scales with a prominent mucro. Seeds, two to four, ovoid.

---

[1] It is not recorded for any of the islands in the Ægean Archipelago except Thasos, where it grows in the pine woods on the coast. Cf. Grisebach, *Veg. der Erde*, 378, 572 (1872). [2] *Veget. Balkanländer*, 152 (1909).

[3] *Holzgewächse Europ. Russlands*, ii. 423 (1889).

[4] In *Gartenflora*, xlvi. 208, t. 26 (1897), and in *Mitt. Deut. Dend. Ges.* 1911, p. 306.

[5] Woodward, *Hortus Arleyensis*, 19 (1907).

[6] Cf. Lace, in *Journ. Linn. Soc.* (*Bot.*) xxviii. 307 (1891). Gamble, *Indian Timbers*, 698 (1902), gives also some particulars concerning this tree, which he considers to be identical with the Himalayan *J. religiosa*.

II. *Juniperus religiosa*, Carrière, *Conif.* 41 (1855) (not Royle[1]).

Juniperus gossainthanea,[2] Loddiges, *Catalogue*, 48 (1836); Loudon, *Trees and Shrubs*, 1090 (1842).
Juniperus chinensis, Parlatore, in De Candolle, *Prod.* xvi. 2, p. 488 (1868 (in part) (not Linnaeus).
Juniperus excelsa, Brandis, *Forest Flora, N.W. India*, 538, t. 68 (1874) (not Bieberstein).
Juniperus macropoda, Hooker, *Fl. Brit. India*, v. 647 (1888) (not Boissier).

A tree, occurring in the inner dry ranges of the north-western Himalayas from Chitral and Kashmir to Nepal, at 5000 to 10,000 ft. altitude. It often attains 50 ft. in height, with a girth of 6 or 7 ft.; but occasionally trees of enormous girth are met with, one at Lahoul measuring $33\frac{1}{2}$ ft. in circumference.

This species has foliage similar to that of *J. macropoda* and *J. chinensis*. Fruit obovoid, widest at the apex, which is depressed, $\frac{1}{4}$ in. in diameter, bluish black with a soft juicy pulp. Seeds, one to three, ovoid, sharp-pointed, with large resin-glands.

(A. H.)

[1] *J. religiosa*, Carrière, the first published name with a description, should be adopted for this species. *J. religiosa*, Royle, *Illust. Himal.* i. 351 (1839), without any description, is possibly *J. squamata*. Cf. p. 1420, note 4.

[2] This name was published without any description, and, moreover, is somewhat doubtful. Cf. p. 1438, note 1.

ATHROTAXIS

Athrotaxis,[1] Don, in *Trans. Linn. Soc.* xviii. 172, tt. 13, 14 (1839); Bentham et Hooker, *Gen. Pl.* iii. 430 (1880); Masters, in *Journ. Linn. Soc.* (*Bot.*) xxx. 21 (1893).

EVERGREEN trees belonging to the division Taxodineæ of the order Coniferæ. Leaves persistent for several years, spirally arranged, homomorphic, crowded, imbricate, spreading or closely appressed, adnate at the base, free at the apex; without scaly buds.

Flowers monœcious, solitary at the apices of the branchlets. Staminate flowers catkin-like, with crowded stamens spirally arranged on an axis; each stamen with a slender stalk and a sagittate connective, which bears two pollen sacs dehiscing longitudinally. Ovuliferous flowers, of ten to twenty-five spirally arranged scales; each scale with an adnate fleshy disc, bearing three to six ovules. Cones ripening in one year; scales, ten to twenty-five, woody, spirally arranged, cuneate and narrow at the base, horizontally spreading, dilated into a clavate or peltate lamina, which bears on the back or at the apex a triangular cuspidate process.[2] Seeds, three to six, pendulous from the thickened part of the scale below the apex; oblong, compressed, with a transverse hilum and two lateral wings. Cotyledons two, longer than the primary leaves.[3]

Athrotaxis is closely related to Cryptomeria; and Kent states that rooted cuttings of *Cryptomeria elegans* are used as stocks for grafting scions taken from the different species of Athrotaxis.

Fossil remains found in various deposits in Europe have been identified, but perhaps erroneously, with Athrotaxis. C. Reid[4] has lately shown that the remains in the Bovey Tracey lignites belong to *Sequoia Couttsiæ*, Heer, and not to Athrotaxis, as had been supposed by Starkie Gardner.

This genus is confined in the living state to Tasmania, where there are three species[5] distinguishable as follows :—

---

[1] Derived from ἀθρόος, crowded, and τάξις, arrangement. Endlicher, *Gen. Suppl.* i. 1372 (1841), and *Syn. Conif.* 193 (1847), gives the erroneous spelling *Arthrotaxis*, which has been followed by several writers.

[2] This process is the extremity of the scale in the flowering stage, which has coalesced almost completely with the ovuliferous disc, the latter having increased much in size during the ripening of the ovules into seed.

[3] Masters, in *Journ. Linn. Soc.* (*Bot.*) xxvii. 235, 237 (1890).

[4] In *Phil. Trans.* series B, vol. 201, p. 171, pl. 15, figs. 40, 41 (1910), where the distinctive characters of the epidermis of the leaves of Sequoia and Athrotaxis are made plain.

[5] *Athrotaxis* (?) *tetragona*, W. J. Hooker, *Icon. Plant.* t. 560 (1843) belongs to a distinct genus, and is *Microcachrys tetragona*, J. D. Hooker, in *London Journ. Bot.* iv. 149 (1845). This is a low rambling shrub, also a native of Tasmania. It is occasionally cultivated in conservatories.

I. Leaves spreading, entire in margin, with two continuous white stomatic bands on the ventral surface, and two lateral stomatic depressions confined to near the base on the dorsal surface.

   1. *Athrotaxis selaginoides*, Don.

   Leaves very spreading, ½ in. long, with a rigid spine-like acuminate apex, and an opaque margin.

   2. *Athrotaxis laxifolia*, W. J. Hooker.

   Leaves slightly spreading, ¼ in. long, obtuse or acute at the apex, and with a translucent margin.

II. Leaves closely appressed, scale-like and apparently in four ranks like Cupressus, but really spirally arranged; margin translucent and denticulate; ventral surface concealed; dorsal surface with indistinct stomatic lines.

   3. *Athrotaxis cupressoides*, Don.

   Leaves rhombic-ovate, obtuse at the apex, ⅛ in. long.   (A. H.)

ATHROTAXIS SELAGINOIDES

Athrotaxis selaginoides, Don, in *Trans. Linn. Soc.* xviii. 172, t. 14 (1839); W. J. Hooker, *Icon. Plant.* t. 574 (1843); J. D. Hooker, in *Lond. Journ. Bot.* iv. 148 (1845), and *Fl. Tasman.* i. 354 (1860); Masters, in *Gard. Chron.* ii. 724, figs. 140, 141 (1887), and iv. 544, fig. 79 (1888); Kent, Veitch's *Man. Conif.* 262 (1900); Rodway, *Tasmanian Flora*, 277 (1903); Baker and Smith, *Pines of Australia*, 303, plates on pp. 304, 305, figs. 217-228 (1910).
Athrotaxis Gunneana, Carrière, *Conif.* 207 (1867).
Athrotaxis Gunniana, Gordon, *Pinetum*, 47 (1875).
Cunninghamia selaginoides, Zuccarini, in Siebold, *Fl. Jap.* ii. 9, note (1844).

A tree, attaining a larger size than the other two species, up to 100 feet in height and 10 feet in girth. Bark described by Baker and Smith, as slightly furrowed and fibrous, but not very rough. Branchlets stout, entirely covered by the decurrent bases of the leaves. Leaves spirally arranged, loosely imbricated, widely spreading but incurved at the apex, rigid, coriaceous, about ½ in. long, subulate, adnate but not jointed at the base, tapering to an acuminate spine-like apex; dorsal surface keeled with two lateral depressions near the base, which are whitened by stomatic lines; ventral surface concave, with two longitudinal white stomatic bands from the apex to the base separated by a green midrib; margin entire, opaque.

Cones ovoid or globose, about 1 in. in diameter, composed of 20 to 24 brown woody scales, which are about ½ in. long, with a narrow cuneate base, and an oval or ovate expanded inflexed lamina, which terminates in a triangular thin process.

The species, which is known in Tasmania as King William Pine, is said by Rodway to occur in the western mountains, extending from Mount Field, Mount Hartz, Adamson Peak, and Mount La Perouse to the west coast. Baker and Smith state that it is common in the immediate neighbourhood of Williamsford, at about 1000 ft. above sea-level. It is a prominent tree in the dense scrub which covers this locality, being associated with *Phyllocladus rhomboidalis*, *Nothofagus Cunninghami*, &c. These authors figure an old tree, said to be typical in habit, which shows a twisted stem, free of branches for three quarters of its height, and surmounted by a small irregular dense crown of foliage. *A. selaginoides* ascends to 3000 ft. or more, as it occurs on the summit of Mt. Reed and other mountains, usually in a much dwarfed and stunted form.

Baker and Smith, who give excellent figures of the structure of the leaves and wood, state that the wood is not unlike that of *Sequoia sempervirens*, both in general characters and in texture, being open and straight in the grain, easy to work, and very light in weight. It is pale reddish when freshly cut, but becomes lighter in colour on exposure. Possessing great durability, and considerable toughness and strength, it is used in Tasmania for cabinet-work, coach-building, and for making oars. Penny[1] states that it occurs in limited quantities; and is apparently never exported.

This species was introduced about the year 1857 by Mr. W. Archer of Cheshunt; but appears to be less common in cultivation than *A. laxifolia*. A thriving specimen at Osborne, Isle of Wight, planted in 1879, was $17\frac{1}{2}$ feet high in January 1912, when it bore both young and old cones. A tree at Lamellen, St. Tudy, Cornwall, which was 26 ft. high, died in 1909. From it Mr. Magor raised a few seedlings, which are still small plants. A specimen at Abbotsbury was killed by drought in the summer of 1911. The finest specimen is at Kilmacurragh, in Ireland, and measured 32 ft. high in March 1912, when it was bearing numerous old cones. A smaller tree is thriving at Rostrevor. (A. H.)

ATHROTAXIS LAXIFOLIA

Athrotaxis laxifolia, W. J. Hooker, *Icon. Plant.* t. 573 (1843); J. D. Hooker, in *Lond. Journ. Bot.* iv. 149 (1845), and *Fl. Tasman.* i. 354 (1860); Masters, in *Gard. Chron.* xxiv. 584, fig. 134 (1885), ii. 724, figs. 142, 143 (1887), and ix. 144, 147, figs. 37, 38 (1891), and in *Journ. Linn. Soc. (Bot.)* xxii. 201, fig. 26 (1886); Kent, Veitch's *Man. Conif.* 261 (1900); Rodway, *Tasmanian Flora*, 277 (1903); Baker and Smith, *Pines of Australia*, 313 (1910).

Athrotaxis Doniana, Maule, *ex* Gordon, *Pinet. Suppl.* 16 (1862).

A tree, attaining about 40 ft. in height. Bark reddish, fibrous, peeling off in long vertical ribbons. Branchlets slender, covered by the decurrent bases of the leaves. Leaves spirally arranged, closely imbricated, slightly spreading, about $\frac{1}{4}$ in. long; incurved at the acute or obtuse, rarely mucronate, apex; dorsal surface keeled, with two lateral depressions near the base, which are whitened by stomatic lines; ventral surface concave, with two longitudinal white stomatic bands; margin entire, thin and translucent towards the apex.

Cones sub-globose, $\frac{3}{4}$ in. in diameter, composed of 15 to 20 brown woody scales,

[1] *Tasmanian Forestry*, xi. 42 (1905). Penny quotes A. O. Green's tests of various Tasmanian timbers. This species is one of the lightest, a cubic foot weighing only 22 lbs.

which are about ⅖ in. long, with a slender cuneate claw, and a thickened oval expanded lamina, bearing on the back a large ovate acute process.

This species was found by Gunn and Archer at the falls of the Meander river, and along rivulets near the summit of the western mountains in Tasmania; and is said by Rodway to occur on Field Range and near Mount La Perouse. It appears to occur at higher altitudes than the other species, reaching 4000 ft.

A. laxifolia was also introduced by Archer in 1857, and appears to be the most successful of the three species in cultivation in this country. The largest specimens are in Cornwall, where Elwes measured in 1911 a fine tree at Scorrier, 38 ft. high and 3 ft. 9 in. in girth; from it three seedlings were raised about twenty years ago, which are now 15 to 20 ft. high. Another fine specimen at Penjerrick was 32 ft. by 2 ft. 8 in. in the same year. One in Mr. R. Gill's nursery at Tremough near Penryn, measured 27 ft. high, and was bearing fruit in January 1912. Another, 17 ft. high, occurs at Trewidden near Penzance.

There is also a good tree at Menabilly, growing at 100 feet above sea-level in an exposed position, which was planted in 1880, and reported to be a perfect pyramid, 12 ft. high in 1891; it was 27 ft. high and 3 ft. in girth in January 1912. A branch of this tree with cones was figured [1] by Dr. Masters.

This species appears to be very hardy, and succeeds as far north as Durris in Kincardineshire, where a tree about 10 ft. high produced cones in 1909, from which numerous seedlings were raised; some of these have been planted out at Bayfordbury and in the Cambridge Botanic Garden.

In Ireland, there is a fine specimen at Kilmacurragh, 28 ft. high in 1912; and a smaller one at Castlewellan.

(A. H.)

ATHROTAXIS CUPRESSOIDES

Athrotaxis cupressoides, Don, in *Trans. Linn. Soc.* xviii. 173, t. 13, fig. 2 (1839); W. J. Hooker,
   Icon. Plant. t. 559 (1843); J. D. Hooker in *Lond. Journ. Bot.* iv. 148 (1845), and *Fl.*
   Tasman. i. 354 (1860); Masters, in *Gard. Chron.* xxiv. 270, fig. 60 (1885), and ii. 725,
   figs. 144, 145 (1887); Kent, Veitch's *Man. Conif.* 261 (1900); Rodway, *Tasmanian Flora*,
   277 (1903); Baker and Smith, *Pines of Australia*, 313 (1910).
Cunninghamia cupressoides, Zuccarini, in Siebold, *Fl. Jap.* ii. 9, note (1844).

A tree, attaining 40 ft. in height and 6 feet in girth, with ascending branches. Branchlets pseudo-opposite or alternate, densely covered with scale-like leaves, which appear to be in opposite decussate pairs, but in reality are in a spiral arrangement. Leaves on the ultimate branchlets, similar to those in Cupressus, homomorphic, densely appressed, closely imbricated, rhombic-ovate, about ⅛ in. long, obtuse at the apex; ventral surface entirely concealed; dorsal surface keeled and marked with white stomatic dots in lines on the two sides; margin denticulate

---

[1] *Gard. Chron.* xxiv. 584, fig. 134 (1885) and *Journ. Linn. Soc. (Bot.)* xxii. 201, fig. 26 (1886). In *Gard. Chron.* xxiv. 660 (1885), it is stated that a tree produced cones in 1875 in Mr. Robert Loder's garden at Whittlebury, Towcester; but this cannot now be found and was probably cut down some years ago.

and translucent. Leaves on the older branchlets increasing in size, becoming dark brown and about $\frac{1}{2}$ in. long in the fifth or sixth year.

Cones, nodding on curved branchlets, $\frac{2}{5}$ in. in diameter; scales 10 or 12, much smaller than in the other species, less than $\frac{1}{4}$ in. long, with a triangular recurved process on the middle of the expanded lamina.

This species is said by Rodway to be a small erect tree, about 40 ft. high, found in the western mountains of Tasmania, near St. Clair and to the west and south-west of Field Range. It rarely exceeds 5 or 6 feet in girth; but Sir J. Hooker mentions one very old and hollow tree, which was 15 ft. in girth at 3½ feet from the ground. Baker and Smith state that the timber resembles in all respects that of *A. selaginoides*.

It was introduced in 1857 by Mr. Archer at the same time as the other species, and resembles them in cultivation; but it is rare and has not attained in this country as great a height as *A. laxifolia*. In Cornwall, a tree 20 ft. high was bearing cones in Gill's nursery, near Penryn, in January 1912; and a smaller specimen, also bearing fruit, occurs at Trewidden. Kent mentioned in 1900 a fine specimen at Upcott, near Barnstaple; which, Mr. W. Harris informs us, has lately lost its leader and is now only 13 ft. high. In Hillier's nursery at Shroner, near Winchester, a tree, planted in 1887, was 12 ft. high in 1910, and had commenced to bear cones in small quantity. There is also a small specimen at Brickendon Grange, Hertford, which was 5 ft. high in 1911.

In Ireland, a fine specimen bearing fruit at Kilmacurragh was 21 ft. high in March 1912; while another at Powerscourt[1] was 22 ft. high. A thriving tree at Castlewellan, 15 ft. high, bore 18° of frost without injury in February 1912.

(A. H.)

[1] Erroneously called *A. Doniana* in *Gard. Chron.* xlix. 219 (1911).

FITZROYA

Fitzroya, J. D. Hooker, *ex* Lindley, in *Journ. Hort. Soc. Lond.*, vi. 264 (1851), and *ex* W. J. Hooker, *Bot. Mag.* t. 4616 (1851); Bentham et Hooker, *Gen. Pl.* iii. 425 (excl. *Diselma*) (1880); Masters, in *Journ. Linn. Soc. (Bot.)* xxx. 17 (1893).

A GENUS belonging to the division Cupressineæ of the order Coniferæ, characterised by the cones, composed of nine scales, in three alternating whorls, the three lower-most scales minute and sterile; those of the intermediate whorl larger, and either empty or each bearing a single two-winged seed; the three upper scales largest and fertile, each bearing two to six seeds, which are partly three- and partly two-winged. The apex of the axis of the cone terminates in three peculiar processes,[1] the precise morphological nature of which is unknown, but possibly they may represent an aborted whorl of scales. Only one species is known, in the description of which below, the vegetative and other characters of the genus are given in detail.

Diselma, a genus founded by Sir J. D. Hooker,[2] was united by Bentham and Hooker[3] with Fitzroya, but appears to be sufficiently distinct. In Diselma, the cones are composed of two pairs of opposite scales, the outer pair small and empty, the inner two scales larger and fertile, each with two seeds, which are three-winged. *Diselma Archeri*, J. D. Hooker, *Fl. Tasman.* i. 353, t. 98 (1860), the only known species, differs greatly from Fitzroya in habit, having minute scale-like appressed leaves, like those of a Cupressus, in opposite decussate pairs; and is a shrub[4] about 6 ft. high, growing between 3000 and 4500 feet elevation in the western mountains of Tasmania. *D. Archeri* was formerly represented in the Temperate House at Kew by a single specimen, which is no longer living. At present it is apparently not in cultivation in England. (A. H.)

FITZROYA PATAGONICA

Fitzroya patagonica, J. D. Hooker, *ex* Lindley in *Journ. Hort. Soc. Lond.* vi. 264 (1851), and *ex* W. J. Hooker, *Bot. Mag.* t. 4616 (1851); Lindley, in Paxton, *Flower Garden*, ii. 115 (1852); Kent, Veitch's *Man. Conif.* 198 (1900); Castillo and Dey, *Jeog. Vej. Rio. Valdivia*, 27, fig. 8 (1908).

An evergreen tree, attaining in Chile 100 to 160 ft. in height, and 10 to 16 ft. in girth; but dwarfed to a low shrub at high elevations. Bark reddish, longitudinally fissured, and peeling off in narrow ribbons. Young branchlets green,

---

[1] These gland-like processes secrete resin and exhale a slight fragrant odour.
[2] *Fl. Tasman.* i. 353, t. 98 (1860).
[3] *Gen. Plant.* iii. 425 (1880).
[4] Cf. Baker and Smith, *Pines of Australia*, 300 (1910).

glabrous, flexile, slender, covered by the decurrent bases of the leaves, which are separated between the whorls by three linear grooves. Older branchlets until the seventh year, stouter, reddish, marked by withered leaves and their remains. Buds ovoid or globose, composed of green scales, which are slightly modified and shortened ordinary leaves. Leaves persistent for several years, in alternating whorls of threes, decurrent by their bases on the branchlets; their free part spreading, spatulate, incurved at the rounded apiculate apex, about $\frac{1}{8}$ in. long; upper surface concave, with two narrow white stomatic depressions extending from the apex to the middle of the leaf or beyond, and separated by a raised green midrib; lower surface convex, with a broad green raised midrib, on each side of which is a narrow white stomatic depression often extending from near the apex to the adnate base of the leaf. On the main axes, the leaves are often $\frac{1}{2}$ in. long, adnate in greater part to the branchlet, and becoming reddish brown in the third and fourth years.

Flowers[1] usually diœcious; sometimes monœcious or hermaphrodite. Staminate flowers solitary in the axils of the leaves towards the apex of the branchlet, cylindrical, subtended at the base by a few scales, composed of 15 to 24 stamens in ternate whorls; anthers 4-celled. Ovuliferous flowers, solitary and sessile on the ends of short leafy branchlets near the apex of a branch, composed of nine scales, in three alternating whorls; the three lowermost scales minute and sterile; the three scales of the intermediate whorl either empty or each bearing a single two-winged ovule; the uppermost three scales always fertile, each bearing a central three-winged ovule and one to five lateral two-winged ovules; oblong tubercles at the summit of the axis of the cone, three, yellowish, translucent, about $\frac{1}{8}$ in. long. Cones, scarcely larger when mature than in the flowering stage, sub-globose when closed and about $\frac{1}{3}$ in. in diameter, ripening in one year, with three minute scales at the base of the six large woody scales, each of the latter with a dorsal process, spreading widely to let loose the seeds, which are variable (nine to sixteen) in number (equalling the number of ovules in the flower). Seed with an oblong compressed body and two or three broad lateral membranous wings; the seed with the wings nearly orbicular and about $\frac{1}{6}$ to $\frac{1}{5}$ in. in diameter. Cotyledons two.

This species is a native of South America, occurring in Chile and northern Patagonia. It extends from the coast range immediately north of Valdivia southward to the island of Chiloe and the mainland opposite in about lat. 42° 40', and reaches inland to the central cordillera of the Andes. It is known to the inhabitants as *alerce*,[2] and covers immense tracts of marshy and peaty ground with extensive woods, which are called *alerzales*. These woods are widely distributed, the best known being in the coast range of Valdivia, around Lakes Llanquihue and Nahuelhuapi, in the neighbourhood of Puerto Montt, and in the valley of the river Maullin.[3]

[1] The flowers of Fitzroya, which are complicated and variable, are being investigated at Cambridge by Mr. R. C. Maclean. Monœcious flowers occur in Chilean specimens, as well as on the tree at Hewell Grange. The hermaphrodite flowers of the latter have several whorls of scales; the scales in the three lowermost whorls bear anthers, those in the upper two whorls bear ovules; the axis ends in the normal three gland-like processes.

[2] *Alerce* is used in Spain as the name of the larch; and is derived from the Arabic, *al-arzah*, signifying cedar.

[3] Cf. Reiche, *Verbreit. Chilen. Conif.* 5 (1900), and *Pflanzenverb. Chile*, 63, 238 (1907).

Castillo and Dey[1] state that enormous trees still occur, up to 160 ft. in height and 16 ft. in girth; and mention veterans, which are said when felled to have shown 4000 annual rings; but this seems scarcely credible. Captain Fitzroy[2] in whose honour the genus was named, and who commanded the " Beagle " between 1828 and 1836, states on good authority, that a tree in the cordillera on the mainland opposite Chiloe had a stem measuring 76 ft. in length to the first branch, and 30 ft. in girth at five feet from the ground. It yielded 1500 planks. W. Lobb saw on the precipices around Valdivia trees 100 ft. in height and 8 ft. in diameter; and states[3] that it ascends the mountains to the limit of perpetual snow, where it is occasionally only a few inches high.

Fitzroya patagonica was first introduced[4] in 1849 by W. Lobb, who sent home seeds from Valdivia; and again by R. Pearce, who collected in Chile for Messrs. Veitch in 1859-1862. It is perfectly hardy in this country, and at Kew[5] bore without injury the severe frost of 1908-1909, when the temperature fell to 10° Fahr.; nevertheless, it is slow in growth, and seems to develop oftener into a bushy shrub than a tree, but this may be due to most of the specimens in cultivation being derived from cuttings.

The finest trees are in Cornwall and Devon. One at Killerton, planted in 1864, was $34\frac{1}{2}$ ft. by 3 ft. 2 in. in 1911. Another at Bicton was 35 ft. high in the same year. An ill-shaped tree at Penjerrick, dividing into several stems at the base, was about 35 ft. high in 1910. At Coldrenick a similar tree with several stems was 25 ft. high in 1911. Masters figured[6] a tree at Pencarrow, planted in 1852 by Sir W. Molesworth, which was reported in 1902 to have been 21 ft. in height, with a spread of branches of 46 ft. At Abbotsbury, a tree was killed by drought in the summer of 1911. At Upcott, Barnstaple, a well-grown and healthy specimen was $28\frac{1}{2}$ ft. high in January 1912. There are smaller trees at Highnam near Gloucester, and at Leaton Knolls, Shrewsbury. In the pinetum at Uplyme, Dorset, there is a fine tree, 30 ft. by 3 ft. 3 in. in 1912.

At Belsay Castle, Newcastle-on-Tyne, there is a fine specimen, which was planted about 1856, and measured $28\frac{1}{2}$ ft. high in January 1912. Sir Arthur E. Middleton, Bart., informs me that it is growing in a sheltered place in an old sandstone quarry, where the soil contains a considerable admixture of clay. This tree has been kept from an early date to one leader; otherwise it would have spread in an irregular way. There are also several bushy specimens at Belsay Castle, where this species has never been touched in the slightest degree by frost.

Nearly all the cultivated trees which we have seen bear only female flowers, and in consequence the seed, which is freely produced, is infertile. A tree, however, at Hewell Grange, Redditch, which is about 28 ft. high, is monœcious, and bore in May 1912 both male and female flowers in profusion, as well as a few which were hermaphrodite.

[1] *Jeog. Veg. Rio Valdivia*, 28 (Santiago, 1908). Reiche gives the maximum as 180 ft. high, 16 ft. in diameter, and 2500 years old.

[2] *Narrative of Voyages of the Beagle*, i. 275, 282, and ii. 391 (1839). Cf. also Cook, in Loudon, *Gard. Mag.* xv. 694 (1839).

[3] *Journ. Hort. Soc.* vi. 262 (1851).

[4] *Hortus Veitchii*, 38, 46, 340 (1906).

[5] *Kew Bull.* 1909, p. 235.

[6] *Gard. Chron.* xxxi. 392, Supply. Illust. (1902).

The only specimen that we know of in Scotland is one at Ardgowan, which was about 25 ft. high in 1909. A tree at Murthly, recorded at the Conifer Conference as 16 ft. high and 30 years old in 1891, was killed in the severe winter of 1893-1894, when the thermometer registered at Murthly 11° Fahr. below zero.

In Ireland a tree at Powerscourt was 30 ft. high in 1910; and another at Fota was 25 ft. high in 1912. There are also good specimens at Kilmacurragh, Rostrevor, and Castlewellan.[1]   (A. H.)

Timber

The wood is very valuable; and is remarkable for the extraordinary straightness of the grain, which makes it very easy to rend into thin boards. These are used as shingles for roofing, which after exposure to the weather turn blue and resemble slates; and also for floors and partitions. Castillo and Dey state that it is very durable in contact with water, is unaffected by heat or humidity, and has lasted in roofs without alteration for over 100 years. Captain Fitzroy[2] says that "it does not shrink or warp; and, though brittle, is of very close grain and well adapted for furniture. Of this wood, the country people make staves for casks; and the bark of the tree is used for caulking the seams of vessels, being extremely durable when constantly wet, though it soon decays when exposed to the sun and air." Spars of alerce, which proved to be very strong, were obtainable 80 or 90 ft. in length. I saw mule-loads of these shingles 8 ft. long and very thin, on my journey from Nahuelhuapi to Puerto Montt, and very large stumps of trees in the country near that port; but I was unable to visit the forest where it grows. The wood is apparently still unknown in Europe; but I bought three boards in London imported from Chile, which are of a rich reddish colour with very fine and close grain.   (H. J. E.)

---

[1] Figured by Earl Annesley, *Beautiful and Rare Trees*, t. 68 (1903), who states that this species should be planted in deep loam and peat, as it is a deep-rooting tree which grows slowly till it is well established.

[2] *Narrative of Voyages of the Beagle*, i. 275, 282, and ii. 391 (1839). Cf. also Cook, in Loudon, *Gard. Mag.* xv. 694 (1839).

SAXEGOTHÆA

Saxegothæa, Lindley, in *Journ. Hort. Soc. Lond.* vi. 258 (1851); Bentham et Hooker, *Gen. Pl.* iii. 434 (1880); Masters, in *Journ. Linn. Soc. (Bot.)* xxvii. 299 (1889), and xxx. 10 (1893); Pilger, in Engler, *Pflanzenreich*, iv. 5, *Taxaceæ*, 42 (1903); Stiles, in *New Phytologist*, vii. 209-222, figs. 28-34 (1908), and *Ann. Bot.* xxvi. 446, 463 (1912); R. B. Thomson, in *Bot. Gaz.* xlvii. 344-354, pl. 22-24 (1909).

A GENUS belonging to the division Podocarpeæ of the order Taxaceæ, mainly characterised by the yew-like foliage with true scaly buds, and by the female cones composed of spirally and loosely imbricated carpellary scales, the uppermost of which are fertile, each bearing internally near the base a cavity from which hangs a single free minute ovule; scales ultimately becoming fleshy, coalescing to form an irregular globose head, only a few of the ovules ripening into seeds, which when mature are set free by the gaping apart of the fertile scales. The genus, which has been studied by W. Stiles at Cambridge, is a remarkable one, all parts of the plant having a simple structure, suggesting that it is a primitive type. *Saxegothæa* is allied to *Araucaria* as well as to *Podocarpus*, resembling the latter in leaf, but the former in the female flowers and in the wingless pollen grains. Only one species is known, in the following description of which the other characters of the genus are given in detail.   (A. H.)

SAXEGOTHÆA CONSPICUA

Saxegothæa conspicua, Lindley, in *Journ. Hort. Soc. Lond.* vi. 258, figs. A and B (1851), and in Paxton, *Flower Garden*, ii. 111, fig. 190 (1852); Masters, in *Gard. Chron.* ii. 684, figs. 130, 131 (1887), and v. 782, fig. 125 (1889); Kent, Veitch's *Man. Conif.* 158 (1900); Pilger, in Engler, *Pflanzenreich*, iv. 5. *Taxaceæ*, 42 (1903); Castillo and Dey, *Jeog. Vej. Rio Valdivia*, 31, fig. 12 (1908).

An evergreen tree, attaining in South America 30 to 40 ft. in height, becoming at high elevations a low dense shrub. Bark greyish brown, scaling off like that of a plane tree, leaving the reddish brown cortex beneath exposed in patches. Branches widely spreading, pendulous at the ends, giving off the branchlets in opposite pairs or in whorls of three or four; young branchlets slender, glabrous, marked by the decurrent bases of the leaves, green with inconspicuous white dots on the lower side. Buds minute, globose, surrounded by three to seven ovate greenish scales, which persist brown and withered at the apex of the branchlet of the second year. Leaves, persistent about five years, arising in spiral order, spreading radially on leading shoots, but thrown into a pectinate arrangement on lateral branches ; linear, straight or curved, ½ to ¾ in. long, narrowed into a petiolate base, decurrent on the branchlet, tapering at the apex, which ends in a sharp cartilaginous point ; upper surface dark green with a raised narrow midrib ; lower surface with a narrow green midrib, on each side of which is a broad stomatic band, composed of about twelve close lines of dots, and wider than the green margins.

Flowers monœcious. Staminate flowers cylindric, solitary or two or three in the axils of the leaves near the end of the branchlet, shortly stalked, subtended by four to six scales ; composed of numerous spirally arranged anthers, each with two cells, dehiscing longitudinally. Ovuliferous flowers solitary on the ends of the branchlets, on short peduncles bearing a few modified leaves ; succeeded by numerous, spirally arranged, densely imbricated, triangular-ovate, pointed carpellary scales, the lower sterile, the upper fertile, each with a depression on the upper surface near the base, in which is borne an inverted ovule. Fruiting head, ripening in the first year, irregularly globose, ⅛ to ½ in. in diameter, with fleshy scales coalescent at the base and free at the apex, containing six to twelve ripened seeds, which are set free by the separating of the fertile scales. Seed, about ⅛ in. in width and length, smooth, shining brown, ovoid, compressed from back to front, with two sharp lateral edges ; base broad, marked by the rough surface of the hilum. Cotyledons two.

This species is a native of Chile and western Patagonia,[1] where it grows in the lower regions of the mountains in the dense forests, composed mainly of evergreen bush and conifers, like *Fitzroya patagonica*, *Libocedrus tetragona*, *Podocarpus chilina*, and *Podocarpus nubigena*. Castillo and Dey say that these two species of Podocarpus and Saxegothæa are known in Chile as *mañiu*, and yield a fine homogeneous wood, yellow in tint, and admirably suited for joiner's work.[2]

Saxegothæa conspicua was discovered[3] by W. Lobb in southern Chile in 1846, and introduced in 1847. It does not appear to have succeeded in our climate, and is now very rare. At Kew, it is perfectly hardy, but grows slowly, and has a stunted appearance.

There are two specimens at Strete Ralegh, near Exeter, the seat of Mr. Imbert Terry, both probably original introductions, and about 30 ft. high in 1909, when the larger was 4 ft. 3 in. in girth at two feet from the ground.[4]

There is a fine specimen at Coldrenick, about 35 ft. high, with a short bole, only a foot long, dividing into about nine stems, 4 to 6 in. in diameter, with wide-spreading branches, which I saw in 1911. Both this tree, and those at Strete

---

[1] Dusen, in Scott, *Princetown Univ. Exped. Patagonia*, viii. 20 (1903), says it grows in the forests of the middle and lower Aysen valley in Patagonia. Reiche, *Verbreit. Chilen. Conif.* 5 (1900), gives the Aysen valley, lat. 45° 10′, as its known southerly limit, while it extends northwards to the Rio Maule in lat. 35° 20′. It occurs on Chiloe, but not in the Guaytecas and Chonos islands.

[2] Capt. Fitzroy, *Narrative of Voyages of the Beagle*, i. 280 (1839), says :—"*Mañu*, a tree of great dimensions, tall and straight, the leaf is like that of a yew : it is a very useful wood for shipbuilding, for planks, and next to *alerce*, is the best for spars that the island of Chiloe produces ; but the large trees have a great tendency to become rotten at the heart owing possibly to the humidity of the climate, and to the very wet soil. Of twenty trees that were cut down, not one was sound at the heart. The wood is heavy, with large knots, which penetrate into the trunk to a great depth. A great deal of this timber grows in the Gulf of Peñas." It is doubtful what species is here referred to ; but Saxegothæa does not now occur so far south as the Gulf of Peñas. [3] *Hortus Veitchii*, 38, 345 (1906). [4] Cf. Dallimore, in *Kew Bull.* 1909, p. 336.

Ralegh, produce flowers and fruit abundantly; but I am not aware that seedlings have been raised. In the pinetum at Uplyme, Dorset, there is a remarkably fine specimen with a single stem, which was 40 ft. high and 3 ft. 4 in. in girth, when I saw it in July 1912. At Abbotsbury this species is represented by a shrubby plant, 4 ft. high and 4 ft. across. At Bury Hill near Dorking there is a tree about 20 ft. high.

Bean[1] saw in 1907 a specimen, about 12 ft. high, at Ochtertyre in Perthshire.

The only specimens that we know of in Ireland are a spreading bush at Rostrevor, where, Sir John Ross-of-Bladensburg informs us, it was planted in 1891; and two bushy trees, about 20 ft. high, which Elwes saw at Kilmacurragh in 1908.

(A. H.)

[1] *Gard. Chron.* xli. 168 (1907)

TORREYA

Torreya,[1] Arnott, in *Ann. Nat. Hist.* i. 130 (1838); Bentham et Hooker, *Gen. Pl.* iii. 431 (1880); Eichler, in Engler and Prantl, *Pflanzenfam.* ii. 1, p. 111 (1889); Masters, in *Journ. Linn. Soc.* (*Bot.*) xxx. 5 (1893); Pilger, in Engler, *Pflanzenreich*, iv. 5, *Taxaceæ*, 105 (1903); Sargent, in *Bot. Gaz.* xliv. 226 (1907).

Tumion, Rafinesque, *Amenities of Nature*, 63 (1840); Sargent, *Silva N. Amer.* x. 55 (1896).

Caryotaxus, Zuccarini, *ex* Endlicher, *Syn. Conif.* 241 (1847).

EVERGREEN trees belonging to the order Taxaceæ, with fissured bark and opposite or whorled branches. Young branchlets green, with linear pulvini, separated by slight grooves. Buds, one terminal, and occasionally two to four lateral, clustered at the end of the branchlet, composed of a few decussately opposite scales. Base of the branchlet marked with scars, left by the fall of the bud-scales of the previous season; occasionally two or three of these persisting brown, unenlarged, and inconspicuous. Leaves spirally arranged, but thrown, by twisting and turning of their bases, into a pectinate arrangement on lateral branches, as in the yew; persistent three or four years; stalked, linear, tipped with a bristle-like cartilaginous point; upper surface green, convex; lower surface with a raised green midrib, and two white stomatic bands, sunk in longitudinal depressions; fibro-vascular bundle undivided, with a solitary resin-canal beneath it.

Flowers diœcious, or monœcious [2] with the sexes on different branches. Staminate flowers solitary in the axils of the leaves of the current year's branchlet, surrounded at the base by several pairs of decussate scales, composed of numerous stamens, in whorls of fours, on a stipitate slender axis; filament expanded into four pollen-sacs; connective truncate or crest-like and dentate.

Pistillate flowers in pairs on rudimentary branchlets,[3] which are solitary in the axils of a few leaves towards the base of the current year's shoot; each flower subtended by four decussate scales and a bract, and consisting of a solitary terminal ovule, surrounded at the base by a small disc, the aril, which grows upwards and ultimately becomes confluent with the succulent testa of the seed. Seed, as only one flower of each pair develops, solitary; ripening in the second year, drupe-like, with an outer succulent resinous coat and an inner woody shell, within which is the ruminate albumen and a minute embryo with two cotyledons. The shell bears a slightly projecting point at the apex, around which there is a dark-coloured oval,

---

[1] According to the Vienna rules of nomenclature, the name Torreya is to be retained for this genus. It had previously been applied to a species of Clerodendron by Sprengel, *Neue Entd.* ii. 121 (1821).

[2] The flowers are usually said to be diœcious, but the trees in cultivation are monœcious.

[3] This occasionally terminates in a third ovule, or produces an extra bract. Cf. Miss Robertson, in *New Phytologist*, iii. 142 (1904).

circular, or saddle-shaped area, representing the outer surface of the integument where it is free from the aril; and below this are two opposite minute shield-like prominences, each with a minute aperture. The seed is subsessile or short-stalked, and subtended at the base by six decussate scales,[1] from which it separates, when it falls after ripening.

Seedling: see under *T. californica*.

All species of Torreya sprout freely from the stump and roots when cut; and bear pruning freely. They are propagated by cuttings[2] or by grafting on the common yew or on species of Cephalotaxus.

During the tertiary period the genus inhabited the Arctic region, and spreading southward existed in Europe. All the living species produce handsome, close-grained, pale yellow wood,[3] useful for cabinet-making, and durable when placed in contact with the soil as posts for fencing.

The species of Torreya superficially resemble those of Cephalotaxus; but the two genera are readily distinguishable.

Torreya.—Leaves with long spines at the apex, and narrow stomatic bands beneath. Buds with few decussate scales, all deciduous or two to four persisting minute and inconspicuous at the base of the branchlet.

Cephalotaxus.—Leaves with short-pointed apices; under surface with broad stomatic bands extending from the midrib almost to the outer margin. Buds with numerous imbricated scales, which persist at the base of the branchlet.

The four living species of Torreya are distinguishable as follows :—

I. *Leaves and branchlets fœtid.*
    1. *Torreya taxifolia*, Arnott. Florida. See p. 1466.
        Leaves linear, $\frac{3}{4}$ to $1\frac{1}{2}$ in. long, $\frac{1}{8}$ in. broad, pale green beneath, with a broad midrib and narrow scarcely depressed stomatic bands; petiole, $\frac{1}{25}$ in.

II. *Leaves and branchlets pungent-aromatic.*
    2. *Torreya californica*, Torrey. California. See p. 1465.
        Leaves linear, $1\frac{1}{4}$ to 3 in. long, $\frac{1}{8}$ in. broad; glaucous beneath with a broad midrib and slightly depressed narrow stomatic bands; petiole, $\frac{1}{12}$ in.
    3. *Torreya nucifera*, Siebold and Zuccarini. Japan. See p. 1463.
        Leaves lanceolate-linear, $\frac{3}{4}$ to $1\frac{1}{4}$ in. long, $\frac{1}{8}$ to $\frac{1}{6}$ in. wide; green beneath with deeply depressed broad stomatic bands, about as wide as the midrib.

III. *Leaves and branchlets devoid of any peculiar odour or taste.*
    4. *Torreya grandis*, Fortune. China. See p. 1464.
        Leaves lanceolate-linear, $\frac{1}{2}$ to 1 in. long, $\frac{1}{8}$ in. wide; thinner in texture than those of *T. nucifera*, but with similar stomatic bands. (A. H.)

[1] These are the four decussate scales and bract of the flower which develops—the sixth scale being the bract and remains of the other flower which does not develop.
[2] Rehder states, in Bailey, *Cycl. Am. Hort.* iv. 1822 (1902), that plants raised from cuttings grow slowly and remain bushy. This may account for the few good specimens of *Torreya* which exist in cultivation. The seeds are difficult to transport, as they soon become rancid.
[3] Figured by Mayr, *Fremdländ. Wald- u. Parkbäume*, 423, pl. x. 23 (1906).

TORREYA NUCIFERA

Torreya nucifera, Siebold and Zuccarini, in *Abh. Akad. München*, iv. 3, p. 234 (1846); Franchet et Savatier, *Enum. Pl. Jap.* i. 475 (1875); Masters, in *Journ. Linn. Soc.* (*Bot.*) xviii. 500 (1881), and xxvi. 546 (1902); Shirasawa, *Icon. Ess. Forest. Japon*, i. text 32, t. 15, figs. 19-34 (1900); Kent, Veitch's *Man. Conif.* 119 (1900); F. W. Oliver, in *New Phytologist*, i. 151, fig. 6 (1902); Pilger, in Engler, *Pflanzenreich*, iv. 5, *Taxaceæ*, 105 (1903).

Taxus nucifera, Linnæus, *Sp. Pl.* 1040 (1753); Thunberg, *Fl. Jap.* 275 (1784).

Podocarpus nucifera, Persoon, *Syn.* ii. 633 (1807); Loudon, *Arb. et Frut. Brit.* iv. 2100 (1838).

Caryotaxus nucifera, Zuccarini, *ex* Endlicher, *Syn. Conif.* 241 (1847); Henkel and Hochstetter, *Syn. Nadelholz.* 366 (1865).

Tumion nuciferum, Greene, in *Pittonia*, ii. 194 (1891).

A tree, attaining in Japan 80 ft. in height, with bright red bark, and pungent aromatic branchlets and leaves. Young branchlets green, glabrous, becoming reddish in the second year. Buds prismatic, about $\frac{1}{8}$ in. long, with six to eight decussate external scales. Leaves, $\frac{3}{4}$ to $1\frac{1}{4}$ in. long, $\frac{1}{8}$ to $\frac{1}{6}$ in. wide, lanceolate-linear, tapering from near the base to the narrow acuminate spine-tipped apex; shining dark or yellowish green above; lower surface with two deeply depressed white stomatic bands, about as wide as the midrib, but not so wide as the marginal green bands; petiole about $\frac{1}{16}$ in. long.

Staminate flowers, $\frac{1}{3}$ in. long; connective crest-like, denticulate. Fruit narrowly obovoid, $\frac{3}{4}$ to 1 in. long, green tinged with purple; flesh thin, resinous; shell light brown, with irregular longitudinal depressions; inner coat dark red, folded for a short distance into the albumen. (A. H.)

This species is a native of Japan, occurring as a rare tree in the southern islands and in the forests of southern and central Hondo, where, according to Sargent, it reaches a height of 80 ft., and is remarkable for the beauty of its bright red bark and lustrous dark green foliage. The only place where I saw the tree wild was in virgin forest on the slopes of the Kireshima volcano in southern Kiusiu. Here there were trees up to about 10 in. in diameter, scattered in a dense forest of *Abies firma* and *Tsuga*, mixed with oaks, chestnut, and other broad-leaved trees. Seedlings were found in dense shade and could be distinguished by the sharp points of their leaves. I was told by the foresters that much larger trees existed, but only in remote and inaccessible places. Where it was planted in Kisogawa and below Koyasan at 1500 ft. elevation, it attained a height of 40 to 50 ft. with a girth of 8 ft. or more; and had very much the habit of a yew, but made a cleaner trunk with less tendency to branch. It is known in Japan as *Kaya*; and an oil is extracted from the seeds, which is used in cooking. The wood, according to Rein, is light yellowish in colour, and is used for making chests and boxes. The wood is also used for making Japanese chess-men.

The Japanese Torreya was described and figured by Kaempfer, *Amœn. Exot.* 815 (1712); and seems to have been introduced[1] into England in 1764 under the

[1] Aiton, *Hort. Kew.* v. 416 (1813).

name *Taxus nucifera*. Loudon gives the date of introduction as 1820, and states that a tree at White Knights was 13 ft. high in 1834. Siebold sent it to Holland about 1840. It is rare in cultivation, and is never seen except as a shrub.

(H. J. E.)

TORREYA GRANDIS

Torreya grandis, Fortune, in *Gard. Chron.* 1857, p. 788, and 1860, p. 170, and in Gordon, *Pinetum*, 326 (1858); Masters, in *Journ. Linn. Soc. (Bot.)* xviii. 500 (1881), xxvii. 323, fig. 28 (1890), and xxvi. 546 (1902), and in *Gard. Chron.* ii. 681, fig. 117 (1884); Franchet, *Pl. David.* i. 292 (1884).

Torreya Fargesii, Franchet, in *Journ. de Bot.* xiii. 264 (1899); Pilger, in Engler, *Pflanzenreich*, iv. 5, *Taxaceæ*, 108 (1903).

Torreya nucifera, Siebold and Zuccarini, var. *grandis*, Pilger, *op. cit.* 107 (1903).

Caryotaxus grandis, Henkel and Hochstetter, *Syn. Nadelholz.* 366 (1865).

Tumion grande, Greene, in *Pittonia*, ii. 194 (1891).

A tree, attaining 80 ft. in height in China, with leaves and branchlets devoid of a disagreeable or pungent odour. Young branchlets green, glabrous, becoming yellowish brown in the second year. Leaves, $\frac{1}{2}$ to 1 in. long, $\frac{1}{8}$ in. broad, linear-lanceolate, similar to those of *T. nucifera* in shape, but thinner in texture, with similar deeply depressed stomatic bands, nearly as wide as the midrib, but narrower than the marginal green bands.

Fruit broadly ellipsoid, $\frac{3}{4}$ to 1 in. long; flesh not disagreeable in odour; shell reddish brown, with irregular shallow depressions over the surface; inner coat only slightly folded into the albumen.

Torreya grandis was discovered by Fortune in 1855 in the coast province of Chekiang in China, in the mountains south-west of Ningpo, at 4000 ft. elevation. Here numerous fine trees were seen, many of which were 60 to 80 ft. in height. It was subsequently collected in the adjoining province of Fukien by Père David. The same tree [1] also exists in the central provinces of Hupeh and Szechwan, in the mountains between 4000 and 6000 ft. elevation; where it occasionally attains a height of 50 ft., but it is more commonly shrubby, bearing fruit when only 8 ft. high.

This tree is known to the Chinese as *fei*; and the kernels, called *fei-shih*,[2] are sold in the drug-shops of most Chinese towns, being considered a valuable remedy in cases of cough, asthma, etc. They are occasionally eaten like hazel nuts, and though reputed laxative, are considered wholesome.

Torreya grandis was introduced by Fortune, who sent seeds in 1855 to Glendinning's nursery at Chiswick, where they germinated freely. This species is not common in collections, the only specimen which we have seen being a small shrub in the Cambridge Botanic Garden, which was obtained from Veitch in 1894.

(A. H.)

---

[1] The Torreya of Central China is considered by Franchet and Pilger to be a distinct species, *T. Fargesii*; but I can see no characters by which it can be separated from the Chekiang species.

[2] Cf. Hanbury, *Sc. Papers*, 233 (1876).

TORREYA CALIFORNICA

Torreya californica, Torrey, in *New York Journ. Pharm.* iii. 49 (1854); J. D. Hooker, in *Gard. Chron.* xxiv. 553 (1885); Masters, in *Gard. Chron.* v. 800, figs. 126, 127 (1889); Kent, Veitch's *Man. Conif.* 117 (1900); Pilger, in Engler, *Pflanzenreich*, iv. 5, *Taxaceæ*, 109 (1903); Jepson, *Silva of California*, 167 (1910).

Torreya Myristica, J. D. Hooker, in *Bot. Mag.* t. 4780 (1854); Murray, in *Trans. Bot. Soc. Edin.* vi. 217, pl. iii. (1860); Masters, in *Gard. Chron.* xxii. 681, fig. 116 (1884).

Caryotaxus Myristica, Henkel and Hochstetter, *Syn. Nadelh.* 368 (1865).

Tumion californicum, Greene, in *Pittonia*, ii. 195 (1891); Sargent, *Silva N. Amer.* x. 59, t. 513 (1896), and *Trees N. Amer.* 98 (1905).

A tree, attaining in California 100 ft. in height, and 12 ft. in girth, usually considerably smaller. Leaves, branches, and wood pungent-aromatic. Young branchlets glabrous, green, becoming brown in the second year. Buds, up to $\frac{1}{4}$ in. long, prismatic, with eight to ten decussate outer scales, those towards the apex elongated. Leaves, $1\frac{1}{4}$ to 3 in. long, $\frac{1}{8}$ in. wide; linear, tapering in the anterior third to an acuminate spine-tipped apex; dark shining green above; lower surface flat, glaucous, with two slightly depressed white stomatic bands (about 0.3 mm. wide), much narrower than the broad midrib (about 1 mm. wide), and the two outer glaucous bands (each about 0.7 mm. wide); petiole stout, $\frac{1}{12}$ in. long.

Staminate flowers $\frac{1}{3}$ in. long; connective truncate, not dentate. Fruit[1] ellipsoid or obovoid, 1 to $1\frac{1}{2}$ in. long; light green streaked with purple; flesh thin, resinous; shell fawn-coloured, smooth or with irregular slight longitudinal ridges; inner coat reddish, deeply folded into the white albumen.

Seedling[2] similar to that of *Ginkgo biloba*, with two thick and fleshy cotyledons, remaining underground; the stem bearing below a few scales, which are succeeded by ordinary leaves, the transition between the scales and the leaves being gradual.

(A. H.)

This species is a native of California, growing on the borders of mountain streams, nowhere common, but widely distributed from Mendocino County to the Santa Cruz mountains in the coast region, and along the western slopes of the Sierra Nevada from Eldorado to Tulare County, at 3000 to 5000 ft. elevation. It is most abundant and of its largest size in the northern coast ranges. Hough, in *American Woods*, pt. vi. p. 50, describes a fine tree overthrown by a flood near the coast in Mendocino County, from which the specimens of wood in his book were cut. It was 85 ft. long to the point, 5 in. in diameter, where its dead top was broken off. Its straight columnar trunk was 12 ft. in girth at eighteen inches, and 8 ft. at thirty-five feet from the ground. Assuming that the growth of this tree had been as uniform as in the section, which shows ten rings to the inch, it would have been from 250 to 300 years old, and the contents of the log about 300 cubic feet.

---

[1] Described from specimens grown at Orton. Cf. F. W. Oliver, in *Ann. Bot.* xvii. 466, pl. xxiv. (1903), for a detailed description of the remarkable structure of the seed of this species.

[2] Miss Chick (Mrs. Tansley) describes seedlings grown from seed produced at Orton, in *New Phytologist*, ii. 83, plates vii. viii. (1903).

Hough states that it is an excellent, light, and durable wood, well suited for boat-building and cabinetmaking, but too rare to be generally known or used. Jepson adds that the wood is strongly odorous, with white sapwood and clear light yellow heartwood, susceptible of a fine polish.

This tree was discovered in 1851 by William Lobb, who sent specimens and seeds in that year to Messrs. Veitch, which were described as *T. Myristica* in 1854 by J. D. Hooker, a short time after it had been published as *T. californica* by Dr. Torrey, who had received specimens from Mr. Shelton.

Though hardy in most parts of Great Britain, this species requires a heavy rainfall together with rich deep soil to grow to any size; and owing to its being usually planted in open situations, instead of in the densely wooded ravines which it likes, it has a tendency to spread and form branches rather than make a trunk.

By far the largest specimen that I know is at Tregothnan, which, when I first saw it in 1905, was 35 ft. by 6 ft.; and in 1911 had increased to 45 ft. high. Its shape is spoilt by three large lower branches; its trunk shows some large nipple-like protuberances resembling those often seen on the deodar.

Another tree at Orton Hall, Peterborough, has produced fruit regularly for the last twenty years, from which numerous seedlings[1] have been raised. It measured about 25 ft. by 4 ft. in 1905. At Poles, near Ware (Herts), a tree, which has never flowered, measured 40 ft. by 4 ft. in 1910. It was planted in 1858.

At Tortworth,[2] there is a handsome specimen, 25 ft. by 4 ft. in 1904, with branches spreading to a diameter of 36 ft.; and at Westonbirt, there is a tree in a shaded position, 29 ft. by 2 ft. 9 in. in 1907. At Chipping Campden, there is a tree about 30 ft. high, which bore staminate flowers in profusion in May 1911.

In Scotland it succeeds as far north as Durris in Kincardineshire, where a healthy specimen is about 20 feet high.

In Ireland, the best example is at Verner's Bridge, near Lough Neagh, where a tree measured 40 ft. by 4 ft. 7 in. in 1904. There are also specimens at Fota and Castlemartyr; the latter, forking near the ground, was 30 ft. high when I saw it in 1908.

(H. J. E.)

TORREYA TAXIFOLIA

Torreya taxifolia, Arnott, in *Ann. Nat. Hist.* i. 130 (1838); W. J. Hooker, *Icon. Plant.* tt. 232, 233 (1840); Kent, Veitch's *Man. Conif.* 119 (1900); Pilger, in Engler, *Pflanzenreich*, iv. 5, *Taxaceæ*, 108 (1903).

Taxus montana, Nuttall, in *Journ. Acad. Sc. Phil.* vii. 96 (1834) (not Willdenow).

Caryotaxus taxifolia, Henkel and Hochstetter, *Syn. Nadelh.* 367 (1865).

Tumion taxifolium, Greene, in *Pittonia*, ii. 194 (1891); Sargent, *Silva N. Amer.* x. 57, t. 512 (1896), and *Trees N. Amer.* 98 (1905).

A tree, attaining in Florida 40 ft. in height and 6 ft. in girth, with fœtid leaves, branches, and wood. Young branchlets green, with occasional minute hairs; older

---

[1] These grow very slowly at Colesborne and are now at eleven years old under two feet high.

[2] A section from another tree which died at Tortworth was sent by the Earl of Ducie to the Cambridge Forestry Museum.

branchlets yellowish red. Buds as in *T. californica*. Leaves $\frac{3}{4}$ to $1\frac{1}{2}$ in. long, $\frac{1}{8}$ in. wide; linear, tapering in the anterior third to an acuminate spine-tipped apex; shining green above; lower surface pale green, with two stomatic bands, scarcely depressed, and narrower than the midrib and the external bands; rounded at the base, with a short petiole, about $\frac{1}{25}$ in. long.

Staminate flowers, $\frac{1}{4}$ in. long; connective minute, rounded, not dentate. Fruit obovoid, 1 to $1\frac{1}{2}$ in. long, dark purple; flesh fœtid, coriaceous; shell smooth, light brown; inner coat brownish and not so deeply folded into the albumen as in *T. californica*.

This species, of which I have seen no living specimen, appears to be very similar to *T. californica*, but has smaller leaves, with very short petioles, and, according to Sargent, is different in odour.

T. taxifolia is restricted to north-western Florida, where it grows on limestone cliffs and in swamps[1] along the banks of the Appalachicola River, from River Junction to near Bristol in Gadsden County. The wood, according to Sargent, is hard, strong, clear bright yellow, with thin lighter-coloured sapwood, and is used locally for fence-posts. Owing to the peculiar odour of the whole tree, noticeable also in the wood when burnt, it is known as "stinking cedar." Dr. Torrey informed Arnott that a blood-red turpentine, of a pasty consistence, flows sparingly from the bark, and is soluble in alcohol.

This species was discovered in 1833 by H. C. Croom, and was introduced[2] into England, in 1840, by A. J. Downing, who sent a living plant to London, which was propagated[3] by Masters of the Canterbury Nursery. It probably proved unsuitable to our climate; and I have seen no specimen which could be identified with this species.

According to Sargent[4] it can be kept alive in eastern Massachusetts in sheltered, well-shaded situations; and occasional individuals have survived a number of years near New York and Philadelphia. (A. H.)

---

[1] Cf. Britton, *N. Amer. Trees*, 126 (1908).   [2] Loudon, *Gard. Mag.* xvi. 658 (1840).
[3] Loudon, *Trees and Shrubs*, 944 (1842), states that it was propagated by grafting on the yew.
[4] Sargent, in *Garden and Forest*, x. 400 (1897).

CEPHALOTAXUS

Cephalotaxus, Siebold and Zuccarini, *ex* Endlicher, *Gen. Suppl.* ii. 27 (1842); Bentham et Hooker, *Gen. Pl.* iii. 430 (1880); Masters, in *Journ. Linn. Soc. (Bot.)* xxx. 4 (1893), and in *Gard. Chron.* xxxiii. 227 (1903); Worsdell, in *Ann. Bot.* xv. 637 (1901); Pilger, in Engler, *Pflanzenreich*, iv. 5, *Taxaceæ*, 99 (1903).

EVERGREEN shrubs or small trees, belonging to the order Taxaceæ, with opposite or whorled branches. Young branchlets green, marked by white stomatic dots, and with linear pulvini, separated by slight grooves. Buds, with numerous imbricated scales, which persist as a conspicuous sheath at the apex of the branchlet of the second year. Leaves spirally arranged, radially spreading on vertical shoots, but on lateral branches thrown by twisting of their bases into a pectinate arrangement; persistent three or four years, very shortly stalked, linear, acute at the apex; upper surface green with a prominent midrib in a depression; lower surface with two whitish broad bands, composed of numerous stomatic lines, separated by a narrow raised green midrib, and bounded on each outer side by a very narrow marginal green band; fibro-vascular bundle undivided, with a single resin-canal beneath it.

Flowers diœcious.[1] Staminate flowers in globose heads, which are solitary in the axils of the leaves of the branchlets of the previous year; each head with six to eleven flowers, each of which is subtended by a bract and has seven to twelve stamens; pollen-sacs two or three, dehiscing longitudinally.[2] Pistillate heads few, each solitary in the axil of a scale-leaf near the base of the branchlet of the current year, and composed of a stipitate axis, towards the end of which are three or four decussate pairs of opposite bracts; each bract is cup-shaped at the base and bears two erect ovules side by side. Usually only one or two of the ovules in a head develops, forming a drupe-like seed, with a fleshly outer covering, and an inner hard woody shell, which encloses the albumen and embryo. The seedling[3] has two long linear cotyledons, immediately above which and decussate with them on the stem is a pair of primary leaves, which are followed at intervals by either whorls or pairs of larger leaves.

Six species[4] of Cephalotaxus are known, one of which is possibly a hybrid,

---

[1] In rare cases, the flowers are monœcious, as in a shrub of *C. Fortuni*, described by Carrière in *Rev. Hort.* 1878, p. 116, fig. 24.

[2] Kerner, *Nat. Hist. Plants*, Eng. Trans. ii. 124 (1898), states that the anthers open and shut periodically.

[3] Cf. Masters, in *Journ. Linn. Soc. (Bot.)* xxvii. 241 (1889).

[4] *C. Mannii*, Hooker, native of the Khasi Mountains; *C. Griffithii*, Hooker, of Assam and Manipur; and *C. Oliveri*, Masters, of Central China, are not now in cultivation. The plants of *C. Griffithii*, formerly in the temperate house at Kew, mentioned by Hooker, *Fl. Brit. India*, v. 648 (1890), died many years ago. The young plants at Coombe Wood, referred to *C. Oliveri* in *Gard. Chron.* xxxiii. 227 (1903), and *Hortus Veitchii*, 338 (1906), are *C. drupacea*.

Podocarpus argotænia, Hance, a peculiar conifer in southern China, is referred to Cephalotaxus by Pilger; but is distinct.

natives of China, Japan, Khasi Mountains, Assam and Manipur. As none of the three species in cultivation forms a tree in this country, they do not properly come within the scope of our work; but are now briefly described, owing to their interest as conifers, which are frequently seen in gardens. All the three species are perfectly hardy; but succeed best in shady situations, sheltered from the wind. When propagated by scions or cuttings, terminal shoots should be selected, as these form regular plants with whorled branches like seedlings; whereas cuttings from lateral branches grow into irregular low spreading shrubs.[1]   (A. H.)

CEPHALOTAXUS DRUPACEA

Cephalotaxus drupacea, Siebold et Zuccarini, *Fl. Jap. Fam. Nat.* ii. 108 (1846), and *Fl. Jap.* ii. 66, tt. 130 and 131 (1870); Franchet et Savatier, *Enum. Fl. Jap.* i. 473 (1875); Masters, in *Gard. Chron.* xxi. 113 (1884), and xxxiii. 228, fig. 94 (1903), and *Journ. Linn. Soc. (Bot.)* xxii. 201 (1886) and xxvi. 544 (1902); Kent, Veitch's *Man. Conif.* 112 (1900); Shirasawa, *Icon. Ess. Forest. Japon*, i. text 31, t. xiv. figs. 1-12 (1900); Pilger, *Taxaceæ*, 100, figs. 19, 20 (1903); Hemsley, in *Bot. Mag.* 8285 (1909).

Taxus baccata, Thunberg, *Fl. Jap.* 275 (1784) (not Linnæus).

A large shrub or small tree, occasionally attaining in Japan 40 ft. in height. Leaves on lateral branches pectinate, but spreading outwards usually in a V-shaped arrangement and not remaining in one plane, linear, $\frac{3}{4}$ to $1\frac{1}{2}$ in. long, straight or falcate, tapering towards the triangular acute apex, which is often tipped with a short spine-like point; stomatic bands beneath, each composed of about thirteen to fifteen lines. Staminate heads $\frac{1}{8}$ in. in diameter, on very short scaly stalks. Fruit brown, $\frac{3}{4}$ to 1 in. long, pyriform, broadest at the rounded apex, which has a circular depression bearing a minute mucro, and narrowed towards the base; kernel light brown, smooth, ellipsoid, $\frac{3}{5}$ in. long, $\frac{2}{5}$ in. wide, rounded at the base, the two sharp lateral edges in the upper half uniting into a slight apiculus at the apex.

This species is a native of Japan and central China. In Japan,[2] it is generally scattered through the mountain forests, extending northward to central Hokkaido, where it grows on low hills as an undershrub, 2 or 3 ft. high, of the deciduous forest; while in Hondo, where it ranges between 2000 and 3000 ft. altitude, it becomes a bushy tree, averaging 25 ft. high, and occasionally attaining 40 ft.[3] It is known to the Japanese as *Inu-gaya*. In China, it has been found in the mountains of Hupeh and Chekiang and in the Chusan Archipelago.

C. drupacea was introduced by Siebold into the Botanic Garden at Leyden in 1829; but does not appear to have been known[4] in English gardens till 1844. It

---

[1] Cf. Rehder, in Bailey, *Cycl. Amer. Hort.* 276 (1900).

[2] A species of Cephalotaxus, not yet determined, occurs in the mountains of Formosa at about 8000 feet altitude. Cf. Hayata, in *Journ. Coll. Sci. Tokyo*, xxv. 215 (1908). Elwes saw this as a bush without flowers or fruit on Arisan in 1912.

[3] Mayr, *Fremdländ. Wald- u. Parkbäume*, 269 (1906), says that in the warmer parts of Japan it becomes a tree, rarely attaining 60 ft. in height.

[4] Nicholson, *Gard. Dict.* i. 294 (1884), gives 1844 as the date of introduction into England. It was mentioned as a cultivated plant by Knight and Perry, *Syn. Conif.* 51 (1850), who gave for it the synonyms *Taxus coriacea* and *Cephalotaxus coriacea*.

always remains a spreading bush in this country, an old specimen at Kew being about 10 ft. high and 15 ft. through, and clothed to the ground with luxuriant dark green foliage.　　　　　　　　　　　　　　　　　　　　　　　　　　(A. H.)

CEPHALOTAXUS FORTUNI

Cephalotaxus Fortuni, W. J. Hooker, in *Bot. Mag.* t. 4499 (1850); Masters, in *Gard. Chron.* xxi. 114, fig. 21 (1884), and xxxiii. 228 (1903), and *Journ. Linn. Soc. (Bot.)* xxvi. 545 (1902); Franchet, *Pl. David.* i. 292 (1884), and in *Journ. de Bot.* 1899, p. 265; Kent, Veitch's *Man. Conif.* 113 (1900); Pilger, *Taxaceæ*, 103 (1903).

A small tree or large shrub, attaining the same dimensions as *C. drupacea*. Leaves on lateral branches pectinate and spreading in one plane, falcate, $1\frac{1}{2}$ to 3 in. long, tapering gradually in the anterior third to an acuminate, usually spine-tipped apex; stomatic bands beneath conspicuously white, each of eighteen to twenty-one lines, covering nearly the whole surface, the midrib and green margins being narrower than in *C. drupacea*. Staminate heads, less than $\frac{1}{4}$ in. in diameter, on scaly stalks, which are $\frac{1}{6}$ in. long. Fruit olive-green, about $1\frac{1}{4}$ in. long and $\frac{5}{8}$ in. in diameter, elongated ovoid, contracted towards the base, and broadest at the rounded apex, which is tipped with a short elevated point, arising from a circular depression; kernel elongated ellipsoid, about 1 in. long and $\frac{2}{5}$ in. wide, light mottled brown, minutely tuberculate on the surface, rounded at the base; upper half with two sharp lateral edges which unite at the apex to form an apiculus.

This species is occasionally monœcious.[1] In the Cheshunt Nurseries,[2] there was a female plant, which bore fruit in 1862 and 1863, though no flowers were borne on a staminate plant close by; and it was supposed to have been pollinated by a yew; but the seeds were not sown, and may have been infertile.

C. Fortuni is a native of China, occurring in the mountain woods of Szechwan, Hupeh, Yunnan, Kiangsu, and Chekiang, where it usually grows as a large bush in the shade of broad-leaved trees. It was discovered in the mountains south-west of Ningpo,[3] in 1848, by Fortune, who sent seeds to the Bagshot Nursery in that year, which germinated freely. It is perfectly hardy in this country, and may be seen in many collections of conifers, forming a spreading shrub, with handsome foliage. The largest that we have seen is about 25 ft. high at Coldrenick. It appears to be little known in America.[4]　　　　　　　　　　　　　　　　(A. H.)

[1] Cf. Carrière, in *Rev. Hort.* 1878, p. 116, fig. 24, where a branch, bearing both male and female flowers, is depicted.

[2] Cf. *Gard. Chron.* 1863, p. 1062.

[3] Bretschneider, *Hist. Europ. Bot. Disc. China*, 502 (1908), points out that the statement usually made, that Fortune found it in North China, is erroneous.

[4] Sargent, in *Garden and Forest*, x. 391 (1897).

CEPHALOTAXUS PEDUNCULATA

Cephalotaxus pedunculata, Siebold and Zuccarini, *Fl. Jap. Fam. Nat.* ii. 108 (1846), and *Fl. Jap.* ii. 67, t. 132 (1870); Franchet et Savatier, *Enum. Pl. Jap.* i. 473 (1875); Franchet, *Pl. David.* i. 292 (1884); Masters, in *Gard. Chron.* xxi. 113, fig. 22 (1884), and xxxiii. 228 (1903), and *Journ. Linn. Soc. (Bot.)* xxii. 201 (1886), and xxvi. 545 (1902); Kent, Veitch's *Man. Conif.* 114 (1900); Diels, *Flora von Central China*, 214 (1901).

Taxus Harringtonia, Knight, *ex* Forbes, *Pin. Woburn.* 217, t. 66 (1839); Loudon, *Gard. Mag.* xv. 273 (1839), and *Trees and Shrubs*, 942 (1842).

Cephalotaxus Harringtonia, Koch, *Dendrologie*, ii. 2, p. 102 (1873).

Cephalotaxus drupacea, Siebold and Zuccarini, var. *pedunculata*, Miquel, *Prol. Fl. Jap.* 333 (1867).

Cephalotaxus drupacea, Siebold and Zuccarini, var. *Harringtonia*, Pilger, *Taxaceæ*, 102 (1903).

A large shrub or small tree. Leaves on lateral branches, pectinate, spreading either in one plane or in a V-shaped arrangement, straight or falcate, $1\frac{1}{2}$ to $2\frac{1}{2}$ in. long, tapering in the anterior third to an acute apex, which is often tipped with a spine; stomatic bands beneath each of sixteen to twenty-one lines, not so white as in *C. Fortuni*. Staminate heads in clusters of two to five, or occasionally solitary, on scaly peduncles, which are $\frac{1}{4}$ to 1 in. in length. Fruit olive-green, $\frac{3}{4}$ to 1 in. long, ellipsoid, not contracted at the basal end, with a circular depression at the apex from which arises a short mucro; kernel similar in size and shape to that of *C. drupacea*, but mottled light brown and slightly tuberculate on the surface.

The following varieties have been described :—

1. Var. *sphæralis*, Masters, in *Gard. Chron.* xxi. 113, fig. 23 (1884), and in *Journ. Linn. Soc. (Bot.)* xxii. 203, plate vii. (1886).

Fruit smaller, globose, not depressed at the apex, which bears a long mucro. This was described by Masters from a specimen growing in the Rev. J. Goring's garden at Steyning; and a shrub at Kew has also borne similar fruit.

2. Var. *fastigiata*, Carrière, *Prod. et Fix. Var.* 44, fig. 1 (1865), and *Conif.* 717 (1867); Masters, in *Gard. Chron.* xxi. 113, fig. 20 (1884).

Cephalotaxus Buergeri, Miquel, *Prol. Fl. Jap.* 333 (1867).

Podocarpus koraiana, Siebold, in *Ann. Soc. Hort. Pays-Bas*, 1844, p. 34; Carrière, in *Rev. Hort.* 1863, p. 349, fig. 36; Maximowicz, *Mél. Biol.* vii. 563 (1870).

A fastigiate form, similar to the Irish yew in habit; branches and branchlets directed vertically upwards; leaves spreading radially on the branchlets.

This handsome shrub appears to have originated in Japan, whence it was introduced in 1861 into England, where it is perfectly hardy.[1] It has never, so far as is known, borne flowers either in Japan or in Europe; and is always propagated by cuttings. It frequently produces near the base lateral branches[2] with normal foliage; and grafts, that are taken from these branches, reproduce the ordinary form of the species.

---

[1] Sargent, in *Garden and Forest*, x. 391 (1897), states that it is not hardy in eastern New England, but there are good plants near New York and Philadelphia.

[2] Figured by Masters, in *Journ. Linn. Soc. (Bot.)* xxvii. 245, fig. 5 (1889), and *Gard. Chron.* xxxiii. 227, fig. 96 (1903). Cf. De Vries, *Mutation Theory*, 110, fig. 16 (1911), who instances these reverted branches as showing the phenomenon of atavism by bud-variation.

C. pedunculata has been long in cultivation in Japan, where it is known as *Chosen-gaya* or *To-gaya*, meaning Korean or Chinese Cephalotaxus; and was introduced there in ancient times from Korea or China by the Buddhist monks. It is unknown in the wild state, and in all probability is a hybrid between *C. Fortuni* and *C. drupacea*, which originated in China, where these two species are both native. It usually resembles more the former species in foliage, and the latter species in fruit; but differs from both in the clustered staminate heads, which is possibly an abnormal condition. There are plants in gardens reputed to be, but not exactly matching *C. Fortuni*, which may be seedlings of *C. pedunculata*. The latter species has leaves of a darker hue than *C. Fortuni* and *C. drupacea*; and is equally hardy, but is scarcely so ornamental as the true *C. Fortuni*, which has the leaves much whiter beneath.

The original *C. pedunculata*, long cultivated in Japan, was always a male plant, no doubt propagated by grafts and cuttings; and it was introduced[1] into England in 1837. So far as can be ascertained, the history of the female plant is as follows:—The seeds of *C. Fortuni*, which were sent by Fortune[2] from China in 1848 to the Bagshot Nursery, produced two kinds of plants; one kind with long leaves, identical with the true wild plant of *C. Fortuni*; and the other kind with shorter leaves, identical with *C. pedunculata*, and comprising individuals which bore fruit.[3]

(A. H.)

[1] Cf. Loudon, *Trees and Shrubs*, 943 (1842). It appears to have been introduced by Siebold into Holland in 1829.

[2] Cf. Fortune, in *Gard. Chron.* 1863, p. 1134.

[3] W. Gorrie, in *Gard. Chron.* 1861, p. 51, points out that the shorter-leaved plants bearing fruit were certainly not *C. drupacea*. Fortune, believing that these plants constituted a new species, sent specimens from Chekiang in 1858, which are now preserved in the Lindley herbarium at Cambridge. These specimens, however, are simply a ♀ branch of *C. Fortuni*, and a ♂ branch of *C. drupacea*; and only show, that as both these species occur in Chekiang, the seed which he sent in 1848 may have been in part of hybrid origin.

KETELEERIA

Keteleeria, Carrière, in *Rev. Hort.* 449 (1866); Pirotta, in *Ann. R. Ist. Roma*, iv. 200 (1889); Masters, in *Journ. Linn. Soc.* (*Bot.*) xxx. 33 (1893).
Abies, Bentham et Hooker, *Gen. Pl.* iii. 442 (in part) (1880) (not Linnæus).

EVERGREEN trees belonging to the division Abietineæ of the order Coniferæ, and closely allied to Abies, from which they differ mainly in the persistent scales of the cones, in the fascicled staminate flowers, and in the leaves with the midrib prominent on both surfaces and ending in a spine on young plants. Branchlets smooth, with circular depressions, from which the leaves arise. Buds, of numerous imbricated scales, persisting after the branch has developed as a conspicuous sheath at its base. Leaves solitary, spiral on the branchlets, but thrown usually into a pectinate arrangement on lateral branches; linear, with the narrow stalk-like base expanded into a circular disc; acute or retuse at the apex (spine-tipped in young plants); upper surface with the midrib prominent in a longitudinal depression; lower surface with a raised midrib, two longitudinal sets of white stomatic lines, and two green marginal bands; fibro-vascular bundle undivided, with two resin-canals close to the epidermis of the lower surface near the outer angles.

Staminate flowers, in umbel-like clusters, each cluster of five to ten flowers either terminal or in the axil of a leaf on the current year's branchlet, arising on a short stalk covered with imbricated scales; each flower with a stipitate axis bearing numerous anthers, each with two pollen sacs; pollen grains with air-vesicles.

Cones erect on the branches, ripening in one year, and very similar to those of Abies, but differing in the persistent bracts and scales, the latter gaping apart to shed the seeds. Seed as in Abies, with resin-vesicles, and a hatchet-shaped wing, but detachable from the latter, which envelops it on one surface and two edges.

About five species of Keteleeria are known in the living state, natives of China, Formosa, and Tonking. They are closely related [1] to the silver firs, but exist in a warmer and drier climate, to which their varnished coriaceous needles are adapted. The leaves resemble superficially those of Cephalotaxus, but are readily distinguishable by the circular disc at the base.

Two species have been introduced into Europe.

---

[1] They have been erroneously supposed by various authors to be allied to *Pseudotsuga* and *Torreya*.

KETELEERIA FORTUNEI

Keteleeria Fortunei, Carrière, in *Rev. Hort.* 1868, p. 132, and 1887, p. 207 ; Pirotta, in *Bull. Soc. Tosc. Ort.* 1889, p. 200 ; Masters, in *Gard. Chron.* ii. 440 (1887), and in *Journ. Linn. Soc. (Bot.)* xxvi. 555 (1902) ; Mayr, *Fremdländ. Wald- u. Parkbäume*, 292, fig. 86 (1906) ; Clinton-Baker, *Illust. Conif.* i. 73 (1909).

Picea Fortuni, Murray, in *Proc. Hort. Soc.* 1862, p. 421.

Abies Fortunei, Murray, *Pines and Firs of Japan*, 49 (1863) ; Hance, in *Journ. Bot.* xx. 32 (1882) ; Masters, in *Journ. Linn. Soc. (Bot.)* xviii. 519 (1881), xxii. 197, figs. 22-25 (1886), and in *Gard. Chron.* xxi. 348, figs. 64-67 (1884), and xxv. 428, figs. 82, 83 (1886).

Pinus Fortunei, Parlatore, in De Candolle, *Prod.* xvi. 2, p. 430 (1868).

Abies jezoensis, Lindley, in Paxton, *Flower Garden*, i. 42 (1850), and *Gard. Chron.* 1850, p. 311 (not Siebold and Zuccarini).

Pseudotsuga jezoensis, Bertrand, in *Ann. Sc. Nat.* xx. 87 (1874).

Abietia Fortunei, Kent, Veitch's *Man. Conif.* 485 (1900).

A tree, attaining in China 80 ft. in height, with thick whitish bark, divided into irregular plates, and somewhat like that of the cork oak. Young branchlets[1] slender, with a scattered short wavy soft pubescence. Buds ovoid, rounded at the apex, with numerous scales. Leaves on adult trees $\frac{1}{2}$ to $1\frac{1}{4}$ in. long, $\frac{1}{8}$ in. wide, linear ; rounded, retuse or acute at the apex ; midrib prominent in a longitudinal depression on the upper surface ; lower surface with twelve to sixteen lines of stomata on each side, extending from the raised midrib nearly to the margin, which shows a very narrow green depressed border.

Leaves on young plants 1 in. long, $\frac{1}{8}$ in. broad, ending in an acuminate apex, tipped by a spine-like point ; upper surface as in adult leaves ; lower surface with a very narrow raised midrib, and two wide bands, each of sixteen stomatic lines, extending to a linear groove just inside the margin.

Cones on pubescent scaly stalks, nearly cylindrical, 4 to 5 in. long, $2\frac{1}{4}$ in. wide, bluish before ripening, brown tinged with purple when mature ; scales about $1\frac{1}{4}$ in. wide, broadly oval with a short claw, concave internally from side to side, with the upper part inflexed, and the margin slightly denticulate ; externally covered with slight pubescence towards the base ; bract $\frac{5}{8}$ in. long, with a linear claw, expanded above into a denticulate lamina, ending in a sharp mucro. Seeds covering the whole of the scale, except its lateral borders.

This species is a native of the coast range of the province of Fukien in eastern China, where it was seen by Maries[2] in 1878, growing wild in quantity and associated with *Pinus Massoniana*. It was discovered in 1844 by Fortune, who found a single tree near the Kushan temple, which is situated in the mountains a few miles to the eastward of Foochow at about 2000 ft. elevation. Fortune sent specimens and seed to Messrs. Standish and Noble, who raised young plants, most of which appear to have been distributed on the continent by Van Houtte. None of the original

---

[1] On the tree at Pallanza the branchlets in November are dark red, with merely traces of pubescence.

[2] *Hortus Veitchii*, 341 (1906).

plants survived in England, but a fine specimen[1] is growing in Rovelli's nursery at Pallanza, which measured 55 ft. by 8 ft. in 1909. This tree produces seed freely, from which numerous plants have been raised since 1884. Mr. Rovelli states that they are very difficult to transplant; and we know of none which have as yet attained any size in England. (A. H.)

KETELEERIA DAVIDIANA

Keteleeria Davidiana, Beissner, *Nadelholzkunde*, 425, fig. 117 (1891); Van Tieghem, in *Bull. Soc. Bot. France*, 411 (1891); Diels, *Flora von Central-China*, 217 (1901); Masters, in *Journ. Linn. Soc. (Bot.)* xxvi. 554 (1902), and in *Gard. Chron.* xxxiii. 84, figs. 37, 38 (1903); Clinton-Baker, *Illust. Conif.* i. 72 (1909).

Keteleeria sacra, Beissner, *Nadelholzkunde*, 426 (1891).

Keteleeria formosana, Hayata, in *Gard. Chron.* xliii. 194 (1908).

Abies Davidiana, Franchet, *Pl. David.* i. 288, t. 13 (1884), and in *Journ. de Bot.* 1899, p. 260; Masters, in *Gard. Chron.* i. 481 (1887).

Abies sacra, Franchet, *Pl. David.* i. 290, t. 14 (1884).

Pseudotsuga Davidiana, Bertrand, in *Ann. Sc. Nat.* xx. 86 (1874).

Podocarpus sutchuenensis, Franchet, in *Journ. de Bot.* 1899, p. 265, *ex* Diels in Engler, *Jahrb.* xxxvi. No. 5, p. 3 (1905).

A tree, attaining[2] in China 100 ft. in height and 16 ft. in girth. Young branchlets slender, with short stiff erect hairs. Buds as in *K. Fortunei*, but with scales not keeled on the back. Leaves on adult trees similar to those of that species, but slightly larger, with a wider prominent midrib beneath, on each side of which are nine lines of stomata.

Leaves on young plants $1\frac{1}{4}$ in. long, $\frac{1}{7}$ in. broad, tapering to an acuminate spine-tipped apex; lower surface with a broad green midrib, elevated in the central line as a narrow ridge, two stomatic bands, each of eight to nine lines, and two marginal green bands, near the edge of each of which is a linear longitudinal groove.

Cones sub-sessile or stalked, nearly cylindrical, 6 to 8 in. long, 2 in. in diameter, brown when ripe: scales about 1 in. long and $\frac{3}{4}$ in. wide, ovate, with a rounded or slightly contracted truncate apex, which is inflexed; concave internally from side to side; outer surface minutely pubescent towards the base: bract $\frac{1}{2}$ in. long, with an oblong claw ending in a denticulate lamina, scarcely broader than the claw, and tipped with an acuminate point. Seeds not extending quite to the apex or lateral margins of the scale.

This species, which was discovered by Père David in 1869, is widely distributed throughout the interior of China, occurring in the provinces of Shensi, Hupeh, Szechwan, and Yunnan; and has recently been found in the mountains of Formosa. In China it is a tree of lower elevations than Abies or Picea, and forms woods in a warm climate on dry hills at about 2000 to 4000 ft. altitude in western Hupeh, and

---

[1] Figured in *Gard. Chron.* xxv. 428, fig. 83 (1886). According to Carrière, *Rev. Hort.* 1887, p. 211, it was planted in 1859. Carrière adds some interesting details of this species, which he states can be multiplied, but with difficulty, by cuttings and layers.

[2] I measured a tree of this size in the Wushan district, north of the Yangtse in eastern Szechwan, in 1888.

at 4000 to 5000 ft. in southern Yunnan. It shows a considerable amount of variation in the wild state; and Franchet described as a distinct species (*Abies sacra*) a form with glabrous branchlets in Shensi. The Formosan tree [1] is distinguished by Hayata [2] as var. *formosana.*

I sent seed from Hupeh in 1888, which was raised at Kew, where there is a specimen in the Temperate House about 4 ft. high. Wilson [3] sent to Coombe Wood in 1901 seeds from western China, which germinated freely, producing hardy plants which are now about 4 ft. high. These appear to constitute two varieties, differing slightly in the length of the foliage.

A log of this species, about 10 in. in diameter, which is now in the Cambridge Forestry Exhibition, was obtained in the mountains of Hupeh by Mr. E. H. Wilson in 1910, during his last voyage in Central China. This was quickly grown, averaging 12 rings to an inch of radius. The wood strongly resembles in appearance that of the common silver fir, and is soft, light, and easily worked. It differs from *Abies* slightly in structure, having resin-canals. (A. H.)

---

[1] I saw this fine tree in North Formosa in March 1912, at about 2000 ft. above sea-level, in a thick virgin forest at Kinkaryo on the Hokusii river, about 6 miles from Heirimbi. It attains a very large size; but I was unable to measure any old trees on account of the difficulty of reaching them. When the Japanese occupied Formosa it was abundant in this district; but most of the large trees were cut down and used in building houses at Taihoku. I saw in the house of S. Nanasumi, chief of police at Heirimbi, very wide boards, cut from this tree, of a rich purplish brown colour, unlike that of any conifer known to me. A specimen of this wood is now in the Cambridge Forestry Museum. *Keteleeria* is known to the Chinese as *Yu-san*, and to the Japanese as *Shima-momi*. The only other part of Formosa where it is known to occur is near Bosan in the southwest.—H. J. E.

[2] *Journ. Coll. Sci. Tokyo*, xxv. art. 19, p. 221 (1908).

[3] *Hortus Veitchii*, 341 (1906).

PSEUDOLARIX

Pseudolarix, Gordon, *Pinetum*, 292 (1858); Bentham et Hooker,[1] *Gen. Pl.* iii. 442 (1880); Masters, in *Journ. Linn. Soc. (Bot.)* xxii. 208 (1886), and xxx. 32 (1893); Eichler, in Engler and Prantl, *Natur. Pflanzenfam.* ii. pt. i. p. 77 (1889); Sargent, in *Garden and Forest*, x. 501 (1897).
Laricopsis, Kent, Veitch's *Man. Conif.* 403 (1900).

A GENUS belonging to the division Abietineæ of the order Coniferæ, similar to Larix in the mode of branching, and in the deciduous needle-like leaves, clustered on the short shoots, and solitary and spiral on the long shoots; differing in the subulate scales of the buds, in the staminate flowers clustered at the tips of leafless short shoots, and in the cones with deciduous scales. Pseudolarix comprises one species, a native of China described in detail below.

PSEUDOLARIX FORTUNEI

Pseudolarix Fortunei, Mayr, *Monog. Abiet. Jap.* 99 (1890), and *Fremdländ. Wald- u. Parkbäume*, 392 (1906); Masters, in *Journ. Linn. Soc. (Bot.)* xxvi. 557 (1902), and xxxvii. 424 (1906); Hemsley, in *Bot. Mag.* t. 8176 (1908).
Pseudolarix Kaempferi, Gordon, *Pinetum*, 292 (1858); Masters, in *Gard. Chron.* xxi. 584, figs. 112, 113, and xxii. 238, fig. 48 (1884), and in *Journ. Linn. Soc. (Bot.)* xxii. 208, fig. 32, and plates ix. x. (1886); Clinton-Baker, *Illust. Conif.* ii. 62 (1909).
Abies Kaempferi, Lindley, in *Gard. Chron.* 1854, pp. 255 and 455 (with figure) (not[2] *Abies Kaempferi*, Lindley, in *Penny Cycl.* i. 34 (1833)); Murray, in *Proc. Hort. Soc.* ii. 644, figs. 172-182 (1862); Fortune, in *Gard. Chron.* 1855, pp. 242, 644, and 1860, p. 170.
Larix Kaempferi, Carrière, in *Flore des Serres*, xi. 97 (1856); Masters, in *Journ. Linn. Soc. (Bot.)* xviii. 523 (1881); Franchet, *Pl. David.* i. 286 (1884).
Pinus Kaempferi, Parlatore, in De Candolle, *Prod.* xvi. 2, p. 412 (not[2] Lambert).
Laricopsis Kaempferi, Kent, Veitch's *Man. Conif.* 404 (1900).

A tree, attaining in China 120 or 130 feet in height. Bark reddish brown, fissured into small narrow scaly plates. Branchlets of two kinds, long shoots and short shoots or spurs, as in the larches (cf. Vol. II. p. 345). Long shoots in the first year glabrous, glaucous, with linear pulvini, separated by slight grooves; in the second year reddish brown with broad white corky fissures between the pulvini.

---

[1] Bentham and Hooker mention Pseudolarix, in a note under Larix, as probably a distinct genus.

[2] *Pinus Kaempferi*, Lambert, *Genus Pinus*, ii. preface, p. v (1824), and the original *Abies Kaempferi*, Lindley, *Penny Cycl.* i. 34 (1833), were names applied to the true larch of Japan, *Larix leptolepis*, which was first mentioned as " *Larix conifera, nucleis pyramidalis foliis deciduis*," by Kaempfer, *Amœn. Exot.* 833 (1712). The name *Abies Kaempferi* was subsequently applied by Lindley in error to the Chinese Pseudolarix, which was quite unknown to Kaempfer or to Lambert; and Gordon's name, *Pseudolarix Kaempferi*, founded on Lindley's erroneous application, was rightly changed by Mayr to *Pseudolarix Fortunei*.

Short shoots with annual zones of growth, each zone marked by a depression and a ring of subulate scales. Buds of three kinds, as in the larches; those terminating (*a*) the long shoots and (*b*) the short shoots, conic, surrounded by acuminate scales, ending in long subulate points; and (*c*) lateral buds, solitary in the axils of a few leaves of the long shoots, globose, with rounded or short-pointed scales.

Leaves deciduous, solitary and spirally arranged on the long shoots, and in clusters of fifteen to thirty at the apices of the short shoots; jointed at the base with the tip of a pulvinus, linear, straight or falcate, $1\frac{1}{2}$ to 2 in. long, $\frac{1}{12}$ in. broad, acute or acuminate, green and slightly convex above; under surface with a raised green midrib, two longitudinal channels covered with white stomatic lines, and a narrow thin outer margin. Fibro-vascular bundle undivided; resin-canals three, all close to the epidermis, one in the median line near the upper surface, and two lateral, near the outer edges of the lower surface.

Flowers monœcious. Staminate flowers, pendulous, twenty-five to thirty in a cluster, at the apex of a leafless short shoot, each subtended by loose scarious scales, and including the slender stalk about $\frac{3}{8}$ in. long; anthers twenty, two-celled, opening transversely; pollen grains winged as in Pinus, and different from the simple pollen of Larix. Pistillate flowers globose, $\frac{3}{4}$ in. in diameter, terminating a short leafy branch, which arises from the apex of a short shoot; ovules, two on each scale, reversed. Cones, erect on the branches, ripening in the autumn of the first year, ovoid, $1\frac{1}{2}$ to 2 in. long: scales numerous, imbricated, coriaceous, reddish brown when ripe, $\frac{3}{4}$ to $1\frac{1}{4}$ in. long, ovate, tapering to a blunt, acute, or notched apex, sagittate at the base, with a claw bent upwards at a right angle, which arises by a narrow linear attachment from the axis of the cone: bract, ovate-lanceolate, $\frac{1}{5}$ to $\frac{1}{3}$ in. long, acuminate, denticulate, adnate to the base of the scale, and deciduous with it. Seeds, two on each scale, which they completely cover with their short body and long wing; wing oval-lanceolate on the outer edge, straight on the inner edge, pale brown, translucent, enclosing the body of the seed on the front and sides in a cavity; body detachable from the wing, white, obovate, with two large resin-vesicles; cotyledons five to seven. As the cone ripens the scales gape apart, showing the wings of the seeds projecting beyond them, and giving them the appearance of a whitish margin. Soon afterwards the scales, bracts, and seeds fall together to the ground, the central axis of the cone being the only part of it left on the branch, as is the case in Cedrus and Abies.

This remarkable conifer is a native of the provinces of Chekiang and Kiangsu in eastern China, where it is known from two localities, both in about lat. 29° 30'. Fortune discovered it in 1853, in the mountains south-west of Ningpo, where there were some fine trees growing near the Tsan-tsin monastery[1] at 1000 to 1500 ft. elevation; and in 1854 he found a plantation, about twenty miles westward, in the vicinity of the Quan-ting monastery, on a mountain slope at about 4000 ft. altitude, one of the trees, standing alone and clothed with branches to near the ground, being 130 ft. in height and 8 ft. in girth. The Rev. G. E. Moule also found

---

[1] This monastery is about a day's journey from Ningpo. Cf. Bretschneider, *Hist. Europ. Bot. Disc. China*, 416 (1898).

some trees[1] in the hills west of Ningpo in 1874. The only other locality where this species has been seen is the Lüshan mountains, south of Kiukiang in Kiangsu, where it was discovered by Abbé David in 1868, and afterwards by Maries,[2] who mentions immense trees; but Wilson[3] only succeeded in finding in these mountains, some wild, and half a dozen planted trees, none of considerable size.

Fortune had been acquainted with this tree for some years previously as a dwarf plant[4] in pots, contrived, though only 1½ to 2 ft. high, to look like an aged Cedar of Lebanon. It appears to be known to the Chinese, as either *chin-sung*, "golden pine," or *chin-lo-sung*, "golden deciduous pine," names applied on account of the beautiful yellow colour of the foliage for a short time before it falls in autumn.

Fortune sent seeds from Chekiang in 1853, and again in 1855 to Glendinning's nursery, Chiswick; but he states[5] that of all the packages of seed, which he sent for several years in succession, only one batch ever germinated; and that the only plants living in England in 1860 were natural seedlings which had been dug up in the woods of China and sent[6] home in Wardian cases in 1854.

The tree is perfectly hardy, as it withstood the severe winter of 1859-1860 at Ambleside[7] and at Hafodunos[7]; and possibly its rarity in collections is due to the small number of plants actually introduced; but it appears to be extremely slow in growth in England. It will not endure lime in the soil, as the seedlings raised and planted at Colesborne soon die. (A. H.)

REMARKABLE TREES

The finest tree that we know of is in a sheltered situation at Carclew, which, when I saw it in 1902, was 35 ft. by 5 ft. In 1910 it was 40 ft. by 5 ft. 2 in. I am told by Mr. Simmons, the head-gardener at Carclew, that he has never found fertile seed on it. There is a healthy tree in the grounds at Hutley Towers near Ryde, which in 1906 was 30 ft. by 2 ft., but it produced no cones either in 1905 or 1906. At Joldwynds, near Dorking, the seat of Sir W. Paget Bowman, Bt., there is a tree, planted about 1879, which is now 27½ ft. by 2½ ft. Though perfectly healthy and branching to the ground it has never produced any seed.

There are several trees at Kew, the largest of which near the main gate is probably one of the original seedlings.[8] It flowered profusely in June 1907, producing fully developed cones with imperfect seed, no embryo being formed.[9] It bore cones freely again in 1910; and measured 31 ft. by 2 ft. 4 in. in 1912.

At Tortworth, a tree measured in 1910, 37 ft. by 4 ft. 1 in. Lord Ducie informs me that it was planted on 3rd November 1858, in a bed of sand overlying carboniferous limestone, deep enough, however, to sustain a heavy growth of

---

[1] Referred by Hance to *Larix dahurica*, but evidently, from the Chinese name "*chin-sung*," used by Moule, he was speaking of Pseudolarix.

[2] Cf. Bretschneider, *op. cit.* 741 (1898). [3] In *Gard. Chron.* xlii. 344 (1907).

[4] Var. *nana*, Masters, in *Journ. Linn. Soc. (Bot.)* xviii. 523 (1881). [5] In *Gard. Chron.* 1860, p. 170.

[6] Fortune, in *Gard. Chron.* 1855, p. 644. Kent says that until the tree coned at Pallanza, plants were obtained by layering. [7] *Gard. Chron.* 1860, pp. 74 and 386.

[8] It is apparently the tree mentioned by J. Smith, *Records of Kew Gardens*, 290 (1888), as 5 ft. high in 1864.

[9] According to Masters, in *Gard. Chron.* ii. 440 (1887) and *Journ. Roy. Hort. Soc.* xiv. 68 (1892), this species first produced fruit in England in 1887 at Lucombe and Pince's nursery, Exeter, but this tree is no longer living.

Rhododendron. The tree is very brittle, and has lost many branches, and is somewhat ragged in look although it is in a very sheltered position. It produced cones on four or five occasions ; but in no case did they contain fertile seed.

The Rev. Hon. W. Ellis informs me that at Bothelhaugh, near Morpeth, Northumberland, a tree planted over thirty years ago is under 6 ft. high. At Coombe Wood, the largest specimen measured 35 ft. by 3 ft. 8 in. in 1910. At Scorrier, near Truro, a healthy tree was 30 ft. by 3 ft. in 1911.

We have seen no trees in either Scotland or Ireland.

The finest specimen in Europe is growing in Rovelli's nursery at Pallanza ; and measured 64 ft. by 6 ft. 10 in. in 1909. This tree produced staminate flowers in 1884 ; and since then has coned regularly every two or three years. Large numbers of natural seedlings appear in prepared soil under the tree ; and the seed is said to germinate better where it falls, than when collected and sown in pans under glass. As the seeds and scales fall together and close to the parent tree, young seedlings probably succeed best with considerable shade.

In Belgium it is said[1] to have attained no less than 46 ft. in height by 3 ft. in girth at the nursery of the Horticultural Society of Calmpthout, near Antwerp, where seedlings grew as fast as those of the common larch ; and Dr. Masters mentioned,[2] in 1883, a fine tree in Linden's nursery at Ghent.

At Verrières, near Paris, one of the original trees[3] is about 35 ft. high and 3 ft. in girth. It produces fruit and fertile seed, but in no great quantity. There is a good specimen[4] at Karlsruhe, about 35 ft. high, which bore cones for the first time in 1896 ; the seed, however, was unfertile.

In the United States this tree thrives well, as it delights in hot summers ; and Sargent states that he never saw a plant which appeared to suffer from heat or cold, fungoid diseases, or the attacks of insects. The largest specimen[5] is growing in Parson's nursery at Flushing, Long Island, which was imported from London in 1859 when it was 3 ft. high. It measured in 1895 55 ft. high, with a stem 2 ft. in diameter, and branches 50 ft. across, and has borne seed frequently. Another specimen in Mr. Hunnewell's pinetum at Wellesley, Mass., measured[6] in 1905 35 ft. in height and 4 ft. in girth, with a spread of branches of 37 ft. This tree has borne seed since 1887, and many seedlings have been raised from it. Sargent reports[6] another large specimen on Mr. Probasco's estate at Cincinnati.   (H. J. E.)

---

[1] *Bull. Soc. Dendr. France*, No. 18, p. 162 (1910).   [2] In *Gard. Chron.* xix. 88 (1883).
[3] Cf. P. L. de Vilmorin, *Hortus Vilmorinianus*, 66, fig. ix. (1906).
[4] Cf. *Mitt. Deut. Dend. Ges.* 1896, pp. 71, fig., and 113.
[5] *Garden and Forest*, 1895, p. 415.
[6] Sargent, *Pinetum at Wellesley in* 1905, p. 10, and in *Garden and Forest*, 1897, p. 317.

CATALPA

Catalpa,[1] Scopoli, *Introd. Hist. Nat.* 170 (1777); Bentham et Hooker, *Gen. Pl.* ii. 1041 (1876); Bureau, in *Nouv. Arch. Mus. Hist. Nat.* vi. 169 (1894); Dode, in *Bull. Soc. Dendr. France*, i. 194 (1907).

DECIDUOUS trees, belonging to the order Bignoniaceæ. Branchlets stout, with thick pith; leaf-scars elevated, orbicular, marked with a circle of dots, which are the tiny scars left by the fibro-vascular bundles of the fallen petiole. Buds minute, globose, immersed in the bark, with two to four external scales; all axillary, no true terminal bud being formed, the top of the branchlet dying in summer, and leaving an elevated circular scar close to the upper axillary bud. Leaves simple, opposite or in whorls of threes, entire or lobed, long-stalked, pinnately-nerved, without stipules.

Flowers perfect, in terminal panicles or corymbs.[2] Calyx gamosepalous, membranous, splitting when the flower opens into two broad ovate entire lobes. Corolla gamopetalous, inserted on a nearly obsolete disc; tube broad, campanulate, oblique, enlarged above into a spreading bilabiate limb, the posterior lip two-partite, the anterior three-lobed. Stamens inserted near the base of the corolla, two, anterior, filaments flattened, anthers bilocular and opening longitudinally. Staminodes similarly inserted, three, posterior, filiform, minute or rudimentary. Ovary sessile, two-celled; style elongated, divided at the apex into two stigmatic lobes[3]; ovules numerous in several series on a central placenta. Fruit, a long nearly cylindrical capsule, tapering from the middle to each end, persistent on the branches during winter, and ultimately splitting into two valves. Seeds numerous, small, oblong, compressed, inserted in two or four ranks near the margin of the woody septum, with broad lateral wings, notched at the base of the seed, and ending in tufts of long coarse hairs.

The leaves of Catalpa show on their lower surface in the axils of the nerves clusters of circular glands which secrete nectar, and are visited by numerous insects, especially ants and bees, the latter getting honey from them as well as from the

---

[1] Catalpa is a corruption of Catawba, the name of an Indian tribe that formerly occupied Georgia and the Carolinas.

[2] Bureau divides the temperate species into two sections :—

(*a*) *Thyrsoideæ*, comprising *C. bignonioides, C. speciosa*, and *C. Kaempferi*; inflorescence a narrow panicle, the secondary axes being branched.

(*b*) *Corymbosæ*, including *C. Bungei* and *C. Fargesi*; inflorescence a corymb, with simple secondary axes.

[3] The stigmatic lobes exhibit sensitive movements, opening and shutting like the leaves of a book, with the visits of bees and other insects. Cf. Masters in *Gard. Chron.* xiii. 651 (1880), and Kerner, *Nat. Hist. Plants*, Eng. Trans. ii. 281 (1898). This phenomenon in *C. bignonioides* has been studied by Meehan, in *Proc. Amer. Assoc. Adv. Sc.* 1873, pp. 72, 73, and in *Bot. Gaz.* viii. 191 (1883); and in the case of *C. speciosa* by Antisdale, in *Bot. Gaz.* viii. 171 (1883).

flowers. These glandular areas, which are large and conspicuous near the base of the leaf, bear no pubescence.[1]

The seedlings[2] of Catalpa have stalked deeply bifid oblate cotyledons raised above ground, and followed on the stem by opposite decussate or ternately verticillate ovate leaves.

Catalpas may be propagated[3] by both stem and root cuttings.

Eight species of Catalpa are known, of which three are natives of the West Indies and not hardy. The West Indian species constitute a distinct section, characterised by lanceolate or elliptic leaves. The remaining five species, with ovate leaves, inhabit the United States and China; and have all been introduced into cultivation in Europe. They may be arranged as follows :—

I. *Leaves glabrous.*

1. *Catalpa Bungei*, Meyer. China. See p. 1489.

    Leaves, with a disagreeable peculiar odour, entire with a long slender acuminate apex, or with one or two long-pointed lateral lobes, or coarsely serrate in margin.

II. *Leaves pubescent with simple hairs.*

* *Branchlets glabrous.*

2. *Catalpa bignonioides*, Walter. United States. See p. 1485.

    Leaves, with a disagreeable peculiar odour, usually entire with a short apex, glabrous above, pubescent beneath, the pubescence not covering the whole surface of the midrib.

3. *Catalpa speciosa*, Warder. United States. See p. 1483.

    Leaves inodorous, usually entire with a long acuminate apex, glabrous above, pubescent beneath, the pubescence covering the midrib entirely.

** *Branchlets with stiff glandular hairs.*

4. *Catalpa Kaempferi*, Siebold and Zuccarini. Wild in Central China, long cultivated in Japan. See p. 1487.

    Leaves inodorous, usually three-lobed; pubescent on the upper surface throughout, and on the lower surface on the midrib and nerves.

III. *Leaves tomentose with branched hairs.*

5. *Catalpa Fargesi*, Bureau. Central China. See p. 1490.

    Leaves entire or with one or two acute lateral lobes; tomentose on the lower surface throughout, and on the upper surface mainly on the nerves.

(A. H.)

---

[1] Cf. Ryder in *Proc. Philad. Acad.* 1879, p. 161, and *Amer. Nat.* xiii. 648 (1879). The glandular areas are greenish in the two species from the United States, and purplish in the three Chinese species. They are almost entirely confined to the base of the leaf in *C. Fargesi* and *C. Bungei*; but are also present in the upper axils of the leaf in *C. Kaempferi, C. speciosa,* and *C. bignonioides.*

[2] Described by Lubbock, *Seedlings,* ii. 335, 339, fig. 571 (1892).

[3] Cf. J. Clarke in *Gard. Chron.* xlvii. 100 (1910).

CATALPA SPECIOSA, WESTERN CATALPA

Catalpa speciosa,[1] Warder, *ex* Engelmann, in *Bot. Gaz.* v. 1 (1880); Sargent, *Silva N. Amer.* vi. 89, tt. 290, 291 (1894), and *Trees N. Amer.* 795 (1905); Bureau, in *Nouv. Arch. Mus. Hist. Nat.* vi. 184 (1894); André, in *Rev. Hort.* lxvii. 136, fig. (1895); Hall and Schrenk, *U.S. Dep. Agric. Bur. Forestry, Bull.* No. 37 (1902); Roberts and Dickens, *Kansas State Agric. College, Bull.* No. 108 (1902); Dode, in *Bull. Soc. Dendr. France*, i. 195 (1907).

Catalpa cordifolia, Jaume, in Duhamel, *Traité des Arb.* ii. t. 5 (1802) (excl. text) (not Moench).

A tree, rarely attaining in America 120 feet in height and 14 feet in girth, usually smaller. Bark thick, deeply furrowed, and roughened with scales. Young branchlets glabrous. Leaves similar to those of *C. bignonioides*, but without their peculiar odour, often larger, up to 10 in. long and 7 in. wide, with longer acuminate points; glabrescent above; lower surface with the pubescence of simple hairs more marked than in *C. bignonioides*, spreading over the whole of the midrib and extending to the petioles.

Flowers appearing two weeks earlier than those of *C. bignonioides*, few in open panicles, which are about 6 in. long and broad; calyx purplish, glandular-pubescent; corolla white, 2 in. long, $2\frac{1}{2}$ in. wide, often spotted externally with purple near the base; marked internally on the lower side with two bands of yellow blotches following two lateral ridges, and a few purple spots on the lobes of the lower lip of the limb. Fruit, 8 in. to 20 in. long, $\frac{1}{2}$ in. to $\frac{3}{4}$ in. in diameter in the middle, with a thick wall, splitting into two concave valves. Seeds 1 in. long, $\frac{1}{2}$ in. wide, light brown, with wings rounded at the ends and ending in a fringe of short hairs.

C. speciosa under favourable conditions differs from *C. bignonioides* in habit, forming a narrow tree with ascending branches; but in the arboretum at Segrez, where there are old trees of both species, they are nearly alike in appearance. They are readily distinguished by their flowers, fruits, and seeds; but when these are absent, the main distinctive character is the odour of the leaves.[2]

This species in its natural range is confined to a limited region, extending from the valley of the Vermilion river, Illinois, through southern Illinois and Indiana, western Kentucky and Tennessee, south-eastern Missouri, and north-eastern Arkansas. It comes in contact with *C. bignonioides* in south-eastern Missouri; and is abundant and of its largest size in southern Illinois and Indiana. It has become naturalised through cultivation in southern Arkansas, western Louisiana, and eastern Texas.

It has been planted in the United States as far north as South Dakota, southern Michigan, and Minnesota, and southern Massachusetts; and westward to eastern

---

[1] Warder, in *Western Hort. Review*, iii. 533 (1853), was the first to distinguish *C. speciosa*, but did not then publish this specific name. It appears to have been first used by Sargent, who, in *Gard. Chron.* xii. 784 (1879), points out that the western Catalpa differs from *C. bignonioides*; and says that if distinct, it should be known as *C. speciosa*.

[2] W. H. Lamb, in *Proc. Soc. Amer. Foresters*, vii. 80, figs. 1, 2 (1912), points out that the septum of the pod (the long wrinkled partition along which the seeds are arranged) is nearly circular in section in *C. speciosa*, and lenticular or narrowly elliptic in *C. bignonioides*.

Nebraska, Kansas, and Oklahoma. It has succeeded on irrigated lands in New Mexico, Utah, and Colorado, at low altitudes and where the soil is free from alkali. The range for economic planting appears to be on the fertile alluvial lands of the middle west, south of lat. 41°.

This species, though only distinguished by Warder as late as 1853, appears to have been introduced early into France, as it was figured in *Nouveau Duhamel* in 1802; but no trees so old as this are now known in Europe. Prof. Sargent sent seeds to Kew in 1880, and probably about the same time to Segrez and Les Barres in France. At Kew *C. speciosa*, though forming a better tree than *C. bignonioides*, is very slow in growth, the tallest example, now thirty years old, from seed, being about 25 feet high. It is perfectly hardy, as it has borne at Kew 0° Fahr. without injury, and does not suffer in the severe winters of New England. Bureau states that on M. André's property in Touraine it did not suffer from a temperature of − 26° Cent.

It came into vogue in America as a tree for planting to produce timber quickly about 1879 to 1883, when large plantations were made by R. Douglas, near Farlington in Kansas, which are now owned by the railway company. One forty-acre tract of these plantations is, however, *C. bignonioides*.   (A. H.)

Cultivation

An immense quantity of literature on this species has appeared in America, mainly by Mr. John P. Brown of Connersville, Indiana, who devoted a great part of his magazine, *Arboriculture*, to advocating the economic value of this tree. Though this publication contains many illustrations and details on the growth of the Catalpa in many localities and under varied conditions, it has more interest for American than for British readers, and our space will not allow me to refer to them in detail. There seems to be little doubt that in the rich alluvial valleys of the Ohio, Tennessee, Arkansas, and other tributaries of the Mississippi, its growth is very rapid when young, and it is one of the most valuable trees for fencing, lumber, railway sleepers, and other purposes,[1] on account of the durability of its timber; but it requires a much longer and hotter summer than any part of our islands afford. The latest account of this species is by Oman,[2] who studied the results obtained by four plantations in Kansas, which were cut in 1902-1906. He gives valuable hints regarding the proper mode of planting, and states that the financial returns on deep fertile porous soil are remarkable. This tree endures inundation, one plantation having been completely submerged for a week without injury. It coppices freely, and can even be propagated by cuttings; but suckers from the roots have not been observed.

Large quantities of seedlings have been raised and distributed in this country on several occasions,[3] but we cannot hear of a single place in which they show any

---

[1] It is also a suitable timber for furniture, as shown by an arm-chair given me by Mr. Brown at Louis in 1904, which has handsome grain, takes a good polish, and has worn well.

[2] *Proc. Soc. Amer. Foresters*, vi. 42-52 (1911).

[3] About 150 trees of this species, which were raised from seed in the Royal Horticultural Society's garden at Chiswick in 1880, were distributed widely to the members; but we have not seen any of these (*Gard. Chron.* xlvii. 245 (1910)).

signs of becoming a timber tree. I agree entirely with the opinion that Mr. Bean has expressed in *Kew Bulletin*, 1907, p. 43, that it is improbable that this tree can be grown anywhere in England with any hope of profit, though as an ornamental tree of small size it may have considerable value in favourable situations.

A plantation of it was made near Tottenham House, Marlborough, by the Marquess of Ailesbury, where the young trees in 1907 were 3 to 6 ft. high, but so far as we can learn they grow slowly and do not ripen their wood in autumn, which is the case with those I have raised myself. (H. J. E.)

CATALPA BIGNONIOIDES, Common Catalpa

Catalpa bignonioides, Walter, *Fl. Car.* 64 (1788); Bureau, in *Nouv. Archiv. Mus. Hist. Nat.* vi. 175 (1894); Dode, in *Bull. Soc. Dendr. France*, i. 194 (1907); Sargent, in *Bot. Gaz.* xliv. 226 (1907).

Catalpa cordifolia, Moench, *Meth.* 464 (1794); Nuttall, *Gen. N. Amer. Pl.* i. 10 (1818).

Catalpa communis, Dumont de Courset, *Bot. Cult.* ii. 189 (1802).

Catalpa syringifolia, Sims, in *Bot. Mag.*[1] t. 1094 (1808); Loudon, *Arb. et Frut. Brit.* iii. 1261 (1838).

Catalpa Catalpa, Karsten, *Pharm. Med. Bot.* 927 (1882); Sargent, *Silva N. Amer.* vi. 86, tt. 288, 289 (1894), and *Trees N. Amer.* 793 (1905).

Bignonia Catalpa, Linnæus, *Sp. Pl.* 622 (1753) (in part).

A tree, rarely attaining in America 60 ft. in height and 10 ft. in girth, usually smaller. Bark separating on the surface into large thin irregular scales. Young branchlets glabrous. Leaves (Vol. III., Plate 204, Fig. 5) emitting when bruised a disagreeable odour, ovate, about 5 to 6 in. long, and 4 to 5 in. wide; cordate, truncate, or cuneate at the base, contracted into a slender acuminate point or rounded at the apex, usually entire or occasionally with one or two slight lateral lobes, glabrous above, pubescent with simple hairs on the nerves and veinlets beneath, the pubescence on the midrib being confined to its edges close to the surface of the blade; glandular areas pale; petioles glabrous.

Flowers numerous in a compact panicle, about 8 to 10 in. long and broad; calyx glabrous, green or light purple; corolla white, $1\frac{1}{2}$ in. long and wide, marked on the inner surface on the lower side by two rows of yellow blotches along two parallel ridges or folds, and on the throat and lower lobes of the limb by numerous conspicuous purple spots. Fruit, 6 to 20 in. long, $\frac{1}{4}$ to $\frac{1}{3}$ in. thick in the middle, with a thin wall, splitting into two flat valves. Seeds about 1 in. long, $\frac{1}{4}$ in. wide, silvery grey, with pointed wings, ending in long pencil-like tufts of white hairs.

The following varieties have arisen in cultivation :—

1. Var. *aurea*, Lavallée, *Arbor. Segrez.* 175 (1877).

Leaves pale yellow, retaining their colour throughout the season. One of the best golden-leaved small trees[2] in cultivation. Its origin is unknown to me.

2. Var. *purpurea*, Rehder, in Bailey, *Cycl. Amer. Hort.* i. 258 (1900).

Leaves purplish, with dark purple glandular spots. It is said by Nicholson to

---

[1] The plate was drawn from a branch of a tree growing in Mr. Granger's garden at Exeter in 1808.

[2] A specimen at Kew is figured in *Gard. Mag.* 1910, p. 709.

have originated in the United States, and is possibly a hybrid. It is cultivated by Simon-Louis at Metz.

3. Var. *variegata*, Bureau, *op. cit.* 183.

Variegated with white or yellow. In var. *Koehnei*, Dode, *op. cit.* 206, the leaves are pale yellow, with irregular angular green patches. Cultivated by Simon-Louis.

4. Var. *erubescens*, Nicholson, in *Woods and Forests*, 1885, p. 52.

Catalpa erubescens, Carrière, in *Rev. Hort.* l. 460 (1869).

This form, which I have not seen, is said to have purplish petioles and glandular spots, with a more compact inflorescence, and a more highly coloured corolla with a less deeply divided limb than the type. It is possibly, as Dode suggests, a hybrid.

5. Var. *nana*, Bureau, *op. cit.* 183.

A low spreading bush, with crowded branches, occasionally grafted high. The leaves are identical in odour and in all other respects with *C. bignonioides*; and there are no grounds for supposing it to be a form of *C. Bungei*, under which name it is commonly known in nurseries and gardens. It has not yet flowered anywhere, and appears to have been first cultivated at Segrez[1] in 1877, where it may possibly have arisen as a sport.

C. bignonioides is a native of the eastern part of the United States; but the exact localities where it is truly native cannot be determined with certainty. It is usually supposed to be indigenous on the banks of rivers in south-western Georgia, western Florida, and central Albania and Mississippi, and to be naturalised throughout the south Atlantic States. On account of its handsome flowers it was extensively planted for ornament; and its dissemination has been aided by its winged seeds, which are borne to a considerable distance by the wind and float on water without injury for a long period. As it bears moderately severe winters it may possibly have been a native of the more northern parts of the Alleghany range, where it is not now met with in the existing forests. It thrives as far north as Philadelphia, but is killed during the winter at Rochester on Lake Ontario, and often succumbs at St. Louis.[2]

(A. H.)

The first account of this species was published in *The Natural History of Carolina* by Catesby, who introduced[3] it into England in 1726.

The largest tree mentioned by Loudon was one at Syon, 52 ft. high and 3 ft. in diameter, of which only the dead stump remains, but there is a spreading tree grown from one of its layered branches on the north side of the lake which was in flower in July 1912 when I saw it last. A tree at Kew, which died in 1907, when it was about sixty years old, was 30 ft. high and 6 ft. 1 in. in girth. A tree in the Terrace Gardens, Richmond, was 35 ft. by 8 ft. 1 in. in 1912. A fine specimen[4] in Mr. Denne Dunn's garden at Canterbury was 32 ft. high in 1876. At Caldrees, Ickleton, near Cambridge, there is a fine tree, which flowers freely every year; it is about 35 ft. high and 7 ft. in girth.

---

[1] Lavallée, *Arb. Segrez.* 176 (1877), where it is named *C. Bungei*, var. *nana* (*pumila*). Cf. also Lavallée, *Icon. Arb. Segrez.* ii. 35 (1880).

[2] R. Douglas, in *Woods and Forests*, 1884, p. 566.　　　[3] Aiton, *Hort. Kew.* ii. 346 (1789).

[4] Figured in *Gard. Chron.* v. 13, fig. 2 (1876). In *Gard. Chron.* xxvi. 257 (1897) mention is made of a large tree at Rosslyn, Stamford Hill.

There are many trees of considerable age, but of no great height, in parks and places[1] in and around London, the best known of which was one at Gray's Inn, which died a few years ago. This Catalpa was reported by tradition to have been brought from America by Sir W. Raleigh, and to have been planted by Bacon; but there is no good authority for this, and the tree is not long-lived in England.

At Ham Manor, near Arundel, I saw a very well-shaped tree (Plate 350) in 1907 which measured 52 ft. by 7 ft. with a clean bole 15 ft. high. There is a tree at Heywood, Wilts, which, in 1906, was about 30 ft. high and 15 ft. in girth below the branches. A photograph sent me by the then gardener, Mr. Robinson, showed it in full flower as a very beautiful tree. Another at Elbridge, as measured by Mr. Furze in 1904, was 41 ft. high, 14 ft. in girth, and had a spread of 61 ft. A fine old tree at Wilton House, Wilts, was, when I saw it in 1906, showing signs of decay, but measured 53 ft. by 6½ ft. (H. J. E.)

CATALPA KAEMPFERI

Catalpa Kaempferi, Siebold and Zuccarini, in *Abhand. Akad. München*, iv. pt. ii. p. 142 (1846); J. D. Hooker, *Bot. Mag.* t. 6611 (1882); Lavallée, *Icon. Arb. Hort. Segrez.* 33, t. 10 (1885); Hemsley, in *Journ. Linn. Soc. (Bot.)* xxvi. 235 (1890); Bureau, in *Nouv. Arch. Mus. Nat. Hist.* vi. 190 (1894).
Catalpa ovata,[2] G. Don, *Gen. Syst.* iv. 230 (1837); Sargent, *Silva N. Amer.* vi. 84, note (1894).
Catalpa Bungei, Decaisne, in *Rev. Hort.* v. 406 (1851) (not C. A. Meyer); Carrière, in *Flore des Serres*, viii. 8 (1852); Jacques, in *Flore des Serres*, x. 188 (1855).
Catalpa Henryi, Dode, in *Bull. Soc. Dend. France*, i. 199 (1907).
Bignonia Catalpa, Thunberg, *Fl. Jap.* 251 (1784) (not Linnæus).

A tree, attaining 70 ft. in height; bark brown, slightly fissured. Young branchlets with numerous sessile glands and scattered stiff glandular hairs.[3] Leaves (Vol. III. Plate 204, Fig. 6) without a disagreeable or peculiar odour, ovate, variable in size, averaging 5 to 6 in. in width and length, cordate at the base, shortly acuminate at the apex; rarely entire, usually with one or two (occasionally three or four) triangular sharp-pointed lateral lobes; upper surface covered with a minute pubescence, the nerves often purple and with scattered long hairs; lower surface pubescent on the nerves and veinlets; petiole with glands and glandular hairs, as on the branchlets.

Flowers numerous in much-branched panicles, which are 4 to 9 in. long; calyx glabrous; corolla pale yellow, about ¾ in. long and broad, marked externally with two orange bands and numerous purple spots. Fruit, 7 to 12 in. long, cylindrical, ⅙ in. in diameter, with a thin wall, splitting into two concave valves. Seeds, ⅓ in.

[1] Mr. Hugh Boyd Watt, in an article on Catalpas in London and neighbourhood, which appeared in *The Field*, Feb. 17, 1912, states that there was abundance of fruit in the autumn of 1911 on the trees in Victoria Embankment gardens, Brunswick Square, Hampstead, Richmond, Kew, and Syon House. Six large trees in Palace Yard, Westminster, bore no fruit, though they flowered in the preceding summer. There was no fruit formed in 1909 and 1910; but there was a good crop in 1906.
[2] This is the oldest name of the species; but it has never been in use.
[3] These peculiar hairs, which are characteristic of this species, are deciduous in the course of the season.

long, $\frac{1}{8}$ in. wide, greyish brown, with pointed wings ending in long pencil-like tufts of white hairs.

This species is much planted in gardens and around temples in Japan; but is not a native of that country, according to the Japanese botanists, who state that it was introduced at an early period by the Buddhist monks. It was found wild [1] in central China by myself and by Wilson in western Hupeh, and by Giraldi [2] in Shensi.

It was first made known to Europeans by Kaempfer, who visited Japan in 1690, and published [3] in 1712 a good description and figure of the tree. It is usually known in Japan as the *Ki-sasage*, or "bean-tree" on account of its peculiar pods. Dupont, in a letter to Lavallée, mentions that he never saw this tree in the forest in Japan, but always planted, and records one 75 ft. in height and 5 ft. in girth growing near a temple.

It was introduced into Europe from Japan by Siebold in 1849, and has probably thriven best at Segrez, where Lavallée mentions a tree, which covered an area over 40 yards in circumference. It appears to be much hardier both in France and at Boston [4] (U.S.) than *C. bignonioides*.

Plants raised from seed [5] sent from central China by Wilson to Coombe Wood, and by Père Farges to Les Barres, appear to be identical in all respects [6] with trees of Japanese origin.

The finest specimen of *C. Kaempferi* known to us in England is a tree at Syon, which measured, when Elwes saw it in July 1912, 62 ft. by 5 ft. 1 in. It was then in full flower, with capsules of the previous year containing ripe seeds. The oldest tree at Kew, about 20 ft. high, was procured from Volxem in 1879.

A remarkable hybrid has arisen between *C. Kaempferi* and *C. bignonioides* :—
Catalpa hybrida,[7] Späth, in *Gartenflora*, lvii. 481, t. 1454 (1898).

Catalpa × J. C. Teas, Sargent, in *Garden and Forest*, ii. 303, fig. (1889); *Catalpa Teasiana*, Dode, in *Bull. Soc. Dend. France*, i. 205 (1907).

A tree, said to be of remarkably vigorous growth, resembling *C. Kaempferi* in foliage, but intermediate in flowers and fruit between that species and *C. bignonioides*. It produces extremely large panicles, 18 to 20 in. long and 10 in. wide, of 200 to 300 fragrant flowers, about 1 in. long; corolla tinged with yellow in the throat and marked with broad purple stripes. Fruit 12 to 15 in. long, $\frac{1}{4}$ in. wide.

This hybrid was raised about 1880 in J. C. Teas' nursery at Baysville, Indiana, from seed out of a peculiar single pod which was observed on a tree of *C. Kaempferi*. The latter grew near trees of both *C. bignonioides* and *C. speciosa*, but the pollen

---

[1] It is figured in the *Chih Wu Ming*, xxxiii. pl. 48 (1848).
[2] Cf. Diels, in Engler, *Jahrb.* xxxvi. heft 4, p. 98 (1905).
[3] *Amœn. Exot.* 842 (1712). [4] Sargent, in *Gard. Chron.* xii. 784 (1879).
[5] Sargent, in a letter to Kew, dated 2nd February 1900, states that seeds received from Shanghai in 1892 of reputed *C. Bungei*, produced plants of *C. Kaempferi*.
[6] A young tree at Kew, 10 ft. high, of Chinese origin, bears leaves identical in odour, shape, pubescence, and colour, with an older tree of Japanese origin. Specimens sent from Les Barres show no differences.
[7] Dode, *op. cit.* 204, identifies *C. hybrida*, Späth, with *C. erubescens*, Carrière, described above, p. 1486. Rehder, however, in Bailey, *Cycl. Am. Hort.* i. 258 (1900), considers Späth's description to refer to Teas' hybrid; and this appears to be correct. Späth does not say, as Dode asserts, that the hybrid originated in his nursery.

appears to have come from the former. As Sargent points out, *C. bignonioides* and *C. Kaempferi* flower at the same time, whereas *C. speciosa* is two to three weeks earlier than *C. Kaempferi*. Moreover, Penhallow,[1] by an examination of the wood of the different species of Catalpa and of the hybrid, has shown conclusively that the latter is a cross between *C. bignonioides* and *C. Kaempferi*, in which the characters of the latter are dominant; and that *C. speciosa* was in no way concerned in its production. Penhallow adds, that some of the seedlings of the hybrid, but not all, revert to *C. Kaempferi*.

"Teas' Japan hybrid," as it is commonly called, was introduced into Kew gardens in 1891, and flowered in 1900 and succeeding years, but has not produced fruit. At Kew it shows no excessive vigour.

C. japonica, Dode, in *Bull. Soc. Dend. France*, i. 200 (1907), said to have been introduced from Japan in 1886 by Simon-Louis, is unknown to me; but from the description appears to be another hybrid between *C. bignonioides* and *C. Kaempferi*. (A. H.)

CATALPA BUNGEI

Catalpa Bungei, C. A. Meyer, in *Bull. Acad. Sc. St. Pétersb.* ii. 51 (1837); Maximowicz, *Prim. Fl. Amur.* 475 (1859); Kurz, in *Journ. Bot.* ii. 193 (1873); Lavallée, *Arbor. Segrez.* 176 (1877); Hance, in *Journ. Bot.* xi. 37 (1882); Franchet, *Pl. David.* i. 229 (1884); Hemsley, in *Journ. Linn. Soc.* (*Bot.*) xxvi. 234 (1890); Bureau, in *Nouv. Arch. Mus. Hist. Nat.* vi. 197, pl. 4 (1894); Bean, in *Kew Bull.* 1907, p. 102.
Catalpa syringæfolia, Bunge, *Enum. Pl. China Bor.* 45 (1835) (not Sims).
Catalpa heterophylla, Dode, in *Bull. Soc. Dend. France*, i. 203 (1907).
Catalpa Duclouxii, Dode, in *Bull. Soc. Dend. France*, 1907, p. 201, and 1909, p. 154.

A tree, attaining in China 40 feet in height. Young branchlets glabrous, covered with minute glands. Leaves with a disagreeable odour, variable in size and shape, about 5 in. long and 4 in. wide, ovate or deltoid, cuneate or truncate at the base, ending in a long slender acuminate apex; entire, or with two lateral long-pointed lobes, or with several irregular acute teeth on each side; upper surface dark green, shining, glabrous; lower surface pale green, glabrous; petiole glabrous.

Flowers, three to nine, in a loose corymb; axis and pedicels glabrous or with a few simple hairs; calyx glabrous, green below, pink above; corolla, $1\frac{1}{4}$ in. long, with a wide campanulate tube, which is tinged with yellow along two projecting ridges and is elsewhere spotted purple, and a white five-lobed limb, with numerous purple dots, which give the whole flower a decidedly pinkish tint. Fruit, described as very long and slender, 25 to 40 in. in length and $\frac{1}{6}$ in. in diameter. Seeds greyish brown, $\frac{4}{8}$ in. long, with narrow pointed wings, ending in pencil-like tufts of pale yellow hairs.

The foliage of this tree is very variable—entire or two- to three-lobed leaves occurring on old trees; whilst those with a dentate margin are characteristic of

[1] In *Amer. Naturalist*, xxxix. 113, figs. 1-8 (1905).

branches ending in an inflorescence, and are normally developed on the ordinary branches of young trees. As both forms occur on the same individual, var. *hetero-phylla*, Meyer, cannot be maintained as a distinct variety.[1] In northern China, the leaves are quite glabrous ; but in the mountains of central China, there are traces of pubescence (simple hairs) on the upper surface of the leaves and on the petioles. Specimens with more numerous flowers in the corymb, which has one or two of the lateral axes branched, have been considered to be a distinct species, *C. Duclouxii*, Dode ; but these are probably trees of greater vigour and not even a distinct variety.

C. Bungei is readily distinguishable by its glabrous shining leaves, with longer and more slender points to the lobes than is the case in the other species.

This tree is widely spread throughout the mountains of China, from Peking in the north to Yunnan in the south-west, and is also recorded from the coast provinces of Shantung and Chekiang. Wilson found it in Szechwan at 8850 feet altitude, as a "tree 40 ft. in height, with white flowers suffused and spotted with pink." In Hupeh, it grows at about 4000 to 5000 feet elevation in the mixed forests of deciduous trees, and, my notes state, "40 feet high, 4 feet in girth, with pinkish flowers, which, together with the leaves, are of a disagreeable odour." It is much planted in temple grounds, at Peking, Shanghai, and elsewhere, and is usually known as the *ch'iu* tree.[2] It flowers at Peking in May.

This species is very rare in cultivation, the plants generally sold under this name by nurserymen being either the dwarf form of *C. bignonioides* or *C. Kaempferi*. Bureau knew of only one living specimen in France, a small tree at Segrez, which had not produced flowers.

It was introduced[3] about 1904 into the Arnold Arboretum by seed procured from Peking ; and a tree sent by Sargent to Kew in 1905 is now about 8 feet high.

(A. H.)

CATALPA FARGESI

Catalpa Fargesi, Bureau, in *Nouv. Archiv. Mus. Hist. Nat.* vi. 195, pl. 3 (1894) ; Dode, in *Bull. Soc. Dend. France*, i. 204 (1907).
Catalpa vestita, Diels, in Engler, *Jahrb.* xxix. 577 (1900).

A small tree, distinct from the other species, in the presence of stellate tomentum or branched hairs on the young branchlets, leaves, axes of the inflorescence, pedicels, bracts, and calyx. Leaves about 5 in. long and 4 in. broad, entire or with one or two acute lateral lobes, subcordate at the base, acuminate at the apex ; tomentose on the lower surface throughout, on the upper surface mainly on the nerves.

---

[1] Cf. Bretschneider, *Bot. Sinicum*, in *Journ. N. China Branch, R. Asiat. Soc.* xvi. 112 (1882), who states that "the leaves on the same tree are very variable, cordate, entire, lobed, laciniate, triangular, sinuate, etc."

[2] Bretschneider gives an account of the Chinese literature of this tree in *Bot. Sin.* ii. 339 (1882) and iii. 478 (1895). The classical name *tze*, with which he identifies the Catalpa, possibly indicated *Sassafras Tzumu*. Cf. vol. iii. p. 515.

[3] Cf. Rehder, in *Mitt. Deut. Dend. Ges.* 1907, p. 76.

Flowers few, seven to ten in a compact corymb; calyx covered externally with stellate tomentum; corolla, similar in size to that of *C. Bungei*, white, spotted with reddish brown dots. Fruit 18 to 20 in. long, ⅛ in. in diameter. Seeds yellowish grey, about ⅔ in. long, with sharp-pointed wings, ending in long silky hairs.

C. Fargesi is a native of western China, where it has been found in the mountains of Szechwan and Shensi by Farges, Giraldi, and Wilson. The latter introduced[1] it in 1901. Young plants show merely slight traces of the branched hairs on the branchlets and leaves, which are so characteristic of wild specimens; but as they grow older this peculiar pubescence may increase in quantity. Plants of this species, raised from seed sent by Wilson (Nos. 636 and 640) in 1905, are growing freely at Colesborne, where they have endured 30° of frost without injury, though the wood does not ripen well in autumn. These are now 4 to 6 ft. high. There are also young plants at Kew and Aldenham. (A. H.)

[1] *Journ. Roy. Hort. Soc.* xxviii. 50 (1902).

PAULOWNIA

Paulownia,[1] Siebold and Zuccarini, *Fl. Jap.* i. 25 (1835); Bentham et Hooker, *Gen. Pl.* ii. 939 (1876); Dode, in *Bull. Soc. Dend. France*, 1908, p. 159; Schneider, *Laubholzkunde*, ii. 618 (1911).

DECIDUOUS trees belonging to the order Scrophulariaceæ. Branchlets with chambered pith, showing in winter large oval raised opposite leaf-scars. Buds axillary, no true terminal bud being formed, minute, covered with two or four pubescent scales. Leaves simple, opposite in decussate pairs, stalked, ovate, cordate.

Flowers in large terminal erect panicles, opening in spring before the leaves; calyx five-cleft, campanulate, persistent at the base of the fruit; corolla gamopetalous, inserted on the base of the calyx, with a long slightly curved tube, and five spreading lobes, the three lower lobes longer than the two upper lobes; stamens four, affixed to the tube of the corolla, didynamous, included, with divaricate anthersacs; ovary superior, two-celled, with numerous ovules; style one, slender, slightly thickened towards the summit, stigmatic on the inner side. Fruit, ripening in one year, a two-celled woody or coriaceous capsule, ovoid, loculicidally dehiscent by two valves; placentæ two, ovate, compressed; seeds numerous, minute, oblong, surrounded by a broad translucent striated wing.

Two species of Paulownia[2] have been clearly distinguished, one of which, little known and not in cultivation in England, may be here briefly described.

I. *Paulownia Fortunei*, Hemsley, in *Journ. Linn. Soc.* (*Bot.*) xxvi. 180 (1890).

Leaves narrowly oval, longer and more acuminate than in *P. tomentosa*, covered beneath with a dense whitish tomentum. Flowers longer and relatively narrower than in *P. tomentosa*; calyx-lobes deltoid, obtuse, usually brown tomentose throughout, occasionally glabrescent except on the borders. Fruit, 3 to 3½ in. long, narrowly ovoid; seeds ¼ in. long, much larger than those of *P. tomentosa*.

[1] Named after Anna Paulowna, Queen of the Netherlands.

[2] Hayata, in *Bull. Congrès Internat. Bot. Bruxelles*, 41, pl. 24 (1910), and in *Journ. Coll. Sci. Tokyo*, xxx. 209 (1911), mentions a possible new species in Formosa. Elwes saw this in 1912 at a village near Horisha in Central Formosa; but it was not in leaf. There is also a supposed new species from Western China, raised at the Arnold Arboretum from seed sent by E. H. Wilson. It is in cultivation at Kew and Aldenham; but the young plants cannot at present be distinguished from *P. tomentosa*. Cf. *Gard. Chron.* xlviii. 275, fig. 116 (1910).—A. H.

I raised seedlings from Mr. Wilson's seeds, No. 769, collected in his journey of 1908 in Western China, which appear at three years old to be hardier and more rapid in growth than those which I have raised from the common species. The seedlings of the latter were killed to the ground for three years after planting out; whilst the West China form is now, at three years old, 14 ft. high, of which 10 ft. is the growth of 1911. This form seems likely to be a most ornamental tree even in cold parts of England; but must be planted in warm sheltered places where its immense juvenile leaves, measuring 21 in. by 23 in., will not be torn by wind.—H. J. E.

This species, which is probably smaller in size than *P. tomentosa*, occurs both in the north of China, where it has been collected at Chefoo, and in the south, where it has been found in the province of Kwangtung.

P. Duclouxii,[1] Dode, in *Bull. Soc. Dend. France*, 1908, p. 162, is apparently a variety with white flowers, tinged with pink, and not spotted as in the type. This is said to have been raised by C. Sprenger[2] of Corfu from seed, which he received from Dr. Dode.

PAULOWNIA TOMENTOSA

Paulownia tomentosa, Koch, *Dendrologie*, ii. pt. 1, p. 299 (1872); Shirasawa, *Icon. Ess. Forest. Japon*, i. text 129, t. 85 (1900).

Paulownia imperialis, Siebold and Zuccarini, *Fl. Jap.* i. 27 (1835); Loudon, *Trees and Shrubs*, 671 (1842); W. J. Hooker, *Bot. Mag.*[3] t. 4666 (1852); Hemsley, in *Journ. Linn. Soc. (Bot.)* xxvi. 180 (1890).

Bignonia tomentosa, Thunberg, *Fl. Jap.* 252 (1784).

A tree, attaining 80 ft. in height and 12 ft. in girth. Young branchlets green, glandular, and pubescent. Leaves broadly ovate, about 7 to 8 in. long, and 6 to 7 in. wide, cordate at the base, shortly acuminate at the apex, undivided or with one or two short lateral deltoid lobes, entire in margin; upper surface dark green, pubescent with short erect hairs; lower surface greyish green, covered with a thin tomentum; petiole 3 to 5 in. long, glandular, pubescent.

Flowers violet, the lower lip marked with dark coloured spots and two yellow bands; calyx with five ovate erect lobes, covered with a dense rusty brown tomentum. Capsules ovoid, about 1½ in. long, and 1 in. broad; seeds minute, about ⅛ in. long.

1. Var. *Fargesii*, Henry (var. *nova*).

Paulownia Fargesii, Franchet, in *Bull. Mus. Hist. Nat. Paris*, 1896, p. 280.

Flowers paler in colour. Adult leaves more glabrous than in the type. This was described by Franchet from a specimen sent by Père Farges from the mountains of north-eastern Szechwan; and appears to be identical with a tree found by me growing on cliffs in the mountains of Hupeh (No. 5346 A). The pentagonal ribbed calyx, noticed by Franchet, appears to be due to drying of the specimen, and not to be specially characteristic of this form, as a similar calyx occurs in some specimens of typical *P. tomentosa*. Var. *Fargesii* was introduced into cultivation in France by M. M. de Vilmorin, who raised it from seed received from Père Farges; and it flowered[4] in M. Boucher's nursery at Paris in 1905.

P. tomentosa is a native of the mountains of central and western China, where it has been found growing wild at altitudes of about 4000 feet by Père Farges in Szechwan, and by myself in Hupeh and Yunnan. It is the *t'ung* tree of the Chinese

---

[1] *P. meridionalis*, Dode, *loc. cit.*, described from a tree growing in Laos in Indo-China, is allied to or perhaps a form of *P. Fortunei*. [2] Cf. *Mitt. Deut. Dend. Ges.* 1910, p. 246.

[3] Figured from a tree in the garden at Bishopstowe, near Torquay, the first which flowered in the open air in England.

[4] Cf. *Journ. Soc. Nat. Hort.* vi. 324 (1905). C. Sprenger of Corfu, states in *Mitt. Deut. Dend. Ges.* 1910, p. 247, that he has raised *P. Fargesii* from seeds received from the mountains of north-west Hupeh in central China.

classics,[1] the wood of which was used in ancient times for making lutes.   It is largely planted as an ornamental tree throughout China, Korea, and Japan; and was probably introduced into the latter country at an early period by the Buddhist monks.[2]

This species often produces root-suckers at a considerable distance from the parent tree; and when cut down, sprouts vigorously from the stool.   The seedling, which has an herbaceous stem, usually dies down at the end of the first year; but in the following spring a permanent and more woody stem arises from a bud close to the ground.   The leaves on young plants (which are very large), on root-suckers, and on coppice shoots differ from the foliage of the adult tree, their margin being furnished with numerous short teeth, while their upper surface is velvety to the touch and very viscid, owing to the presence of numerous glandular hairs and sessile glands.

The Paulownia may be raised from seed sown in spring, or from root-cuttings; and may also be propagated by stem-cuttings under glass, or even from leaf-cuttings, At Kew this species is very effective as a foliage plant in beds out of doors.   The plants, when about three years old, are cut down in early spring to within 6 inches of the ground; and when they start to grow, all the buds except one on each stem are removed.   Watered in dry weather and mulched with manure, stems are produced about 6 ft. high, which bear enormous leaves, 12 to 18 in. in diameter.

This tree was introduced[3] into Europe in 1834, by seeds sent from Japan to Paris by M. de Cussy; and one of the original trees in the Jardin des Plantes measured in 1904 about 60 ft. in height, and 12 ft. in girth.   It succeeds well in France and Italy, where it attains a considerable size, and regularly produces flowers and fruit.

It was introduced[4] into England by seeds from Japan in 1838, and was cultivated in the Horticultural Society's garden at Chiswick; but it has never attained large dimensions in this country, and probably many of the older trees were killed in the severe winters of 1860 and 1866.   It flowers frequently in the south of England, but rarely produces fruit, and very seldom fertile seed.   (A. H.)

The largest tree I know of in England is one standing near the entrance lodge at Westonbirt, Gloucestershire, which as measured in August 1911, by Mr. A. Chapman, was 56 ft. by 7 ft. at 3 ft. from the ground.

The next is at Wilton House, which in 1906 measured 53 ft. by 6 ft. 8 in., but being in a damp situation it has only flowered twice in fifty years, and Mr. Challis informed me that it suffered much in the hard winters of 1860-61, and 1879-80.   A fine tree at Linton Park, Maidstone, in 1911 measured 45 ft. high and 9 ft. in girth at 6 ft. from the ground.   At Caldrees, Ickleton, Cambridgeshire, there is a tree about 25 ft. high and 3½ ft. in girth, which produces flowers and fruit nearly every year.   From its seeds, plants were raised in 1902, one of which was planted at Ickleton Grange, and is now 12 ft. high.

---

[1] Cf. Bretschneider, *Bot. Sinic.* ii. 348 (1892).   It is colloquially known in China as the *pao-t'ung*, in order to distinguish it from the *wu-t'ung* (*Sterculia platanifolia*), and the *t'ung-yu* (*Aleurites cordata*).

[2] It is figured by Kaempfer, *Amœn. Exot.* 860 (1712).

[3] *Actes Premier Congrès Internat. Bot., Paris*, 536 (1900).   Cf. also *Gard. Chron.* 1841, pp. 349, 701, where it is stated that a single plant was raised in the Royal Garden in Paris, from Japanese seed, in 1834.

[4] Loudon, *Gard. Mag.* xvi. 635 (1840).   It flowered in a greenhouse at Oakfield near Cheltenham in 1843, according to Loudon, *Gard. Mag.* 1843, p. 649.

At Swanmore Cottage, Hants, Mr. Molyneux showed me a well-shaped tree, 35 ft. by 7½ ft., in 1906, which flowers occasionally but never ripens seed. At Ashstead Park, Surrey, the seat of Mr. P. Ralli, there was in 1892 a fine tree, 45 ft. by 7½ ft., which flowers freely in warm seasons.[1] At Whitbourn Court, near Worcester, Sir R. Harrington has a tree which he raised from seed gathered in the Vatican gardens at Rome in 1888, which in 1905 was 23 ft. by 4 ft., and with a head 25 ft. in diameter. In Cornwall, where the climate does not suit it so well, the largest I have seen, at Scorrier, was about 25 ft. high in 1911.

There are specimens 20 to 30 ft. high in the Kew and Cambridge Botanic Gardens; at Grayswood, Haslemere; in several gardens in Kent; at East Cowes Castle in the Isle of Wight; at Hursley Park, Winchester; at Abbotsbury[2]; at Bicton[3]; at Trevarno,[4] Cornwall; and at Singleton Abbey, Swansea.

In Scotland, the only places where I have seen it are at Castle Kennedy, where in 1906 I saw a poor-looking tree about 25 ft. high, which evidently does not like the climate; and at Tyninghame, East Lothian, where Mr. Brotherston measured a tree 3 ft. 9 in. in girth at 3 ft. from the ground, which has been in bad health for years, so we may conclude that the summers of Scotland are too short and cold for it.

In Ireland, there are two old but not very thriving trees at Glasnevin; but at Mount Usher, there is a fine tree nearly 40 ft. high, and over 6 ft. in girth.

In America,[5] it does not flower regularly north of New York; but is fairly hardy in sheltered positions as far north as Massachusetts, where the flower-buds are killed every winter; and it can be cultivated as a foliage plant even in Montreal.

TIMBER

In Japan, it is known as *kiri*; and produces a very light, dull white, shining wood, which is used for making boxes, musical instruments, linings of safes, clogs, doors, and in cabinet work. As large planks are not usually obtainable, Paulownia boards are made by joining small pieces together with paste and bamboo pegs, as shown at the Anglo-Japanese Exhibition at Shepherd's Bush, London. On my last visit to Japan I saw well-made wardrobes of this wood, which is preferred to all others for this purpose on account of its resistance to damp. Clothes kept in drawers of this wood are said to remain free from mould during the rainy season. These wardrobes are sometimes framed in the wood of *Diospyros Kaki*, the heart-wood of which is black mottled with grey, and very handsome. The large braziers, called *Hibashi* in Japan, are often made of sections of the trunk of Paulownia turned in such a way as to show its beautiful grain. (H. J. E.)

[1] *Gard. Chron.* xii. 440 (1892).   [2] Fruit was sent to Kew from Abbotsbury in January 1902.

[3] *Quart. Journ. Forestry*, i. 54 (1907).

[4] Cf. *Gard. Chron.* xxvi. 211 (1899). In *The Field*, 1908, p. 233, the tree at Trevarno, said to be twenty years old and 30 ft. high, is reported to have produced a few years previously a good crop of fruit.

[5] Rehder, in Bailey, *Cycl. Amer. Hort.* 1223 (1901). Britton and Brown, *Illustrated Flora Northern U.S.* iii. 157 (1898), state that it has escaped from cultivation in southern New York, in New Jersey, and in the Southern States.

ROBINIA

Robinia, Linnæus, *Gen. Pl.* 220 (1737) and *Sp. Pl.* 722 (1753); Bentham et Hooker, *Gen. Pl.* i. 499 (1865); Schneider, *Laubholzkunde*, ii. 79 (1907).
Pseudacacia, Moench, *Meth.* 145 (1794).

DECIDUOUS trees and shrubs belonging to the division Papilionaceæ of the order Leguminosæ. Leaves unequally pinnate, alternate, stalked : leaflets opposite, rarely sub-opposite or alternate, stipellate, stalked, entire, penninerved. Stipules in pairs, at first setaceous, ultimately either deciduous or developing into persistent spines. No true terminal bud is formed, the tip of the branchlet falling off early in summer and leaving a scar at the apex of the twig. Buds minute, multiple, three to five superposed vertically, not apparent[1] in summer and autumn, being concealed by the enlarged base of the petiole; in winter, embedded in a projection on the branchlet between the stipules, and covered by three scales, which are united together and form the leaf-scar. Usually only the uppermost bud develops—the scales, which are very tomentose within, bursting open, and afterwards persisting at the base of the new branchlet during the following season.

Flowers in pendulous racemes, arising from the axils of the leaves, with long pedicels and caducous bracts and bracteoles. Calyx campanulate, unequally five-toothed. Corolla papilionaceous ; petals with short claws, inserted on a tubular glandular disc, connate with the base of the calyx-tube; standard large, reflexed, obcordate ; wings oblong, curved, free ; keel-petals incurved, obtuse, united below. Stamens ten, inserted with the petals ; nine inferior united into a tube ; upper stamen free at the base. Ovary stalked ; style subulate, inflexed, pubescent, ending in a small stigma ; ovules numerous, hanging in two rows from the ventral suture. Pods in pendulous racemes, linear-oblong, compressed ; valves two, thin, membranous ; seed-bearing suture with a narrow wing. Seeds numerous, reniform, oblique, with a persistent incurved stalk.

About eight species of Robinia are known, confined to the United States and Mexico. Four, all natives of the United States, occur in cultivation :—

I. *Branchlets without glands.*

1. *Robinia Pseudacacia*, Linnæus. See p. 1497.

Branchlets at first slightly pubescent, soon becoming glabrous. Leaflets

---

[1] Occasionally a supra-axillary bud is formed, visible above the insertion of the leaf; it develops into a short-lived feeble branch.

pubescent, minutely apiculate. Stipules persisting as glabrous hard woody spines.

2. *Robinia neomexicana*, Gray. See p. 1506.

Branchlets densely pubescent. Leaflets pubescent, tipped at the acute apex with a long mucro; stipels mostly persistent. Stipules persistent as slightly pubescent hard woody spines.

II. *Branchlets glandular.*

3. *Robinia viscosa*, Ventenat. See p. 1507.

Branchlets with short-stalked glands, exuding a viscid matter. Leaflets pubescent, shortly apiculate. Stipules usually deciduous, occasionally persistent as slender short spines.

4. *Robinia hispida*, Linnæus. See p. 1508.

Branchlets with glandular bristles and dense pubescence. Leaflets glabrous, except for slight pubescence on the midrib beneath. Stipules usually deciduous, occasionally persistent as minute blunt spines. (A. H.)

ROBINIA PSEUDACACIA, False Acacia, Locust

Robinia Pseudacacia, Linnæus, *Sp. Pl.* 722 (1753); Cobbett, *Woodlands*, Nos. 322-398 (1825); Loudon, *Arb. et Frut. Brit.* ii. 609 (1838); Withers, *The Acacia Tree* (1842); Wilkomm, *Forstliche Flora*, 930 (1887); Sargent, *Silva N. Amer.* iii. 39, tt. 112, 113 (1892), and *Trees N. Amer.* 572 (1905); Mathieu, *Flore Forestière*, 119 (1897); Schneider, *Laubholzkunde*, ii. 82 (1907).

Robinia fragilis, Salisbury, *Prod.* 336 (1796).

Pseudacacia odorata, Moench, *Meth.* 145 (1794).

A tree, attaining about 80 feet in height and 15 feet in girth. Bark thick, brownish, with broad and deep longitudinal fissures, separated by scaly ridges. Young branchlets, at first slightly pubescent, soon becoming glabrous. Leaf-rachis slightly pubescent. Leaflets, nine to nineteen, elliptic or oval, rounded at the base; rounded, truncate or emarginate at the apex, which terminates in a minute mucro; upper surface covered with minute appressed pubescence, lower surface paler with scattered hairs; petiolule, $\frac{1}{8}$ in., slightly pubescent; stipels linear, $\frac{1}{16}$ in., early deciduous.

Flowers in loose pubescent non-glandular racemes; pedicels slender, $\frac{1}{2}$ in. long, pubescent; calyx gibbous, pubescent, the lowest lobe acuminate and longer than the others; petals white, with a greenish yellow patch on the middle of the inner surface of the standard. Pod, 2 to 4 in. long, glabrous, persistent on the leafless branches late in winter; seeds usually four to eight.

The stipules,[1] at first linear, subulate, membranous, pubescent, and about $\frac{1}{2}$ in. long, are either deciduous or develop into spines, which persist for several years and occasionally attain $\frac{3}{4}$ in. in length. These spines are glabrous, triangular, com-

[1] Cf. Colomb, in *Ann. Sc. Nat.* vi. 65 (1887).

pressed, straight or turned up at their sharp points; and in winter, together with the concealed buds and the stout glabrous angled twigs, serve to identify this species.

The flowers are fragrant,[1] and are visited by bees, as they contain much honey. In southern Italy,[2] the flowers are developed before the leaves; but north of the Alps the leaves unfold earlier or at the same time as the flowers. The leaves droop at night.

Seedling:[3]—The cotyledons, raised above ground on a glabrous caulicle, are oblong-oval, obtuse, entire, shortly stalked, light green, thick in texture, obscurely veined, and about $\frac{1}{2}$ in. long. The slightly pubescent stem bears alternate leaves; first leaf always unifoliolate, rounded, obtuse, and broader than long; second leaf with three leaflets, the terminal orbicular one larger than the two lateral elliptic leaflets; succeeding leaves with three to five leaflets.

This species produces numerous root-suckers,[4] arising singly or in groups of two or three at a considerable distance from the tree. Leaves on root-suckers smaller than on adult trees, but with the usual number of leaflets, which are cordate at the base; stipules usually aborted or very short.

The tree, when cut down, produces coppice shoots freely; and on this account, and because of its facility of reproduction from root-suckers, it is much cultivated in the warmer parts of Europe as coppice with a short rotation.

VARIETIES

A large number of varieties have arisen in cultivation in Europe. Some have been successfully raised from cuttings; but they are usually propagated by grafting.

I. *Leaves as in the type; habit peculiar.*

1. Var. *pyramidalis*, Pépin, in *Rev. Hort.* iv. 240 (1845) (var. *fastigiata*, Nicholson).

Narrowly pyramidal, resembling the Lombardy Poplar in shape; branches directed vertically upwards. This was introduced by M. Leroy of Angers, and a fine specimen in the Jardin des Plantes at Paris was planted by Pépin in 1843. The best example of this we know is in the public gardens at Le Mans, in France, where Elwes measured one in 1908 nearly 60 ft. high. There are four specimens in the public garden at Genoa, one about 40 ft. in height, which are more broadly pyramidal in habit, the branches not being so upright. These correspond to the description of var. *stricta*, Link, *ex* Loudon, *Arb. et Frut. Brit.* ii. 610 (1838). Both these varieties are suitable for planting in streets; but seem to be scarcely known in England, except at Kew, where there is a good example of var. *pyramidalis* about 35 ft. high.

2. Var. *Ulriciana*, Reuter, *ex* Dippel, *Laubholzkunde*, iii. 702 (1893). Branches pendulous.

[1] Cf. Mesnard, in *Ann. Sc. Nat.* xviii. 341 (1893).
[2] Kerner, *Nat. Hist. Plants*, Eng. Transl. i. 562 (1898).
[3] Cf. Lubbock, *Seedlings*, i. 422, fig. 275 (1892), and Ledoux, in *Ann. Sc. Nat.* xviii. 369, fig. 39 (1903).
[4] Cf. Dubard, in *Ann. Sc. Nat.* xvii. 167 (1903).

A tree at Kew, called var. *pendula nova*, has slightly pendulous branches with small leaflets.

3. Var. *tortuosa*, De Candolle, *Cat. Pl. Monspel.* 136 (1813).

A tree, with wide-spreading twisted branches.[1] The habit is well figured by Schneider, *Dendrol. Winterstudien*, 89, fig. 90 (1903).

4. Var. *umbraculifera*, De Candolle, *Cat. Pl. Monspel.* 137 (1813).

A compact, rounded bush, giving dense shade; branches without spines; seldom if ever flowering.[2] Suitable for rock gardens, and often called Parasol Acacia. It is said[3] to have originated from a burr on a tree of the typical form. There are two good specimens at Kew, about 15 ft. high, grafted at about 10 ft. from the ground.

5. Var. *Bessoniana*, Nicholson.

A compact, dense, round-headed tree, without spines. This resists wind, and is much used for avenues and street planting. It is readily propagated by cuttings.

6. Var. *inermis*, De Candolle, *Cat. Pl. Monspel.* 136 (1813).

Branches without spines; otherwise as in the type.

II. *Leaves different from those of the type.*

7. Numerous varieties with differently coloured foliage are known, the peculiarities of which are indicated by their names, as var. *aurea*, Kirchner; var. *aureo-variegata*, Schneider; var. *argenteo-variegata*, Kirchner: and var. *purpurea*, Dippel.

8. Var. *crispa*, De Candolle, *Prod.* ii. 261 (1825).

Leaflets curled; branches without spines. On a tree of this variety in the Jardin des Plantes at Paris, only the upper leaflets were abnormal, the lower three or four pairs of each leaf being of the ordinary form.

9. Var. *bullata*, Koch, *Dendrologie*, i. 56 (1869).

Leaflets crowded and puckered with swellings.

10. A series of forms, with remarkably small leaflets, have arisen in cultivation, as var. *amorphaefolia*, Link; var. *myrtifolia*, Koch; var. *tragacanthoides*, Kirchner; var. *sophoraefolia*, Kirchner (*R. coluteoides*, Koch). In var. *linearis*, Kirchner, and var. *dissecta*, Koch, the leaves are remarkably narrow.

11. Var. *monophylla*, Kirchner, *Arb. Musc.* 377 (1864).

Leaves with only one leaflet, which is much enlarged, or with, in addition, three to five leaflets, all larger than in the type. This appeared in 1855 as a single plant in a bed of seedlings in M. Deniaux's nursery at Brain-sur-Authion (Maine-et-Loire). The original plant was transplanted into the Jardin des Plantes at Paris, where it flowered and gave seeds in 1865. The seeds when sown were said to have yielded about one-quarter of the variety, the remainder being normal. The one-leaved Robinia is also said to be liable to a petaloid alteration of the stamens, which impairs fertility.[4]

A tall, slender tree of var. *monophylla* at Brocklesby Park, 57 ft. high and

---

[1] This variety occasionally develops remarkably thick corky bark. Cf. *Mitt. Deut. Dend. Ges.* 1911, p. 404, fig.

[2] Koch, *Dendrologie*, i. 57 (1869) states that he saw a flowering branch, which had been produced by an unclipped bush of this variety. The flowers were white, and not yellow, as stated by Dumont de Courset, *Bot. Cult.* vi. 140 (1811).

[3] Carrière, *Prod. et Fixat. Vars.* 54 (1865).   [4] De Vries, *Plant Breeding*, 617, 664 (1906).

2 ft. 8 in. in girth in 1908, showed no sign of having been grafted, and had produced three suckers with normal leaves, one being 36 ft. distant from the parent tree and 8 ft. in height.   There are two trees about 30 ft. high at Kew, differing in habit— one, narrow with ascending branches; the other wider, with pendulous branches.

III. *Flowers differing from those of the type.*

12. Var. *Decaisneana*, Carrière, in *Rev. Hort.* 1863, p. 151, with coloured plate.

Flowers pale pink, and larger than in the type; otherwise similar.   This originated in the nursery of M. Villevielle at Manosque (Basses-Alpes), and flowered for the first time in 1862.   This variety is said to come fairly true from seed.   It is well known in English nurseries, and seems to be a very vigorous grower.

13. Var. *lutea*, Schneider, *Laubholzkunde*, ii. 83 (1907).

Flowers pale yellow.   This variety is mentioned by Dumont de Courset, *Bot. Cult.* vi. 140 (1811) as var. *flore luteo*, but it is doubtful if it is now in cultivation.

14. Var. *semperflorens*, Carrière, in *Rev. Hort.* xlii. 502 (1871) and xlvii. 191, with coloured plate (1875).

A tree of great vigour, producing annual shoots of abnormal length (up to 6 ft. long), which bear flowers continually from June to September.   This originated as a single seedling in M. Durousset's nursery at Genouilly (Seine et Loire), and has been in commerce since 1875.   The original tree was transplanted in 1874 into the Jardin des Plantes at Paris, and was remarkable for its constant production of numerous flowers.   Carrière counted on one branch no less than 145 racemes.

HYBRIDS

1. *Robinia dubia*, Foucault, in Desvaux, *Journ. de Bot.* ii. 204 (1813); Loudon, *Arb. et Frut. Brit.* ii. 627 (1838).

Robinia ambigua, Poiret, in Lamarck, *Encyc. Suppl.* iv. 690 (1816).
Robinia intermedia, Soulange-Bodin, in *Ann. Soc. Hort. Paris*, ii. 43 (1828).

This hybrid between *R. Pseudacacia* and *R. viscosa* is a small tree, differing mainly from the first named species in the young branchlets and flowering peduncles being slightly viscid-glandular and in the flowers being pale pink.   It has been in cultivation about 150 years.[1]

2. *Robinia bella-rosea*, Nicholson, *Dict. Garden.* iii. 310 (1887).

This is similar to the last, but the branchlets are more plainly viscid-glandular, whilst the flowers are larger and deep pink in colour.

3. *Robinia Holdtii*, Beissner, in *Mitt. Deut. Dend. Ges.* xi. 117 (1902); Koehne in *Gartenflora*, lii. 272 (1903).

A hybrid between *R. Pseudacacia* and *R. neomexicana*, which originated[2] about 1890 in Mr. Von Holdt's garden at Alcott in Colorado.   Leaflets larger than those of *R. neomexicana*, darker green than those of *R. Pseudacacia*.   Flowers varying in colour, white flushed with pink or deep pink.   A few stalked glands occur on the flowering peduncle and on the pod.

---

[1] It is possibly *R. echinata*, Miller, *Dict.* ed. 8, No. 2 (1768).
[2] Beissner, *op. cit.* 118, states that a similar plant also originated in Späth's nursery.

4. *Robinia coloradensis*, Dode, in *Bull. Soc. Bot. France*, lv. 650 (1908).

A peculiar tree[1] raised at Les Barres from seed, sent by M. Berthoud from Golden in Colorado. It is probably a seedling of *R. Holdtii*.

5. *Robinia Holdtii britzensis*, Späth, in *Gartenflora*, lii. 557 (1903).

This is a vigorous plant which was raised in Späth's nursery from a seed produced by a tree of *R. neomexicana* in 1893. It, however, closely resembles *R. Pseudacacia*, only differing from this species in having traces of purple on the standard of the flower.

DISTRIBUTION

R. Pseudacacia is considered by American botanists to be indigenous in the Alleghany Mountains from Pennsylvania to Georgia, but it is now widely naturalised throughout most of the United States east of the Rocky Mountains and in Nova Scotia and Ontario. Dame and Brooks say[2] that it is thoroughly at home in Maine, forming wooded banks along streams, and fairly abundant in thickets and along roadsides and fences in the New England States generally. Plate 353, reproduced from a photograph sent by Miss Cummings, shows a tree at Boston (Mass.). A shrubby form, which occurs in Kansas, Arkansas, and Indian Territory, is perhaps a true native and not an escape from cultivation. The tree attains its largest size[3] in West Virginia. It is found in the forest in mixture with other trees, growing on the slopes with oaks, chestnut, hickories, and maples, while along the streams it is associated with ash and black walnut. It often spreads by its root-suckers, forming thickets of small trees. In the mountains of Pennsylvania and West Virginia, burnt and cut-over forest lands are speedily covered with seedlings,[4] which often grow up into pure stands of considerable extent.

On account of the excellence of its timber for fence posts and for all uses requiring contact with the soil, and also for fuel, the Robinia was much planted[5] formerly throughout the United States, and succeeded well south of the 38th parallel; but at the present time planting is restricted or rendered useless on account of the ravages of the locust-borer[6] (*Cyllene Robiniæ*, Forster). The grubs of this beetle bore holes through the bark deep into the wood, which becomes completely honeycombed with galleries. Young plantations are often attacked, especially in the States east of the Rocky Mountains. The Robinia thrives in the dry prairie regions, and has been planted with success on alkaline soil in the San Joaquin valley in California.[7]

INTRODUCTION

This species was introduced into France by Jean Robin, who received seeds from America in 1601 ; and the oldest tree known was planted in 1636 by Vespasian

---

[1] Cf. Vilmorin, *Frut. Vilmor.* 54, fig. (1994), whose description does not agree with that of Dode. The individuals that were raised may have differed considerably.   [2] *Trees of New England*, 131 (1902).

[3] Ridgway measured a cultivated tree in Illinois, 95 ft. high and 11½ ft. in girth.

[4] The seed is retained in the pods on the trees till late in the year, and is distributed to great distances by the strong winter winds.

[5] Directions for planting in the United States are given in the *United States Forest Circular*, No. 64 (1907).

[6] Cf. *United States Entom. Comm.*, *Fifth Report*, 355 (1890).   [7] Hilyard, *Soils*, 480 (1906).

Robin in the Jardin des Plantes at Paris, where the stump, largely made up by cement, still remains, and produces a few shoots and flowering branches every year.[1]

The date of introduction[2] into England is uncertain ; but Parkinson, in his *Theatre of Plants*, published in 1640, mentions the tree as having been grown by Tradescant " to an exceeding height." During the seventeenth and eighteenth centuries the Robinia was occasionally planted ; but it only came prominently into vogue by the vigorous advocacy of Cobbett, who began to write about it in 1823.

(A. H.)

CULTIVATION

No tree except the oak has probably been so much written on in England as the Robinia or Locust, as William Cobbett, its great advocate, always called it, though now it is more commonly, though incorrectly, called Acacia. That great Englishman was, during his exile in America, so much impressed with the valuable qualities of its wood, that, though suffering under a grievous and unjust political persecution, " nevertheless I did not forget my country and the duty I still owed to her "; and when he returned to England in 1819 brought with him a parcel of seed, which he had no means of sowing till 1823. He then began sowing it on a very small scale at first, but later he raised it in large quantities, and sold more than a million plants. In *Woodlands*, Cobbett devoted many pages to an account of this species, which he considered was going to supplant all other trees in England ; but Cobbett's enthusiasm in arboriculture, as in politics, often outran his discretion, and though many of his trees still remain, mostly long past their prime and in a more or less decayed condition, the Robinia has never realised his predictions. It has always been obnoxious to the British woodman, partly on account of its thorns which tore his clothes and hands, partly on account of the hardness of its wood which blunted his tools, and partly because of its liability to be broken by the wind, and to reproduce from suckers when cut.

Cobbett had many admirers, one of whom, Lord Folkestone, afterwards Earl of Radnor, bought 13,600 locust trees, which were one-year seedlings one year transplanted ; and carried them in a large waggon from Sussex to Coleshill, Berkshire, in the latter part of March 1824. These were planted in clumps of one to several hundreds in a large plantation in Coleshill Park ; and when Cobbett saw them in 1826 they averaged more than 12 ft. high, and, as he says, " looked like clumps of trees which had been planted many years previous to the planting of the trees of the rest of the plantation." He calculated the cost of trees, trenching, and planting, at £134, and thought that, at six or seven years old, they would be fit to cut for hop-poles, and then be worth 1s. each, leaving a profit of £529 on the five acres which he supposed them to cover. Of course if this profit or anything like it had been realised here or elsewhere, the tree would not have been neglected as it has been

[1] Cf. *Actes Premier Congrès Internat. Bot. Paris*, 536 (1900). This old tree is figured by Sargent in *Garden and Forest*, iii. 305, fig. (1890). When Elwes saw it in 1905 it had a branch about 40 ft. high producing good foliage.

[2] Dr. Yule, in *Mem. Caled. Hort. Soc.* ii. 413 (1819), states that it was first planted in Scotland by Mr. Cockburn at Ormiston Hall, Haddingtonshire.

almost everywhere in England. When I visited Coleshill in June 1905, I could find no one who remembered them, and the late Mr. Pleydell-Bouverie, uncle of the present Lord Radnor, who then lived at Coleshill, thought that they had all disappeared with the exception of a few near the house, which measured 60 ft. by 8 ft. 3 in. and 65 ft. by 9 ft. 3 in., and two or three in the park.

Neither at Botley, where Cobbett lived for some years and planted American trees, nor in the surrounding district, where the soil and climate suit this tree as well as any in England, could I find any trace of locust plantations.

The only plantation which we have seen and of which any exact details are on record, is the Brickhills plantation on the estate of Sir Hugh Beevor at Hargham, Norfolk. This has been described [1] by its owner in the *Quarterly Journal of Forestry*, ii. 301-303 (1908), and the results are so favourable that they should be read by everyone. Sir Hugh sums up by saying, " For its own sake I shall continue to grow it for estate use, having always utilised and refused to sell any. The chief use I put it to now is as stakes for rabbit-proof fencing ; and these when taken up at the end of ten years will be used again for a similar term. To supply stakes no tree is so well fitted, and it would be worth planting for this purpose only."

In 1842 W. Withers of Holt, Norfolk, compiled a work [2] of over 400 pages on this tree, in which he prints a great many communications, among which one written by F. Blaikie from Holkham in 1828 states, " I found great difficulty in protecting young locust trees from hares and rabbits. These animals prefer it to any forest tree. They could not have been so plentiful as they are now, at the time Mr. Coke reared the innumerable locust trees growing at this place. Those trees thrive on our most inferior sandy soils where other forest trees barely exist. We do not succeed in raising plants from home-grown seed, though it appears sound and well grown." Here also I failed to find any evidence that this species was now valued ; and it may be said generally that though in almost every southern county large trees may be found, which are often decayed at the heart, yet Robinia is no longer looked on as useful for general planting ; and that unless pruning of the branches is carefully attended to, this tree is more likely to suffer from wind when young than other trees.

All my observations go to show that Robinia is essentially a lover of a hot, dry, and sandy soil, though it only attains a large size and age on a good sandy loam ; and while it tolerates lime, grows much better without it. It is easy to raise from seed and grows rapidly from the first ; but the shoots often fail to ripen and are liable to freeze when young. It should be transplanted in the spring when a year old, as its root system is not naturally fibrous, and be planted out when two or three years old in a sunny position. Though Robinia trees may be seen drawn up to a great height in mixed plantations,[3] where their rapid growth has given

---

[1] Sir Hugh Beevor gives me the following notes on this plantation, which was made in 1829 and measured in 1901 :—" Where pure the trees cover not quite an acre ; on the best half-acre there are 40 trees per acre with a volume of 1000 cubic ft. ; canopy incomplete on account of windfalls ; largest tree with a volume of 80 cubic ft."

[2] *The Acacia Tree* (1842).

[3] Paeske of Brunswick, in *Mitt. Deut. Dend. Ges.* 1911, p. 77, recommends the planting of Robinia as scattered solitary trees in birch woods, which will provide shelter against the wind, and yet give enough sunlight for it to develop well. Amidst birch, the Robinia cleans its stem perfectly, being often free from branches up to 40 ft. in fairly good soil.

them the lead, this species will not bear much shade; and if grown for economic purposes it ought to be cut when twenty to thirty years old, or as soon as it is fit to make a good gate-post. On any suitable soil the wood is then at its best; whereas if left to grow into larger timber, the trunk becomes deeply furrowed or burry, and usually begins to decay inside at fifty to sixty years old. It would perhaps make a good mixture if planted in alternate lines five feet apart with sweet chestnut and cut out as soon as large enough; leaving the chestnuts to form a clean crop dense enough to suppress the stool shoots of the Robinia by its shade, and in their turn to be clean felled at fifty to sixty years old. This opinion is shared by Mr. Braid, forester to the Earl of Dudley, who finds Robinia a valuable tree when properly treated, on dry sandy land near Kidderminster.

REMARKABLE TREES

Perhaps the finest tree in England is the one growing in front of Frogmore House, near Windsor (Plate 351), which, when I measured it in 1908, was 88 ft. high by 14 ft. 7 in. in girth. President Roosevelt told me in 1904 that he hardly thought that such a tree could be now found in the United States. It is probably over a hundred years old, as I am told by Mr. Nutt, Clerk of Works at Windsor Castle, that Frogmore was bought in 1748 by Sir E. Walpole; and leased to Queen Charlotte in 1809. Another fine tree at The Mote, Maidstone (Plate 352), which I saw in 1902 measured about 80 ft. high.

At Kew a very old tree was probably planted by Aiton about 1760.

At Pains Hill there are some very large trees, one of which in 1904 was about 60 ft. high by 17 ft. 3 in. in girth; another was 70 ft. by 12½ ft. These were probably planted about 1750.

At Burwood House, Surrey, Col. Thynne in 1900 measured a tree 86 ft. by 13 ft. 3 in., which I have not seen. At Bowood, Wilts, I found, in 1908, a tree nearly 90 ft. high by 8½ ft. in girth. At Arley Castle there is a tree 85 ft. by 9 ft. in 1905 which was planted in 1820. At Stanway, Gloucestershire, there is a very old tree whose top is dying, and has been taller; in 1911 I made it 65 ft. by 11 ft. 10 in. At Audley End, in 1908, there was a healthy tree close to the house 80 ft. by 11 ft. 2 in. At Hatfield, Herts, there are several good trees, the largest, near the conservatory, measuring, in 1911, 67 ft. high by 12 ft. 8 in. in girth at three feet from the ground. In Tree Court of Caius College, Cambridge, there is an old tree about 60 ft. high, and 11 ft. 7 in. in girth, dividing at 10 ft. from the ground into two main stems. At Kenwood, near London, a very old tree, split nearly to the ground, with a decayed top, measured 45 ft. by 16 ft. in 1909. In an old gravel pit at Hitchin, Herts, drawn up in a thick wood among elms, I saw in 1905 a tall slender tree 80 to 90 ft. high, with a clean bole over 50 ft. high. At Holly Dale, near Keston, Kent, Mr. Webster records[1] a tree 78 ft. by 11 ft. 7 in., containing 110 ft. of timber and with a spread of 54 ft. At Chilham Castle, Kent, there is a fine tree, 75 to 80 ft. high and of moderate girth, on the lawn.

[1] *Trans. Roy. Scot. Arb. Soc.* xii. 312 (1890).

Sometimes the trunks of old trees are covered with burrs, the most remarkable that I have seen being a tree at Coolhurst, Sussex, which had a short trunk 14½ feet round; and another at Henley Park, Surrey, the seat of Sir J. Roberts, which was 65 ft. by 12½ ft.

In the north-west of England the tree rarely attains any great size; but Clark[1] records one at Dovenby Hall, Cumberland, 55 ft. by 9 ft. in 1887, which was planted in 1795.

In Wales the best I have seen is at Golden Grove, which in 1906 was 65 ft. by 8½ ft.

In Scotland it only thrives in the drier climate of the east, the best I know of being at Gordon Castle, a sound healthy tree which in 1904 was 56 ft. by 9 ft. At Murthly Castle a tree, forked near the ground, was in 1906 about 50 ft. by 6 ft. The largest trees measured by Mr. Renwick are two at Cordale House, Dumbartonshire, 64 ft. by 7 ft. and 53 ft. by 6 ft. 7 in., and one at Mauldslie Castle, 60 ft. by 6 ft. 1 in., all taken in 1911.

In Ireland, Henry has seen very few large trees, the finest being probably one at Doneraile, Co. Cork, which is about 12 ft. in girth, with a *Pyrus Aucuparia* arising from the stem at 9 ft. from the ground. At Glenomera, Co. Clare, a young healthy tree in a wood measured, in 1907, 59 ft. by 5¹ ft. 9 in. At Woodstock, Kilkenny, I saw in 1909 a tree near the Lodge, about 60 ft. by 6 ft. 9 in.

TIMBER

Such a mass of details are found in the writings of Loudon, Cobbett, and Withers as to the durable properties of this timber, that I need not repeat them at length; but I can say, from personal experience that for gate- and fencing-posts it is unrivalled for strength and durability by any native timber except that of the yew. Mr. H. Clinton-Baker[2] showed me the hanging-post of his lodge gate, which is now absolutely sound at the ground level, with the date 1849 cut on it. I was told that the spurs put up to support the oak posts of a fence at Pains Hill had remained sound for eighty years. I have myself proved that a tree planted on fair loam will in twenty-five years produce three good gate-posts, whilst oak at the same age would not make one.

On the Continent it is considered by wheelwrights to be superior to all other timber for spokes, and is now used extensively for wheel spokes of motor cars. For such purposes it must be cut before the wood has become old and lost its elasticity; and I believe that it is best when 9 to 12 inches in diameter. From one short log of this size, for which I paid 5s., the spokes of a heavy timber carriage wheel were made. Notwithstanding these good qualities the timber has no recognised value among English timber merchants, and "Acorn," in *English Timber*, says that it is so little known in home timber yards that there are no specific outlets for its disposal. Therefore I should advise that it be kept for home use only.

---

[1] *Trans. Eng. Arb. Soc.* 1887, p. 143.

[2] Two fencing stakes made of Robinia, which had been put down in 1848 in the pinetum at Bayfordbury, were perfectly sound in October 1911, when they were presented by Mr. H. Clinton-Baker to the Cambridge Forestry Museum.

The Robinia has proved very useful in the afforestation of the steppes of Russia and Hungary. The plantations in Hungary,[1] mostly pure and growing on light dry soil, covered an area of 70,000 hectares in 1899. These plantations are usually treated as coppice with a rotation of twenty years. Illes[2] states that in Hungary, a plantation, fifty years old, produced 8800 cubic feet of timber per acre, the trees averaging 90 ft. in height and 10 in. in diameter. In Roumania, the tree has been used with great success in fixing the moving sands on the Danube, and 5000 to 6000 hectares have been planted.

(H. J. E.)

ROBINIA NEOMEXICANA, Western Locust

Robinia neomexicana, Gray, in *Mem. Amer. Acad.* v. 315 (1853); Sargent, *Silva N. Amer.* iii. 43, t. 114 (1892), and *Trees N. Amer.* 573 (1905); Wittmack and Brettschneider, in *Gartenflora*, li. 649, t. 1385 (1892); Robinson, *Flora and Sylva*, 1904, p. 57, with coloured plate.

A small tree, attaining 25 ft. in height and 2 ft. in girth, with thin scaly slightly furrowed bark. Young branchlets densely covered with appressed pubescence, some of which is retained in the second year. Leaflets nine to twenty-three, elliptical, rounded at the base, acute or rounded at the apex, which ends in a long slender mucro; pubescence as in *R. Pseudacacia*; stipels, $\frac{1}{10}$ in. long, often persistent; petiolules $\frac{1}{8}$ in. long, pubescent. The leaflets are somewhat smaller and bluer in tint than those of *R. Pseudacacia*.

Flowers[3] in short compact glandular long-peduncled racemes; calyx-lobes all acuminate, appressed pubescent; corolla pink or white tinged with pink; standard narrow with a yellow blotch on the inner surface. Pod about 3 in. long, glandular pubescent.

This species differs little in appearance from the common Robinia, but is readily distinguished by the densely pubescent branchlets and the persistent stipels at the base of the leaflets. In winter the branchlets retain some of the pubescence, and are often covered with a glaucous bloom. The stipules become spinescent, as in *R. Pseudacacia*, but retain traces of pubescence, and are not quite glabrous as in that species.

This species forms hybrids with *R. Pseudacacia*, which are described on p. 1500.

R. neomexicana grows on the banks of mountain streams, and is distributed from the valley of the Purgatory river in southern Colorado through northern New Mexico to the Santa Catalina and Santa Rita Mountains in Arizona, whence it extends northward to southern Utah. It ascends to 7000 feet altitude.

It was introduced[4] into cultivation in the Harvard Botanic Garden, U.S.A., in

[1] Cf. Booth, *Einführ. Ausländ. Holzarten*, 65 (1903).
[2] Quoted by Unwin, *Future Forest Trees*, 45 (1905). A plantation in France, fifty years old, yielded about 4300 cubic ft. per acre. Cf. *Bull. Soc. Forest. Franche-Comté*, x. 18 (1911).
[3] A form with white flowers is said to have been found wild in Arizona. Cf. *Mitt. Deut. Dend. Ges.* 1911, p. 423.
[4] Cf. W. J. Bean, in *Gard. Chron.* xxxv. 229 (1904). It appears to have been introduced on the Continent by Dieck, who states in his catalogue, *Neuheiten Offerte Nat. Arb. Zöschen*, 1889-1890, p. 13, that he received it from Prof. Sargent. It was soon afterwards obtained from the same source by Späth.

1882, and was sent from there to Kew in 1887. It is perfectly hardy at Kew, ripening seed, and has already attained a height of 25 feet. It is an ornamental small tree, producing beautiful pale pink flowers in June, and occasionally a second time in autumn. It flowers best in hot dry summers.   (A. H.)

In my trial ground at Colesborne a seedling procured from Kew has endured winter frosts below zero without injury, and ripens its wood in a young state better than the common Robinia. At Aldenham House a young tree 15 ft. high bore fruit in 1911, and evidently liked the hot dry summer. Mr. Vicary Gibbs considers its flowers more ornamental than those of any variety, except *R. Pseudacacia* var. *Decaisneana*.   (H. J. E.)

ROBINIA VISCOSA, Clammy Locust

Robinia viscosa, Ventenat, *Hort. Cels.* 4, t. 4 (1800); Loudon, *Arb. et Frut. Brit.* ii. 626 (1838); Sargent, *Silva N. Amer.* iii. 45, t. 115 (1892), and *Trees N. Amer.* 574 (1905).
Robinia glutinosa, Sims, in *Bot. Mag.* xvi. t. 560 (1803).

A tree, attaining 40 ft. in height and 3 ft. in girth, with thin smooth dark brown bark. Young branchlets covered with sessile and short-stalked glands, exuding a viscid matter. Leaflets nine to twenty-five, ovate or oval, rounded at the base, acute or rounded at the apex, which ends in a short mucro; pubescence similar to that of *R. Pseudacacia*; stipels slender, mostly deciduous; petiolules $\frac{1}{8}$ in. long, pubescent.

Flowers, about ten to fifteen, inodorous, in pubescent glandular short racemes; pedicels pubescent; calyx-lobes all subulately pointed, pubescent; corolla pale pink, with a narrow standard marked on the inner surface with a pale yellow blotch. Pod about 3 in. long, glandular-hispid.

This species is readily distinguished in winter by the viscid glandular twigs, which look as if oiled. The stipules are subulate, ultimately either deciduous or developing into slender short spines.

No varieties[1] of this species are known; but it forms hybrids with *R. Pseudacacia*, which are described on p. 1500.

The Clammy Locust, as it is called in America, is confined in the wild state to the mountains of North and South Carolina, where it occurs as a small tree or large shrub, often spreading by root-suckers and forming considerable thickets. It has been widely planted in the eastern United States, and is now naturalised in many places east of the Mississippi and as far north as the Canadian frontier.

It was introduced into England in 1797, and first flowered, according to Sims, in Mr. Whiteley's garden at Old Brompton in 1800. It is occasionally seen in botanic gardens, as at Kew, where there are several old specimens 25 to 35 ft. in height, which are probably the finest in Europe. This species is much slower in growth than *R. Pseudacacia*, but is ornamental on account of its pink flowers, which are produced in June and July.   (A. H.)

---

[1] In Loudon, *Gard. Mag.* xvii. 391 (1841), it is stated that Vilmorin found at Verrières seedlings of this species with branchlets, which were not viscid.

ROBINIA HISPIDA, Rose Acacia

Robinia hispida, Linnæus, *Mantissa*, 101 (1767) ; Curtis, *Bot. Mag.* t. 311 (1795) ; Loudon, *Arb. et Frut. Brit.* ii. 627 (1838) ; Sargent, *Silva N. Amer.* iii. 37 (1892).
Robinia hispida-rosea, Loiseleur, in *Nouv. Duham.* ii. 64 (1804).
Robinia rosea, Loiseleur, *op. cit.* t. 18 (1804).
Pseudacacia hispida, Moench, *Meth.* 145 (1794).

A shrub, attaining about 10 ft. in height. Young branchlets covered with dense pubescence, interspersed with glandular bristles, which persist in the second year. Leaf-rachis with similar pubescence and bristles. Leaflets, seven to eleven, larger than in the other species, 1 to 2 in. long, oval, rounded at the base, rounded or occasionally slightly cuspidate at the apex, which is tipped with a long pubescent mucro ; glabrous, except for slight pubescence on the midrib beneath ; stipels persistent ; petiolule, $\frac{1}{8}$ to $\frac{1}{4}$ in. long, pubescent and glandular-hispid.

Flowers few, five to nine, in loose pendulous racemes, pink, without odour ; calyx-lobes all ending in long subulate teeth ; peduncles, axis, pedicels, and calyx covered with white pubescence interspersed with glandular bristles. Fruit,[1] about 2 in. long, slightly constricted between the seeds, covered with glandular bristles. The stipules are usually deciduous, but occasionally develop on the older branchlets as minute blunt spines.

Two forms of this species and a probable hybrid are known :—

1. Var. *typica*, Schneider, *Laubholzkunde*, ii. 81 (1907), described above. This is characterised by numerous glandular bristles on the branchlets and leaf-rachis.

2. Var. *macrophylla*, De Candolle, *Prod.* ii. 262 (1825).

> Var. *inermis*, Kirchner, *Arb. Musc.* 373 (1864).
> *Robinia macrophylla*, Schrader, *ex* De Candolle, *loc. cit.*

Glandular bristles few or none on the branchlets and leaf-rachis, but present on the peduncles and calyx of the flower and on the pod. Leaves and flowers larger than in the typical form. Var. *macrophylla* appears to have arisen in cultivation in Europe.[2]

3. *Robinia Kelseyi*, Cowell, in Bailey, *Cycl. Amer. Hort.* 1538 (1902) ; Hutchinson and Bean, in *Bot. Mag.* t. 8213 (1908).

A small shrub, which originated[3] in H. P. Kelsey's nursery at Boston, U.S.A.

---

[1] Mr. T. Meehan, who studied this plant in the wild state in Tennessee, says that it produces fruit very rarely, and is usually reproduced in the forests by root-suckers. He procured two pods in 1894, one of which he sent to the Kew herbarium. Cf. *Kew Bulletin*, 1893, p. 341, and Nicholson, in *Gardeners' Magazine*, 1894, p. 118. Carrière, in *Rev. Hort.* xxxix. 431, fig. 38 (1867), describes imperfect fruits which were produced on a tree at Paris.

[2] Cf. Carrière, *Prod. et Fixat. Vars.* 54 (1865), and in *Rev. Hort.* liv. 109 (1972).

[3] Mr. Kelsey in a letter to Kew says that it came up spontaneously in his nursery ; but supposes that it may have come into his collection with seed of other plants from the southern Alleghany Mountains.

With *R. Kelseyi* should be compared *R. Boyntoni*, Ashe, in *Journ. Elis. Mitch. Soc.* xiv. part ii. p. 53 (1897), which I have not seen. This is identified by Schneider, *Laubholzkunde*, ii. 82, note (1907) with *R. hispida*, var. *rosea*, Pursh. *Fl. Amer. Sept.* ii. 488 (1814) ; and is said to occur in North Carolina, Tennessee, Georgia, and Alabama. *R. Boyntoni*, according to Ashe's description, is very similar to *R. Kelseyi*, but with glabrous fruit. Small, *Flora South-eastern U.S.* 613 (1903), however, says that the pod is hispid.

It is probably a hybrid between *R. hispida* and *R. Pseudacacia*, resembling the latter in the slightly pubescent and usually non-glandular[1] branchlets, and in the small size of the leaflets, which are seven to eleven in number, $\frac{1}{2}$ to $1\frac{1}{2}$ in. long, glabrous, lanceolate, and tipped with a conspicuous mucro. The flowers are similar to those of *R. hispida*, but smaller; and are pink in colour, few (about five to eight) in the raceme,—with the peduncle, axis, pedicels, and calyx more or less covered with stalked glands, all the calyx-lobes being subulately pointed. Fruit reddish, $1\frac{1}{2}$ to 2 in. long, densely covered with glandular hairs; seeds three to four. Stipular spines are present on the branchlets, but are feebly developed.

This was introduced into commerce about 1901, and shrubs planted at Kew in 1903 have flowered and produced fruit.[2] It is not so rank in growth as *R. hispida*, and is less liable on this account to be injured by the wind. According to Mr. Bean, it can be readily propagated by grafting on the roots of *R. Pseudacacia*.

R. hispida, commonly known as the rose acacia, occurs in the mountains of Virginia, Kentucky, Tennessee, North and South Carolina, Alabama and Georgia.

It was introduced[3] into cultivation in England by Sir John Colliton at Exmouth, in 1741. It is a handsome shrub with beautiful flowers, produced in June and July. It is usually grafted on the common Robinia; but Nicholson says that as it is readily propagated by root-cuttings, this mode of reproduction is preferable.

(A. H.)

[1] The branchlets bear occasionally one or two stalked glands.
[2] Cf. *Gard. Chron.* xlvii. 391, fig. 177 (1910), where the plant is figured. It obtained an award of merit at the Royal Horticultural Society on June 7, 1910.
[3] Loudon, *Gard. Mag.* 1841, p. 637.

GLEDITSCHIA

Gleditschia, Clayton, *ex* Linnæus, *Gen. Pl.* 480 (1742); Willdenow, *Berl. Baumz.* 132 (1796);
Spach, *Hist. Veg.* i. 90 (1834); Bentham et Hooker, *Gen. Pl.* i. 568 (1865); Maximowicz,
Mél. Biol. xii. 450 (1886); Rehder, in Bailey, *Cycl. Amer. Hort.* 650 (1900).
Gleditsia, Linnæus, *Sp. Pl.* 1056 (1753); Schneider, *Laubholzkunde*, ii. 8 (1907).

DECIDUOUS trees belonging to the division Cæsalpineæ of the order Leguminosæ.
Stem and branches often armed with stout rigid sharp long-pointed spines, which
are simple or branched. These spines are abortive branches, which arise above the
axils of the leaves, being developed from buds embedded in the bark. Buds of the
leafy branches, axillary, no true terminal bud being formed, concealed in the tissue
of the twig, multiple, three or four superposed in a vertical line, the two or three
lower without scales and covered by the base of the petiole in summer and by the
leaf-scar in winter; the uppermost bud, larger than the others, covered with minute
scales.

Leaves alternate or fascicled, long-stalked, equally pinnate or bipinnate, some
of the pinnæ in the latter often reduced[1] to simple leaflets. Leaflets membranous, opposite or alternate, crenulate in margin, without stipels. Stipules leafy, early
deciduous.

Flowers polygamous, diœcious, or perfect; regular, minute, greenish white, in
axillary simple or spicate racemes, with scale-like caducous bracts. Calyx campanulate, lined with a disc, three- to five-lobed. Petals as many as the lobes of the calyx.
Stamens, six to ten, inserted with the petals on the margin of the disc, exserted;
filaments free, erect; anthers abortive in the pistillate flowers; ovary usually one-
celled, rudimentary or absent in the staminate flowers; style short, with a dilated
terminal stigma; ovules two or many.

Pod coriaceous, compressed; either elongated, containing pulp, many-seeded,
and indehiscent; or short, without pulp, one- or two-seeded, and ultimately dehiscent.
Seed oval or orbicular, more or less compressed, attached by a long slender
funicle.

About a dozen species of Gleditschia are known, natives of North America,
Caucasus, North Persia, Japan, China, Formosa, the Philippines, Celebes, Tropical
Africa, and Argentina.

The following key comprises the species cultivated in Europe, which I have
been able to identify. The leaflets described are those of the simply pinnate leaves
on adult trees.

---

[1] Leavitt, in *Bot. Gaz.* xlvii. 49, fig. 14 (1909), describes and figures these peculiar leaves, occurring in *G. triacanthos.*

Gleditschia 1511

I. *Pods elongated, many-seeded, indehiscent.*
* *Leaflets, eight to twelve, prominently reticulate beneath.*
(a) *Branchlets pubescent.*
1. *Gleditschia Delavayi*, Franchet.   South-western China.   See p. 1513.
Leaflets ovate-oblong, 2 to 2½ in. long, ¾ to 1½ in. wide, almost entire in margin.   Pod, 16 to 20 in. long, flattened, twisted, not dotted with pits.
(b) *Branchlets glabrous.*
2. *Gleditschia sinensis*, Lamarck.   North China.   See p. 1513.
Leaflets ovate-lanceolate, 1¼ to 2½ in. long, ⅝ to 1 in. wide, crenulate.   Pod, 6 to 10 in. long, convex on both surfaces, dotted with minute pits.
3. *Gleditschia macracantha*, Desfontaines.   Central China.   See p. 1514.
Leaflets ovate-oblong, 2 to 3 in. long, 1 to 1½ in. wide, crenulate.   Pod, 6 to 12 in. long, slightly convex on both surfaces, dotted with minute pits.

** *Leaflets twelve to twenty-eight, not prominently reticulate beneath. Branchlets glabrous.*[1]   *Pods flattened, averaging 12 in. long, not dotted with pits.*
4. *Gleditschia triacanthos*, Linnæus.   North America.   See p. 1517.
Leaf-rachis pubescent on all sides.   Leaflets lanceolate, 1 to 1½ in. long, ⅜ to ⅝ in. wide, with dense conspicuous pubescence on the midrib beneath.
5. *Gleditschia caspica*, Desfontaines.   South-west coast of the Caspian Sea.   See p. 1512.
Leaf-rachis pubescent only on the outer edge of the groove.   Leaflets narrowly elliptical or lanceolate, 1¼ to 1½ in. long, ⅝ to ¾ in. wide; glandular on the midrib above; shining and quite glabrous beneath.   Young branchlets green.
6. *Gleditschia japonica*, Miquel.   Japan, Manchuria, North China.   See p. 1516.
Leaf-rachis as in *G. caspica*.   Leaflets similar, but narrower and not glandular on the midrib above.   Young branchlets purplish.

II. *Pods short, one- to two-seeded, dehiscent.*
7. *Gleditschia aquatica*, Marshall.   North America.   See p. 1520.
Leaf-rachis pubescent only on the outer edge of the groove.   Leaflets, ten to twenty, lanceolate, about 1 in. long and ⅓ in. wide, distinguishable from the other species by their small size, and by being quite glabrous on both surfaces and only slightly ciliate on the margin.   Young branchlets glabrous.

In addition to the species enumerated in the key, mention may be made here of a peculiar Gleditschia, growing in the nursery of Messrs. Simon-Louis at Plantières, which has been identified by Schneider, *Laubholzkunde*, ii. 10 (1907), with *Gleditschia ferox*, Desfontaines, *Hist. Arb.* ii. 247 (1809).   This tree has remarkably stout, compressed, reddish brown spines, which attain on the trunk 10 in. in length, with branches 3 to 5 in. long; on the branchlets, the spines are about 1½ in. long, bearing two short lateral opposite smaller spines.   The foliage on the specimen which I have seen, is mostly bipinnate, with six to ten pinnæ, each bearing twenty or more

[1] The branchlets of *G. triacanthos* are usually pubescent close to the base; but are glabrous elsewhere.

VI

2 A

small leaflets. This tree has never borne fruit, and cannot be identified with any species known in the wild state. Whether it is a hybrid, which originated at an early period in France, or an unknown species, is uncertain.[1] A specimen is growing at Colesborne, where it is quite hardy, but does not seem likely to attain a tree-like habit. (A. H.)

GLEDITSCHIA CASPICA

Gleditschia caspica, Desfontaines, *Hist. Arb.* ii. 247 (1809); Spach, *Hist. Vég.* i. 97 (1834); Loudon, *Arb. et Frut. Brit.* ii. 655 (1838); Boissier, *Fl. Orient.* ii. 631 (1872).
Gleditschia horrida, Makino, var. *caspica*, Schneider, *Laubholzkunde*, ii. 12 (1907).

A tree, attaining about 35 ft. in height. Young branchlets glabrous. Leaves simply pinnate or bipinnate with six to eight pairs of pinnæ; leaflets of the pinnate leaves, fourteen to twenty, narrowly elliptical or lanceolate, about $1\frac{1}{4}$ to $1\frac{1}{2}$ in. long, and $\frac{5}{8}$ to $\frac{3}{4}$ in. wide, rounded at the base, truncate or emarginate with a short mucro at the apex; upper surface with scattered short hairs on the midrib and nerves, and numerous minute glands on the midrib; lower surface glabrous, shining; margin crenulate, ciliate; stalklet very short, broad, pubescent above, glabrous beneath; rachis pubescent on the outer edge of the broad glandular often winged groove, elsewhere glabrous.

Flowers sub-sessile in slightly pubescent racemes or spikes; ovary glabrous. Pod thin and flattened, indehiscent, 9 to 12 in. long, 1 to $1\frac{1}{4}$ in. broad, with a sweet edible pulp,[2] straight or falcate, dark chestnut brown, smooth without dotted pits, slightly pubescent. Seeds numerous, close to the upper suture of the pod, oval, compressed, about $\frac{1}{2}$ in. long, shining, dark brown, marked on the surface with transverse raised lines.

The thorns on the branches in this species are chestnut brown, flattened or compressed in section, usually bipinnately branched, the primary branches coming off in different planes and bearing two or three lateral thorns. The thorns on the stem are much branched, and often very formidable.

G. caspica is limited to the wooded districts along the south-west coast of the Caspian Sea in Ghilan and Talysch. It is a rare tree, never attaining a large size, Medwedew[3] giving its height as 30 to 35 ft. It ascends in the valleys to about 500 ft. altitude. The wood is used in the construction of mills, as it is durable under water. Cattle and wild pigs are fond of the sweetish pods.

G. caspica was introduced into England about 1822; and is usually seen as a small tree in botanic gardens,[4] where its remarkable spiny trunk renders it an object of interest. It flowers, and occasionally produces fruit at Cambridge, where there are two trees of different sexes. (A. H.)

---

[1] Rehder, in Bailey, *Cycl. Amer. Hort.* 650 (1900), states that *G. ferox* is often cultivated; but all the specimens which we have received from the Continent with this name, except the one from Simon-Louis, are referable to *G. caspica*.

[2] Cf. Hohenacker, in *Bull. Soc. Mosc.* iv. 351 (1838).

[3] Cf. Köppen, *Holzgewächse Europ. Russlands*, i. 236 (1888), and Radde, *Pflanzenverb. Kaukasusländ.* 185, 189, 198 (1899).

[4] Bean, in *Kew Bull.* 1908, p. 400, records a handsome tree, 40 ft. high, in the Schönbrunn Botanic Garden near Vienna.

GLEDITSCHIA DELAVAYI

Gleditschia Delavayi, Franchet, *Pl. Delav.* 189 (1889); Schneider, *Laubholzkunde*, ii. 11 (1907).

A tree, attaining in Yunnan 40 ft. in height. Young branchlets covered with a dense minute pubescence. Leaves on adult trees simply pinnate; leaflets ten to fourteen, oblong-ovate, 2 to 2½ in. long, and ¾ to 1½ in. broad; rounded or emarginate, and minutely apiculate at the apex; almost entire in margin, the crenations being few and indistinct; upper surface slightly pubescent on the midrib and veins; lower surface prominently reticulate, slightly pubescent on the midrib, elsewhere glabrous; stalklets short, stout, pubescent; rachis covered on all sides with a dense minute pubescence.

Flowers with short pedicels, perfect, in simple loose pubescent racemes; ovary glabrous. Pod indehiscent, 16 to 20 in. long, and 2 to 2½ in. wide, flattened, twisted, glaucous, dark brown, slightly pubescent, not dotted with pits on the surface; seeds numerous, close to the upper suture.

The spines on the branchlets are terete, pubescent, simple or two- to three-forked.

G. Delavayi is readily distinguishable from all the other species by its pubescent branchlets and very large pods. It is a native of south-western China, where it was found in Yunnan by Delavay and myself at elevations of 4000 to 6000 ft. It was introduced into England by E. H. Wilson, who sent seeds in 1900 from Yunnan to Messrs. Veitch; but the plants raised from this consignment in the Coombe Wood nursery died in the winter of 1905-1906. Other plants raised two or three years afterwards from seed sent by Wilson from the Min valley in Szechwan still survive at Coombe Wood, as well as small plants at Kew and Cambridge.

This species is also in cultivation at Verrières near Paris.   (A. H.)

GLEDITSCHIA SINENSIS

Gleditschia sinensis,[1] Lamarck, *Encycl.* ii. 465 (1786); Bunge, *Enum. Pl. Chin. Bor.* 21 (1831); Loudon, *Arb. et Frut. Brit.* ii. 654 (1838); Hemsley, in *Journ. Linn. Soc.* (*Bot.*) xxiii. 209 (1887); Schneider, *Laubholzkunde*, ii. 9 (1907).
Gleditschia horrida, Willdenow, *Sp. Pl.* iv. 2, p. 1098 (1806) (not Makino).
Gleditschia xylocarpa, Hance, in *Journ. Bot.* xxii. 366 (1884).
Gymnocladus Williamsii, Hance, *loc. cit.*

A tree, attaining in China 40 feet in height. Young branchlets glabrous. Leaves usually simply pinnate; leaflets eight to twelve, ovate-lanceolate, 1¼ to 2½ in. long, ⅝ to 1 in. broad; tapering to a rounded or truncate apiculate apex; upper surface minutely pubescent, but usually becoming glabrescent except on the midrib;

[1] *G. sinensis*, Bentham, *Fl. Hongk.* 100 (1861), and Maximowicz, in *Mél. Biol.* xii. 453 (1886), is not Lamarck's species; and is referable to *G. australis*, Hemsley.

lower surface, with scattered minute pubescence, which usually soon disappears, and with prominent reticulate venation; margin crenulate and ciliate; rachis pubescent on the edges of the groove; petiolule pubescent.

Flowers, on pedicels about $\frac{1}{8}$ in. long, in loose pubescent racemes; ovary pubescent on the borders. Pod indehiscent, 6 to 10 in. long, $\frac{3}{4}$ to $1\frac{1}{4}$ in. broad, swollen and convex on both surfaces, glaucous, glabrous, dark purplish brown, marked on the surface with dot-like pits. Seeds numerous, placed in the centre of the pod, oval, about $\frac{2}{5}$ in. long, convex on both surfaces, pointed at the apex, smooth, shining, brown, marked with transerve lines.

The spines on the branchlets are terete, usually with one or two lateral smaller spines. A variety[1] without spines is said to have arisen in Camuzet's nursery in 1823.

G. *sinensis* was described in 1786 by Lamarck from a tree, which had been cultivated for nine years in the Royal Garden at Versailles. It was said to have been raised from seed[2] received from China; and Lamarck adds that the pods were somewhat cylindrical and contained globose seeds. In Gay's herbarium at Kew, there is a specimen, labelled G. *sinensis*, Lamarck, which was gathered in 1822 from a tree, probably the original one, at the Trianon; and this specimen may be regarded as an undoubted type of Lamarck's species.[3] It is identical with a species of Gleditschia which occurs wild on the mountains near Peking, and is cultivated farther south, as at Shanghai and Ningpo. The pods, which are called *tsao-chia*, are used as soap for washing clothes by the Chinese.

Loudon reported large trees of this species at Syon and the Mile End Nursery in 1838; but it is doubtful if these were accurately named; and they cannot now be found. We have not seen a single example in Britain. It is not common even in France, though there are large specimens at Verrières[4] near Paris, and at Montpellier.　　　　　　　　　　　　　　　　　　　　　　　　　　　　(A. H.)

GLEDITSCHIA MACRACANTHA

Gleditschia macracantha, Desfontaines, *Hist. Arb.* ii. 246 (1809); Loudon, *Arb. et Frut. Brit.* ii.
　　654 (1838).
Gleditschia Fontanesii, Spach, *Hist. Vég. i.* 95 (1834).
Gleditschia officinalis,[5] Hemsley, in *Kew Bulletin*, 1892, p. 82.

A tree, attaining in central China 50 ft. in height. Young branchlets glabrous. Leaves simply pinnate, with six to twelve leaflets, which are variable in size; those

---

[1] De Vries, *Plant Breeding*, 617 (1906).

[2] Bretschneider, *Hist. Europ. Bot. Disc. China*, 49, 52, 77 (1898), supposes that the seed was sent from Peking by Père D'Incarville; but the latter died in 1757, and Lamarck's account, though not quite clear, seems to indicate that the seeds were received about 1777.

[3] G. *sinensis* has been much confused with other Chinese species by various authors. There are good specimens in the British Museum, gathered by Bretschneider and Williams near Peking.

[4] Cf. *Hortus Vilmorinianus*, 19 (1906), where a tree of this species, wrongly identified as G. *caspica*, is reported to be 50 ft. high and 5 ft. in girth. It was probably planted in 1825.

[5] This species was founded on specimens, bearing unripe and aborted pods, which were gathered on 26th July 1888.

towards the base of the shoot, very large, ovate-oblong,[1] 2 to 3 in. long, and 1 to 1½ in. broad ; those towards the apex of the shoot, smaller, ovate-lanceolate, 1½ in. long and ¾ in. broad ; glabrous, except on the midrib above, which is pubescent throughout, and on the midrib beneath, which bears a few scattered hairs ; margin crenulate ; lower surface prominently reticulate ; rachis pubescent on the edges of the groove ; petiolules pubescent.

Flowers in loose pubescent simple racemes ; pedicels slender, pubescent, ¼ to ½ in. long ; ovary pubescent. Pod indehiscent, 6 to 12 in. long, 1 to 1½ in. broad, straight or curved, not twisted, slightly convex on both surfaces, not so swollen as in *G. sinensis*, and not so flattened as in *G. triacanthos*, dark purplish brown, glabrous, shining, often glaucous, dotted with minute pits. Seeds numerous, placed towards the centre of the pod, oval, about ½ in. long, convex on both surfaces, shining brown, marked with transverse raised lines, becoming deeply pitted when dried in the immature state.

The spines on the branchlets are terete, conical, reddish brown, simple or with one or two small lateral thorns.

This species is closely allied to *G. sinensis*, but appears to differ in the shape of the leaflets, which are larger, broader, and more prominently reticulate ; and has pods, which can be readily distinguished. In the Kew herbarium, there is a branch with a pod, gathered by Gay in 1814 from a tree in M. Morel's garden near Paris. This tree is stated by Gay to have been identical with a tree of *G. macracantha* in the Jardin des Plantes, doubtless the one on which Desfontaines founded this species. Gay's specimen may then be accepted with certainty as a type of *G. macracantha*.

G. macracantha occurs in the wild state in the mountains of central China,[2] where it was found by Dr. Faber and myself in the provinces of Hupeh and Szechwan. Young unripe and aborted pods are gathered by the Chinese, and used as a drug, known as *ya-tsao*.[3]

G. macracantha was probably introduced into France by seed sent from China by the missionaries towards the end of the 18th century. A large tree of this species is said by Loudon to have been growing at Syon in 1838 ; but little reliance can be placed on his identification. It is apparently now unknown in cultivation in England ; but it thrives well in the south of France and in Italy, whence we have obtained specimens from Montpellier and La Mortola. (A. H.)

---

[1] On one of the trees at Montpellier, the large leaflets are occasionally obovate, a shape not seen in any other species.

[2] A specimen gathered by Ford's collector in 1888 on Lantao Island, near Hong-Kong, is probably this species.

[3] Cf. Hanbury, *Science Papers*, 248, 1876. These pods, of which there are specimens from China in the London Pharmaceutical Museum, appear to be aborted and contain no seeds. Similar pods have been produced on a solitary tree at La Mortola, which bears only pistillate flowers, the anthers being absent or sterile.

GLEDITSCHIA JAPONICA

Gleditschia japonica, Miquel, in *Ann. Mus. Bot. Ludg. Bat.* iii. 54 (1867); Franchet et Savatier,
  Enum. Pl. Jap. i. 114 (1875) and ii. 327 (1879); Maximowicz, *Mél. Biol.* xii. 452 (1886);
  Sargent, in *Garden and Forest*, vi. 163, t. 27 (1893), and *Forest Flora Japan*, 35, t. 11 (1894);
  Shirasawa, *Icon. Ess. Forest. Japon*, i. text, 87, t. 51, figs. 15 to 30 (1900).
Gleditschia horrida,[1] Makino, in *Tokyo Bot. Mag.* xvii. 12 (1903); Schneider, *Laubholzkunde*, ii. 11
  (1907).
Fagara horrida, Thunberg, in *Trans. Linn. Soc.* ii. 329 (1794).

A tree, attaining in Japan 60 to 70 ft. in height, with a trunk occasionally 3 ft. in diameter. Young branchlets glabrous, purplish. Leaves similar to those of *G. caspica*, but with the leaflets apparently narrower, always lanceolate, tapering to a rounded or acute (not emarginate) apiculate apex, and without glands on the midrib of the upper surface; lower surface slightly pubescent on the midrib.

Pods thin, flattened, about 10 to 14 in. long, falcate, often twisted, shining or glaucous, brown, glabrescent, without dot-like pits on the surface. Seeds numerous, situated towards the upper suture of the pod, oval, flattened, about $\frac{1}{3}$ in. long, marked with transverse lines.

This species is closely allied to, if not really identical with *G. caspica*; but the material examined has been insufficient for determination. The purplish colour of the young branchlets in cultivated specimens of *G. japonica* may not be a constant character.

This species is a native of Japan, Manchuria, Korea,[2] and north China. In Japan, it is rare towards the north, where it is occasionally seen on the banks of rivers at no great altitude. It is more abundant and attains its largest size on the banks of the Kisogawa and other streams in the centre of the main island at elevations of about 2000 ft., where it thrives best in rich alluvial soil, though it is often seen on gravelly slopes. The pulp of the pods is much used for washing cloth by the Japanese, who call the tree *Saikachi*.

In Manchuria, it is said by Komarov[3] to be wild in the neighbourhood of Mukden; and there are specimens in the Kew herbarium collected by Ross and Webster in other districts. Bretschneider also sent this species to Kew from the neighbourhood of Peking.

The only specimen which we have seen in England is a tree at Kew, about 15 ft. high, obtained from Yokohama in 1894. A young tree at Verrières, raised from Chinese seed, appears to belong to this species.   (A. H.)

---

[1] This name is inadvisable, as *G. horrida*, Willdenow, *Sp. Pl.* iv. 2. p. 1098 (1806), is a synonym of *G. sinensis*, Lamarck.

[2] Cf. T. Nakai, in *Journ. Sci. Coll. Tokyo*, xxvi. 142 (1910), who records a variety of this species without spines from Korea. His *G. caspica* from the same region is doubtless *G. japonica*.

[3] In *Act. Hort. Petrop.* xxii. 566 (1904).

GLEDITSCHIA TRIACANTHOS, Honey Locust

Gleditschia triacanthos, Linnæus, *Sp. Pl.* 1056 (excl. β) (1753); Loudon, *Arb. et Frut. Brit.* ii. 650
(1838); Sargent, *Silva N. Amer.* iii. 75, tt. 125, 126 (1882), and *Trees N. Amer.* 556 (1905);
Schneider, *Laubholzkunde*, 12 (1907).
Gleditschia spinosa, Marshall, *Arb. Amer.* 54 (1785).
Gleditschia meliloba, Walter, *Fl. Carol.* 254 (1788).
Gleditschia elegans, Salisbury, *Prod.* 323 (1796).
Gleditschia brachycarpa,[1] Pursh, *Fl. Amer. Sept.* i. 221 (1814).
Gleditschia heterophylla, Rafinesque, *Fl. Ludovic.* 99 (1817).

A tree, attaining in America 70 to 140 ft. in height and 10 to 20 ft. in girth. Bark divided by deep fissures into narrow longitudinal scaly ridges. Young branchlets green, slightly pubescent at the base below the insertions of the lower leaves, elsewhere glabrous; shining brown and glabrous in the second year. Leaves simply pinnate or bipinnate, with four to seven pairs of pinnæ; leaflets of the pinnate leaves, 14 to 28, lanceolate, averaging 1 to $1\frac{1}{2}$ in. long, and $\frac{3}{8}$ to $\frac{5}{8}$ in. wide, minutely apiculate at the rounded or acute apex; margin crenulate and ciliate; both surfaces pubescent with scattered curved hairs, which are dense and conspicuous on the yellowish green midrib beneath, and on the short stout stalklets; rachis with a narrow groove, and densely pubescent on all sides; venation inconspicuous and not prominent on the lower surface.

Flowers with short pedicels; staminate flowers numerous in usually clustered pubescent racemes; pistillate flowers few in solitary racemes. Pod, 12 to 16 in. long, 1 to $1\frac{1}{2}$ in. broad, indehiscent, thin and flattened; straight or falcate, often twisted; shining dark brown without dot-like pits, pubescent; inner coat thin and papery. Seeds numerous, placed close to the upper suture of the pod, separated by a succulent pulp, oval, about $\frac{1}{3}$ in. long, compressed, marked with a few transverse lines, and becoming pitted on drying.

The spines on the branchlets are conical, terete, unbranched or three-forked, the branching being simply pinnate in one plane; their colour is reddish at first, becoming ultimately shining dark brown.

The seedling[2] has oval-oblong sessile cotyledons, and bears in the first year pinnate leaves with about 20 very small leaflets.

VARIETIES AND HYBRID

1. Var. *inermis*, Pursh, *Fl. Amer. Sept.* i. 221 (1814).
Spines entirely absent or feebly developed.[3] According to Loudon, abundance

---

[1] Said to differ from the type in bearing short pods. The pods, however, are variable in size on individual trees; and this cannot be maintained even as a distinct variety.

[2] Cf. Tubeuf, *Samen, Früchte u. Keimlinge*, 127 (1891).

[3] Sargent, *Silva N. Amer.* iii. 75, note 3 (1882), states that trees growing under conditions where they have been freely exposed to light most frequently develop spines; while those growing in the forest, shaded by other trees, are often unarmed.

of plants without spines may be selected from beds of seedlings of the ordinary form.

2. Var. *Bujoti*, Rehder, in Bailey, *Cycl. Amer. Hort.* 650 (1900).

Gleditschia Bujoti, Neumann, in *Rev. Hort.* iv. 205 (1845).

Branches pendulous; leaflets narrower than in the typical form.

3. Var. *nana*, Hort. Kew.   A small round-headed tree of compact habit, with dark green foliage, and leaflets shorter and broader than in the type.   This, of which there is a living example at Kew, appears to be identical with *G. ferox*, var. *nana*, Rehder, in Bailey, *Cycl. Amer. Hort.* 650 (1900); and is possibly *G. sinensis*, var. *nana*, Loudon, *Arb. et Frut. Brit.* ii. 654 (1838).

4. This species probably hybridizes with *G. aquatica*.   Schneck, in *Plant World*, vii. 252 (1904), states that he found several trees of evidently hybrid origin growing on the edge of a swamp on Mt. Carmel, Illinois.   These bore pods about 5 in. long and $1\frac{1}{2}$ in. broad which were entirely destitute of pulp.

G. texana, Sargent, in *Bot. Gaz.* xxxi. 1 (1901), *Silva N. Amer.* xiii. 13, t. 627 (1902), and *Trees N. Amer.* 557 (1905), bears similar pods, and is said to be only found as a single grove of large trees, growing in alluvial soil along the Brazos river, near Brazoria in Texas.   As both *G. triacanthos* and *G. aquatica* are found in the valley of the Brazos river, it is probable that *G. texana* is a hybrid between these two species.   There are similar pods, without foliage, in the Kew Herbarium, which are labelled " *Gleditschia brachyloba*,[1] Mississippi Banks, Nuttall."

A small plant of *G. texana*, obtained from the Arnold Arboretum in 1900, is now growing at Kew.   It differs from native specimens in having spines on the branchlets.   (A. H.)

DISTRIBUTION AND CULTIVATION

G. triacanthos is a native of North America, extending from the western slope of the Alleghany Mountains in Pennsylvania westward through Ontario and Michigan to Minnesota, Nebraska, Kansas, and Indian Territory, and southward to Alabama, Mississippi, and the valley of the Brazos river in Texas.   It usually grows on the borders of streams or in valleys in moist fertile land, either singly or in rare cases covering considerable areas almost exclusively, and occasionally occurs on dry gravelly hills.   Sargent states that it is now often found naturalised in the region east of the Alleghany Mountains.   It attains its largest size in Indiana and Illinois, where Ridgway[2] states that it was formerly one of the most majestic trees of the forest.   Many were 120 to 140 ft. high, with straight trunks clear of branches to 50 or 70 ft., and 4 to 6 ft. in diameter.   No trees except the Cypress and Catalpa had a more characteristic appearance, its tall straight but usually inclined trunk, of a dark grey or nearly black colour, being conspicuous at a distance, while the delicate foliage made its top contrast with other species.   He[3] gives 156 ft. as its maximum height, on the authority of Dr. Schneck, who showed me the remains of

---

[1] *G. brachycarpa*, Pursh, is probably meant, the specimens being collected by Nuttall.
[2] *Proc. U.S. Nat. Mus.* 1882, p. 64.   [3] Ridgway, *Additional Notes*, 419.

these wonderful woods in 1904, when I could find no tree of this species larger than about 110 ft. by 9 ft.

G. triacanthos is a very ornamental tree, with fine foliage, which turns a bright golden yellow colour in autumn. It was first cultivated in England at Fulham, where a tree planted by Bishop Compton about 1700 bore fruit[1] in 1729. Though the seeds rarely if ever ripen in England, I have easily raised seedlings from French seed by soaking them in water before sowing. The seedlings do not ripen their wood in the open, and should be kept in a frame for two or three years. This species does not appear to be long lived in our climate, as most of the old trees mentioned by Loudon in 1838 cannot now be found. He figures one at Syon 57 ft. high, with a trunk 3 ft. in diameter; but the largest tree[2] there now was only 46 ft. by 2 ft. in girth in 1910.

In England one of the finest trees is growing near the Palm House at Kew. It measures 54 ft. by 6 ft. 10 in., and bears fruit occasionally. At Fawley Court a fine tree with flaky bark, like that of a tree at Verrières, and with very few spines, was 74 ft. by 5 ft. 10 in. in 1905. An old tree at Arley Castle[3] measured 51 ft. by 5 ft. 5 in. in 1904. At Belton, Grantham, Miss Woolward reports a tree 47 ft. by 4 ft. in 1907.

Other large trees measured by me are at Bisterne Park, Hants, 62 ft. by 8 ft., with a bole of 30 ft., decaying in 1906; at Ham Manor, Sussex, 60 ft. by 4 ft. 7 in. in 1907; at Stowe, Bucks, 53 ft. by 4 ft. 10 in. in 1905; in the grounds of Wadham College, Oxford, 53 ft. by $4\frac{1}{2}$ ft. in 1907; and one of the same size at Bibury Rectory, Gloucestershire, in 1908.

In Wales the only large tree I have seen is a very fine one at Golden Grove, 65 ft. by $4\frac{1}{2}$ ft., a healthy tree with very few spines.

In Scotland and Ireland we have not seen any trees of note.

This species attains a large size, and produces fruit with good seed regularly in the south of France and in Italy. In the neighbourhood of Parma it is used for making formidable hedges. At Montpellier there is a fine specimen, narrowly pyramidal in habit, very different in aspect from the usual form seen in cultivation. In the Bishop's garden at Beauvais I measured in 1908 a tree 80 ft. by 7 ft.; and at the Château de Geneste, near Bordeaux, in 1909, one 80 ft. by 8 ft. At Schloss Dyck, in Germany, an even larger tree is reported,[4] measuring about 90 ft. by 7 ft.

TIMBER

The wood is heavy, hard, and strong, with a coarse open grain, reddish brown at the heart with yellowish white sapwood. Being durable in the soil, it is used in the United States for posts and rails and to a small extent for wheel hubs; but I have never seen or heard of it in the English timber market, and it is not likely to have

[1] Cf. *London Catalogue of Trees*, 87, plate 21 (1730).
[2] A. B. Jackson, *Cat. Trees Syon House*, 15 (1910).
[3] R. Woodward, jun., *Cat. Trees Arley Castle*, 15 (1907).
[4] *Mitt. D. D. Ges.* 1904, p. 18.

any special commercial value.   The best specimen of it which I have seen was cut down at Wretham Rectory, Norfolk, in 1884, when it produced a log 30 ft. long, containing 69 cubic feet of timber.   From it a bookcase of very nice appearance was made.   (H. J. E.)

GLEDITSCHIA AQUATICA

Gleditschia aquatica, Marshall, *Arb. Amer.* 95 (1785); Sargent, *Silva N. Amer.* iii. 79, t. 127 (1892), and *Trees N. Amer.* 558 (1905).
Gleditschia inermis, Miller, *Dict.* ed. 8, No. 2 (1768); Schneider, *Laubholzkunde*, ii. 13 (1907).
Gleditschia caroliniensis, Lamarck, *Encyc.* ii. 465 (1786).
Gleditschia monosperma, Walter, *Fl. Carol.* 254 (1788); Loudon, *Arb. et Frut. Brit.* ii. 653 (1838).

A tree, attaining in America 50 to 60 ft. in height, with a short trunk occasionally 8 ft. in girth.   Young branchlets glabrous.   Leaves simply pinnate, or bipinnate with three or four pairs of pinnæ; leaflets of the simple pinnate leaves, ten to twenty, lanceolate, about 1 in. long and $\frac{1}{3}$ in. wide, gradually narrowing to an obtuse apex, which is usually without any distinct mucro; margin crenulate, slightly ciliate; both surfaces shining, glabrous; stalklet slender, slightly pubescent; rachis pubescent on the edge of the narrow groove, elsewhere glabrous.

Pod thin and flattened, without pulp, dehiscent; either one-seeded, rhomboid, 1 to $1\frac{1}{2}$ in. long, or two-seeded, with the edges nearly parallel, about $2\frac{1}{4}$ in. long; reddish brown, glabrous, and without dotted pits on the surface.   Seed orbicular, flattened, about $\frac{1}{3}$ in. in diameter, orange-brown, marked on the surface with transverse wavy lines.

The spines in this species are shining dark reddish brown, compressed and not terete in section, usually unbranched, or rarely with short branches arising pinnately in one plane.

G. aquatica is a native of the coast region of the United States from South Carolina, Georgia, and Florida westward through the Gulf States to the Brazos river in Texas; and extends northward in the Mississippi basin, through Louisiana, Arkansas, Kentucky, and Tennessee to southern Illinois and Indiana.   East of the Mississippi river, it is rare and only found in river swamps; but westward it is very abundant on rich alluvial land, occupying in Louisiana and Arkansas extensive tracts, which are inundated for a considerable period every year.

It is said by Loudon to have been introduced into England in 1723 by Catesby; but it is extremely doubtful if the large trees considered to be this species by Loudon were correctly named, as in all probability they were simply *G. triacanthos*.

The only specimen which we have seen in this country, is a small tree at Kew, about 10 ft. high.   (A. H.)

LABURNUM

Laburnum, Linnæus, *Syst. Nat.* 4 (1735); Medicus in *Vorl. Churfürstl. Ges.* ii. 363 (1787); Grisebach, *Spicil. Fl. Rum.* i. 7 (1843); Bentham et Hooker, *Gen. Pl.* i. 481 (1865); Schneider,
Laubholzkunde, ii. 37 (1907).
Cytisus, Linnæus, section *Laburnum*, Wettstein, in *Oesterr. Bot. Zeitschrf.* xl. 435, t. iv. (1890), and
xli. 127, 261 (1891).

DECIDUOUS small trees or shrubs, belonging to the division Papilionaceæ of the order
Leguminosæ. Leaves alternate, compound, with three stalked leaflets, and more or
less persistent stipules. Flowers yellow, in terminal racemes; calyx campanulate,
two-lipped, the upper lip with two short teeth, the lower lip with three short teeth;
standard ovate or orbicular, erect; wings obovate; keel incurved, shorter than the
wings; stamens monadelphous, five short versatile anthers alternating with five basi-
fixed anthers; ovary stalked, ovules numerous; style glabrous, incurved, with a
terminal stigma. Pod linear, flattened, stalked. Seeds without an excrescence at
the hilum.

This genus comprises three species, one of which *L. caramanicum*, Bentham
and Hooker, a native of Greece and Asia Minor, is a low shrub, which need not be
further alluded to. (A. H.)

LABURNUM ALPINUM, ALPINE LABURNUM

Laburnum alpinum, Berchtold and Presl, *Opir. Rost.* iii. 99 (1830-1835); Grisebach, *Spicil. Fl.
Rum.* i. 7 (1843); Schneider, *Laubholzkunde*, ii. 39 (1907).
Cytisus alpinus, Miller, *Dict.* ed. 8, No. 2 (1768); Loudon, *Arb. et Frut. Brit.* ii. 591 (1838).
Cytisus angustifolius, Moench, *Meth.* 145 (1794).

A tree similar in size, bark, and mode of branching, to *L. vulgare*. Young
branchlets glabrous. Leaves similar to those of *L. vulgare*, but with glabrous
petioles; leaflets more variable in shape, oval, oblong, or even obovate, acute
and mucronate at the apex, entire in margin; dark green and glabrous above;
light green and glabrous beneath, except for a few long hairs on the base of
the midrib.

Flowers in long and slender racemes,—smaller, paler yellow, and opening later
than those of *L. vulgare*; calyx and pedicels glabrous. Pod, about $1\frac{1}{2}$ to 2 in. long,
glabrous, with the upper suture winged; seeds brown.

This species is variable in the wild state :—

1. Var. *pilosum*, Koehne, *Deut. Dendr.* 326 (1893), has leaflets with a few scattered hairs on the under surface.

2. Var. *microstachys*, Koehne, *loc. cit.* Racemes shorter and less pendulous than in the type. This occurs in sunny arid localities.

3. Loudon mentions a pendulous form in Loddiges' nursery, which, from the description, appears to have been a hybrid between this species and *L. vulgare*.

This species has much the same distribution as *L. vulgare*, but is found at a higher elevation in the mountains of central Europe, being common in the Jura, the Alps, and the Carpathians.

The two species were confused ; but *L. alpinum*[1] appears to have been as early introduced as the other. It is more hardy, and is cultivated 5° of latitude farther north in Norway and Sweden. It thrives in the Highlands of Scotland,[2] and hence is often, but erroneously, called Scotch Laburnum. The largest specimens, which we have seen of this species in Britain, are the fine old trees in the Edinburgh Botanic Garden, which were planted about 1820 and are now showing signs of old age. One of these in 1911 was over 40 ft. high and nearly 5 ft. in girth.

(A. H.)

LABURNUM VULGARE, Laburnum

Laburnum vulgare, Berchtold and Presl, *Opir. Rost.* iii. 99 (1830-1835) ; Grisebach, *Spicil. Fl. Rum.*
  i. 7 (1843).
Laburnum anagyroides, Medicus, in *Vorl. Churfürstl. Ges.* ii. 363 (1787) ; Schneider, *Laubholzkunde*,
  ii. 37 (1907).
Cytisus Laburnum, Linnæus, *Sp. Pl.* 739 (1753) ; Loudon, *Arb. et Frut. Brit.* ii. 590 (1838).

A small tree, seldom exceeding 30 ft. in height, with smooth greenish bark, and few long shoots, most of the lateral buds developing only short shoots, on which the leaves are borne in fascicles. Young branchlets covered with appressed silky pubescence. Leaves on long pubescent stalks. Leaflets three, elliptic, the terminal one larger than the two lateral ; tapering at both ends, mucronate at the apex ; upper surface dark green, glabrous ; lower surface pale, more or less covered with appressed silky pubescence ; margin entire ; venation pinnate ; stalklets silky pubescent. Buds with two to three silky external scales, a true terminal bud being present, which is surrounded in winter by acuminate silky stipular leaf bases.

Flowers in long pendulous racemes ; the axis, long slender pedicels, and calyx being covered with silky appressed hairs. Corolla large, yellow ; standard veined with dark purple lines. Pods about 2 in. long, often contracted in the middle, at first silky, then becoming glabrescent, thickened on both sutures, dehiscent. Seeds reniform-orbicular, depressed, black.

---

[1] Loudon, *Arb. et Frut. Brit.* ii. 917 (1838) says that the Alpine Laburnum, though actually a much lower tree than the mountain ash, will, when drawn up in woods, attain twice the height of the latter tree. This species was called Tree Laburnum by Sang, Nicol's *Planter's Kalendar*, 91 (1812) to distinguish it from the common Laburnum, which was said to be more shrubby in growth in Scotland.

[2] See Plate 370, where in the accompanying letterpress are given further particulars.

This species is variable in the wild state, three sub-species having been established by Wettstein; but Schneider only recognises the typical form and the following variety :—

1. Var. *Alschingeri*, Reichenbach, *Icon. Flor. Germ.* xxii. 30 (1869).

    Cytisus Alschingeri, Visiani, *Fl. Dalmat.* iii. 262 (1852).

Lower lip of the calyx much longer than the upper lip, not approximately equal in length as in the type. Leaflets greyer and more pubescent beneath.

A considerable number of varieties have arisen in cultivation, the most important of which are :—

2. Var. *quercifolium*, Loudon. Leaflets, three or five, deeply lobed.

3. Var. *bullatum*, Koch, *Dendrologie*, i. 17 (1869). Leaflets curled, and puckered with swellings.

4. Var. *sessilifolium*, Kirchner, *Arb. Musc.* 399 (1864). Leaflets crowded, sessile.

5. Var. *aureum*, Simon-Louis, *Cat.* 1880, p. 51; Van Houtte, *Flore des Serres*, xxi. 2242, 2243 (1875). Foliage yellow.

6. Var. *pendulum*, Koch, *loc. cit.* Branchlets pendulous.

7. Var. *Carlieri*,[1] Kirchner, *Arb. Musc.* 398 (1864). Leaflets smaller than in the type. Flowers small, in short, more or less erect racemes.

HYBRIDS

L. vulgare is one of the parents in each of the following hybrids :—

I. *Laburnum Watereri*, Dippel, *Laubholzkunde*, iii. 673 (1893).

    Laburnum vulgare, vars. *Watereri*, *intermedium*, and *Parksii*, Kirchner, *Arb. Musc.* 399, 400 (1864).
    Cytisus Watereri, Wettstein, in *Oesterr. Bot. Zeitschrf.* xli. 129 (1891).

This is intermediate between *L. vulgare* and *L. alpinum* in the characters of the branchlets, foliage, and pods, and in the time of flowering; but it approaches the former species in the colour of the flowers, and exceeds it in the length of the racemes, which bear numerous flowers on long pedicels. The young branchlets, petiole, and under surface of the leaflets, have a few scattered appressed hairs.

This hybrid has been noticed in the wild state,[2] as at Bozen in the Tyrol, where Hausmann gathered it in 1856. It appears to have originated several times in cultivation; and was first noticed in 1842, when it was offered for sale by T. D. Parkes,[3] a nurseryman at Dartford. He states that it was raised from seed, and had flowers of a deeper colour than *L. vulgare*, borne in racemes 15 in. long.

It afterwards appeared in a bed of seedlings in Waterer's nursery. Darwin[4] found that 20 per cent of the pollen grains were ill-formed and useless, and that in most seasons it yielded no fruit.[5] In 1865 his tree produced a few pods, some of

---

[1] Several plants appear to be known under this name, and possibly may be different forms, arising from the hybrid in the second generation.

[2] Wettstein, in *Oesterr. Bot. Zeitschrf.* xli. 170 (1891).   [3] Cf. *Gard. Chron.* 1842, pp. 365 and 705.

[4] *Variation of Animals and Plants*, i. 416 (1890). A shrub at Aldenham, named *L. Alschingeri*, agrees in foliage and long racemes with *L. Watereri*; but produces every year abundance of pods, which apparently contain good seed. Mr. Vicary Gibbs raised seedlings from it some years ago.

[5] Wettstein states that 42 per cent of the pollen grains are useless; and in 1890 made the following observations on three trees of similar age in the Vienna Botanic Garden :—*L. vulgare* bore 654 pods, with about 3000 seeds; *L. alpinum* produced 562 pods, with about 2500 seeds; while *L. Watereri* bore only 7 pods, with 21 seeds, of which only 5 were fertile.

which contained one or two apparently good seeds.　Some of these germinated ; and he raised two trees, one of which resembled the parent, but the other was remarkably dwarf in habit with small leaves.

This hybrid, which is usually sold as *L. Watereri* or *L. Parksii*, is very ornamental, occasionally bearing racemes 16 or 18 in. long, and is said to be very rapid in growth.[1]

II. *Laburnum Adami*, Kirchner, *Arb. Musc.* 397 (1864).

> *Cytisus Adami*, Poiteau, in *Ann. Soc. Hort. Paris*, vii. 501 (1830), and in Loudon, *Gard. Mag.* xvii. 59 (1841).
>
> *Cytisus Laburnum*, Linnæus, var. *coccineum*, Lindley, in *Bot. Reg.* xxxiii. t. 1965 (1837).
>
> *Cytisus Laburnum*, Linnæus, var. *purpurascens*, Loudon, *Arb. et Frut. Brit.* ii. 590 (1838).

This remarkable hybrid, between *Laburnum vulgare* and *Cytisus purpureus*, Scopoli, has normally glabrous branchlets, petioles, and leaves, with leaflets smaller than those of *L. vulgare* ; and bears dingy red small flowers, in short pendulous racemes, which never set seed.　It is usually reproduced by grafting on the common Laburnum, and is remarkable in often producing stray branches which revert back to one or both of the parents.　A single tree is thus often seen bearing three kinds of foliage and flowers : (*a*) the hybrid foliage and flowers ; (*b*) branches with the yellow flowers and leaves of *L. vulgare* ; and (*c*) branches with the leaves and small purple flowers of *Cytisus purpureus*.[2]　The yellow flowers produce pods which yield fertile seed, from which Darwin[3] raised seedlings resembling in most respects the common Laburnum.　The branches with purple flowers are said by Darwin to be not quite the same as those of *Cytisus purpureus*, and are not perfectly fertile ; but he raised seedlings from their seed which differed in no respect from pure *Cytisus purpureus*.

This hybrid is said to have originated in 1825 in the nursery of M. Adam at Vitry, near Paris.　He inserted on a stock of *L. vulgare* a shield of the bark of *Cytisus purpureus*, which produced in the following year several shoots, one of which was more vigorous and had larger leaves than the others.　This shoot was propagated, and as soon as it bore the peculiar flowers intermediate between the two species, was recognised as a hybrid which had been produced by grafting.　This account has been disputed by some botanists ; and De Vries,[4] who has made a study of the plant, states that Camuzet, a contemporary of M. Adam, maintained that he had seen the tree from which the latter had taken his graft, and that it was not *Cytisus purpureus*, but *L. Adami* itself, so that the latter must have originated earlier and been an ordinary chance hybrid from seed.　Camuzet's statement seems very improbable ; and against this view may be stated the fact that Reisseck, Caspary, and Darwin tried in vain to cross the flowers of *C. purpureus* with those of *L. vulgare*.

---

[1] Cf. Harrison Weir, in *Gard. Chron.* xxvi. 83 (1899).

[2] Mixed flowers also occur.　Braun, *Rejuvenescence* (Ray Soc.), 317, plate v. (1853), gives an instance of a raceme which bore 21 flowers of *L. Adami*, 3 flowers of *L. vulgare*, and 8 mixed flowers.　In the latter, half of the corolla was reddish like *L. Adami*, the other half yellow like *L. vulgare* ; similarly, half of the calyx was reddish brown and glabrous, as in *L. Adami*, the other half green and pubescent, as in *L. vulgare*.

[3] *Variation of Animals and Plants*, i. 414 (1890).

[4] *Mutation Theory*, ii. 622, figs. 139, 140 (1911).　Cf. also Bateson, in *Nature*, lxxxviii. 37 (1912).

The literature concerning *L. Adami* is very extensive, and need not be cited[1] here; but the reader may refer to an article by R. P. Gregory in *Gard. Chron.* l. 162, 185 (1911) on graft-hybrids, in which is given the anatomical evidence that *L. Adami* resembles *Cytisus purpureus* in the structure of the epidermis, whilst its internal structure is that of *L. vulgare*.

There are good specimens[2] at Kew, and in several private gardens.

DISTRIBUTION

L. vulgare is a native of Central Europe, occurring in the east of France from Lorraine, through the Jura to the Dauphiné, where it is common in woods on calcareous soil. It is rare in Switzerland, but is widely spread through Lower Austria, Styria, Carinthia, Carniola, and Hungary south of the Danube, attaining its southerly limit in northern Italy, Dalmatia, Servia, and Bulgaria.

All parts of the plant are said to be poisonous; but this is doubtful, as we have heard no case of animals[3] being affected from browsing on the leaves alone. The seeds, however, contain a highly poisonous alkaloid,[4] *cytisine*, discovered by Husemann and Marmé in 1865; and ten seeds have proved fatal to a child.

The flowers are produced regularly in May; but in some seasons, as in 1908, a few trees flower a second time[5] in October.

In old trees of Laburnum, decayed at the base, adventitious roots are occasionally thrown out at some distance above the ground, and act as props to the stem. An instance of this in the Edinburgh Botanic Garden was described and photographed by Dr. Borthwick.[6]   (A. H.)

CULTIVATION

It was cultivated[7] by Tradescant in England in 1596, and is largely planted everywhere as a small ornamental tree.

The common Laburnum is very hardy. It was not hurt in Suffolk by the severe winter of 1860-1861, and flowered well in the following May. The blossoms,

[1] The early history is given by Loudon, *Gard. Mag.* xii. 225 (1836), and xv. 122 (1839); and by Dillwyn, in *Gard. Chron.* 1841, pp. 325, 366, and 1842, p. 397. Cf. also Kerner, *Nat. Hist. Plants*, Eng. Trans. ii. 570 (1898).

[2] Cf. Bean, in *Gard. Chron.* xxxv. 371 (1904). Cf. also *The Garden*, lxix. 333 (1906), and *Gard. Chron.* xxxvi. 217 (1904), where an illustration is given of a tree of *L. Adami* in Pennick's nursery at Delgany, Co. Wicklow.

[3] In the case reported by Dr. Stark in *Gard. Chron.* xvi. 666 (1881), where cattle died at Huntfield, Lanarkshire, after browsing on Laburnum, the pods were eaten as well as the leaves and branches. In *Gard. Chron.* l. 310 (1911) a case is recorded of two cows which died after eating the tops of a felled Laburnum tree; but these bore seed-pods as well as foliage. The flowers are poisonous, according to Taylor, *Med. Jurisp.* 730-733 (1905), who considers a recorded case of poisoning by the bark at Inverness to be doubtful.

[4] Flückiger and Hanbury, *Pharmacographia*, 172 (1879). The alkaloid *laburnine* here mentioned proved to be only impure *cytisine*. Cf. Kunkel, *Toxicologie*, ii. 847 (1901).

[5] Cf. *Gard. Chron.* xxxii. 253, 271, 290 (1902), and xliv. 313, 363 (1908). Autumnal flowers, seen in October at Kingston-on-Thames, and on 1st November at Antwerp, were noticed as borne in short racemes.

[6] *Notes R. Bot. Garden, Edinburgh*, xviii. 121, plate 24 (1907).

[7] By Act V, Ed. IV. cap. 4 (Ireland), every Englishman was required to have a bow of his own height made of yew, wych hazel, ash, or awburne. The *awburne* was possibly the Laburnum, as it is identical with the Scottish name for this tree, *hoburn saugh*, which is given in Jamieson's *Dictionary*. Cotgrave's *Dictionary* (1611) gives the French word, "*Aubourt*, a kind of tree, Latin *alburnus*; it bears long yellow blossoms which no bee will touch." The flowers of the Laburnum are visited by bees; but Cotgrave here repeats an erroneous statement of Pliny. Cf. the correspondence on this subject in *The Phytologist*, iv. 191, 255 (1860). Matthiolus mentions that in his time the wood of the Laburnum was considered to make the best bows.

however, are sometimes completely killed by late frosts in April, as in 1859 at Mildenhall.[1]   It is planted in Norway as far north as Trondhjem, and in Sweden as far as Stockholm.[2]

No tree is more easily raised from seed, or more easily transplanted than the Laburnum; and its rapid growth when young, and ability to thrive on the poorest, driest, and most chalky soils, makes it a valuable addition not only to pleasure grounds, where it is most generally seen, but also for plantations.

Boutcher[3] recommends planting it in plantations where hares are numerous, because as long as they can get Laburnum they will touch no other tree, and though eaten to the ground every winter they will spring up again every summer.   I have proved the truth of this to some extent at Colesborne, and planted Laburnums in quantity on the edge of plantations.   But as the tree has a tendency to branch freely, and the branches are easily broken by the wind, pruning must be done for some years if a clean bole is desired.   There is much variation in the habit; some, having a weeping tendency, should be selected for ornament; whilst for timber, clean stems of 10 to 12 ft. should be aimed at; and on good soils I believe that, though the wood is hardly known to modern cabinetmakers, it will have a considerable value at an age when few other hardwoods have any.

REMARKABLE TREES

One of the finest I have seen in the south, is near the East Lodge at The Mote, near Maidstone, in a sheltered dell.   This in 1911 was 38 ft. high, with a bole 8½ ft. in girth and 8 ft. high, but the stem was partly decayed and patched with lead and plaster.

At Ickleton Grange, near Cambridge, Mr. G. W. H. Bowen showed Henry, in June 1912, a remarkable plantation of Laburnum (*L. vulgare*), which is said to have been made about 1790 by Mr. Percy Wyndham.   It is four acres in extent, and is situated on the south slope of a hill, the soil being very thin and resting on impermeable chalk.   The plantation was coppiced at an early period, most of the trees being bushy, with several wide-spreading limbs, arising from stools of considerable size.   The largest specimen has eight great stems, 3 to 5 ft. in girth, and about 30 ft. high.   Though close together and almost forming pure canopy, the trees are thinly clad with foliage; and the soil beneath is covered with thick grass. Seedlings, as soon as they appear, are eaten by rabbits; but in the same soil in a garden at some distance, which is surrounded by a Laburnum hedge, seedlings are very numerous.   Mr. Bowen has also some thriving Laburnums, which were originally put down as stakes for fencing, but speedily developed roots and grew in this chalky soil.

At Coolhurst, near Horsham, there is a very old tree which leans over the road and has been propped up to keep it from falling.   In 1906 it was about 35 ft. high by 3 to 4 ft. in girth.   At "The Laburnums," near Stroud, there is a group of three trees, the stems of which are close together at the base, girthing respectively

---

[1] Bunbury, *Arb. Notes*, 6 (1889).   [2] Willkomm, *Forstl. Flora*, 918 (1887).   [3] *Forest Trees*, 109 (1775).

8 ft. 3 in., 7 ft. 5 in., and 6 ft. near the ground, with a height of about 30 ft., and a spread of foliage of about 18 yards.

In the laundry yard at Alnwick Castle there is an immense spreading tree, dividing at the ground into two main trunks about 43 ft. high and 11 ft. 2 in. round at the ground. This is probably one of three trees mentioned by Loudon,[1] the largest of which in 1835 measured 6 ft. 11 in. in girth, and contained 46 ft. of timber. Mr. Gillanders believes it to be over one hundred years old, and says that it shows signs of decay.

At Inveraray it also grows to a large size, as I saw a log lying in the Duke of Argyll's yard in 1905 which was about 8 ft. in girth, and from which some very handsome furniture was afterwards made. At Dalkeith Palace[2] there is a remarkable Laburnum, low and spreading in habit, but covering a piece of ground 60 ft. across.

TIMBER

The wood of Laburnum is hard and heavy, weighing about 53 lbs. to the cubic foot when dry, according to Loudon, and is very durable when exposed to wet; and Mr. Vicary Gibbs tells me that a clothes-line prop of laburnum outlasted one of yew.

In colour the heartwood is dark when old, but the wood varies in colour with age, and rapidly becomes darker when exposed to light, becoming dark olive or red brown, and showing small medullary rays. When thoroughly dry it makes remarkably good joints, and is a first-class cabinetmakers' wood, taking a fine surface and good polish. Sang says[3] that in his time (1812) it was the most valuable timber grown in Scotland, and that a quantity of it was sold in 1809 at Brechin Castle and Panmure at 10s. 6d. per foot. It was then used for cabinetmaking, musical instruments, handles, and chairs;[4] and I can say from personal experience that it is a most excellent wood for furniture. I have seen old cabinets made in Scotland in which this wood was used in transverse sections, like the so-called oyster-shell cabinets made from slices of walnut wood. I believe it would also be well adapted for parquet.
(H. J. E.)

[1] *Arb. et Frut. Brit.* iv., Suppl. 2551 (1883).   [2] Bean, in *Gard. Chron.* li. 168 (1907).
[3] Nicol, *Planter's Kalendar*, 91 (1812).
[4] Cf. *Gard. Chron.* xxxvii. 397 (1905), where Mr. Coomber mentions a beautiful set of chairs made by Mr. Ross from trees which grew near Peterhead.

CASTANOPSIS

Castanopsis, Spach, *Hist. Vég.* xi. 185 (1842); Bentham et Hooker, *Gen. Pl.* iii. 409 (1880).
Castanea, Endlicher, *Gen. Pl.* 275 (in part) (1836); Prantl, in Engler and Prantl, *Nat. Pflanzenfam.*
    iii. pt. 1, p. 55 (1894).
Callæocarpus, Miquel, *Pl. Jungh.* i. 13 (1851).

TREES or shrubs belonging to the order Fagaceæ, differing mainly from Castanea in the leaves being evergreen, and the buds covered with numerous imbricated scales; moreover, a true terminal bud is formed. Leaves coriaceous, five-ranked, entire or dentate. Catkins erect. Staminate flowers usually in clusters of threes, in simple or branched catkins; calyx, five- or six-partite; stamens six to twenty. Pistillate flowers, one to three, in an involucre, in simple or branched catkins or scattered at the base of the staminate catkin; calyx six-cleft, with abortive stamens; ovary three-celled, with two ovules in each cell; styles three, each ending in a minute stigma. Fruit, ripening in the second year, the spiny involucre enclosing one to three nuts. Nuts ovoid or globose, angled, usually containing one seed, which bears at its apex the aborted ovules.

About thirty species of Castanopsis have been described, one of which is a native of California and is the only species in cultivation. The others occur in south-eastern Asia and the Malay Archipelago, extending as far north as the eastern Himalayas, southern China, and Formosa.

CASTANOPSIS CHRYSOPHYLLA

Castanopsis chrysophylla, A. de Candolle, in *Journ. Bot.* i. 182 (1862), and *Prod.* xvi. 2, p. 109
    (1864); Sargent, *Silva N. Amer.* ix. 3, t. 439 (1896), and *Trees N. Amer.* 223 (1905); Earl
    of Ducie, in *Gard. Chron.* xxii. 411, fig. 120 (1897); Masters, in *Gard. Chron.* xxxvi. 152,
    fig. 59 (1904); Jepson, *Silva of California*, 239, plate 74 (1910).
Castanea chrysophylla, Douglas, *ex* W. J. Hooker, *Comp. Bot. Mag.* ii. 127 (1836), *Fl. Bor. Amer.*
    ii. 159 (1839), *London Journ. Bot.* ii. 496, t. 16 (1843), and *Bot. Mag.* t. 4953 (1856);
    Jepson, *Fl. W. Mid. Calif.* 145 (1901).

A tree, rarely attaining in America 120 ft. in height and 20 ft. in girth, but usually smaller. Bark, 1 to 3 in. thick, and deeply divided by longitudinal fissures into rounded scaly ridges. Young branchlets covered with yellow scurfy scales. Leaves (Vol. III, Plate 202, Fig. 10), persistent two or three years, oblong, averaging 3 to 4 in. in length, and 1 to 1½ in. in breadth, tapering at the base and apex, often contracted at the apex into a short broad point ; thick, coriaceous, entire in margin ; dark green, shining and glabrous above ; under surface bright yellow, coated with minute scales ; lateral nerves pinnate, twelve to sixteen pairs, looping just before they reach the margin ; petiole $\frac{1}{4}$ to $\frac{1}{3}$ in. long.

Staminate catkins simple or branched ; stamens six to ten. Pistillate flowers borne on short separate catkins or scattered at the base of the staminate catkin. Fruit-involucre spiny, like that of the common chestnut, irregularly four-valved, containing one or occasionally two nuts ; nuts $\frac{1}{3}$ in. long, angled, with a hard shell and a sweet kernel.

The typical form described above is a tree, with large leaves, which gradually passes at high elevations into a shrub, distinguished as follows :—

1. Var. *minor*, A. de Candolle, *Prod.* xvi. 2, p. 110 (1864).

> Var. *pumila*, Vasey, in *Rept. Comm. Agric.* 1875, p. 176 (1876).
> *Castanea chrysophylla*, var. *minor*, Bentham, *Pl. Hartweg.* 337 (1857).

A shrub 2 to 15 ft. high. Leaves similar to those of the type in shape, but smaller, 1 to 2 in. long, very golden beneath.

This variety occurs in the Santa Cruz mountains and northwards as far as the south fork of the Salmon river. It is often gregarious on chaparral slopes and on the pine barrens of the Mendocino coast.

2. Var. *sempervirens*, Henry.

> *Castanopsis sempervirens*, Dudley, in Merriam, *Biol. Survey Mt. Shasta*, 142 (1899).
> *Castanea sempervirens*, Kellogg, in *Proc. Calif. Acad.* i. 71 (1855).

A shrub, 1 to 8 ft. high. Leaves elliptical, 1 to 2$\frac{1}{2}$ in. long, variable at the base, rounded or obtuse at the apex, pale yellow beneath. This is retained by Jepson, *Silva of California*, 241 (1910), as a distinct species, on account of the usually more numerous (ten to seventeen) stamens ; but he acknowledges that there is no difference in the fruits. It grows on arid mountain slopes from Mt. Jacinto northwards through the Sierra Nevada and coast ranges to Mt. Shasta at 3000 to 8000 ft. elevation.

This species, in its typical form as a large tree, occurs in the Redwood belt of Mendocino and Humboldt counties of California, and northwards in the Siskiyou, Coast Range and Cascades of Oregon, as far as the valley of the Columbia river. Southward and eastward in California, only the shrubby forms occur. As a tree it appears to be rare, fruiting very sparingly, so that good seed can scarcely ever be procured. Jepson states that the largest trees now existing scarcely exceed 115 ft. in height ; but are occasionally 4 to 6 ft. in diameter.

C. chrysophylla was discovered in 1830 by David Douglas on the Grand Rapids of the river Columbia in Oregon, at Cape Orford, and near Mt. Hood. Seeds were sent to Kew, by the collector Burke,[1] in 1845, from which a solitary plant was raised, which produced fruit when only 5 ft. high in 1856. It was reintroduced by William Lobb, who[2] sent seeds to Messrs. Veitch in 1853 ; but, so far as I can judge, the

---

[1] Cf. *Bot. Mag.* t. 4953. A MS. note at Kew states that Burke collected in North America in 1845.
[2] *Hortus Veitchii*, 393 (1906).

typical form has never been introduced, all the specimens which I have seen being var. *minor*.
(A. H.)

The finest specimen in England is probably that at Tortworth,[1] which was procured from Messrs. Veitch about 1854-1856, and had attained 20 ft. by 1 ft. 5 in. in 1879, and 27 ft. by 3 ft. in 1897; and on 1st August 1911 measured 29 ft. by 3 ft. 5 in., with a circumference of branches as much as 75 ft. This tree is growing in a sheltered position on the side of a hill in a sandy loam, resting on the Old Red Sandstone formation. Up to 1882 it bore sterile fruits; but since that date has produced fertile seed, from which many seedlings have been raised. One of these in the late Sir J. D. Hooker's garden at Sunningdale was 10 ft. high in 1897. I have several times endeavoured to grow seedlings from this tree at Colesborne, but in every case they have died before or after turning out of their pots. I attribute this to excess of lime, and it is clear that *Castanopsis* is intolerant of lime in the soil.

At the Heatherside Nursery, Farnborough, there is a fine specimen with a single stem, but with very wide-spreading branches, which was 35 ft. high and bearing fruit in July 1910. There is said to be a good specimen at Pencarrow in Cornwall.

A shrub[2] only 2 ft. high produced fruit in 1904 at Kaimes Lodge, Murrayfield, Midlothian.

In Ireland the finest specimen, about 30 ft. high, was at Old Connaught House, Bray; but it died a short time before 1905, when Henry saw the dead stump still remaining.
(H. J. E.)

[1] Cf. *Gard. Chron.* xiv. 435 (1880), xviii. 716 (1882), and xxii. 411, fig. 120 (1897).
[2] Cf. *Gard. Chron.* xxxvi. 152 (1904).

UMBELLULARIA

Umbellularia, Nuttall, *Sylva*, i. 87 (1842); Bentham et Hooker, *Gen. Pl.* iii. 162 (1880).
Oreodaphne, sub-genus *Umbellularia*, Nees ab Esenbeck, *Syst. Laurin.* 462 (1836).
Drimophyllum, Nuttall, *Sylva*, i. 85 (1842).

A GENUS belonging to the order Lauraceæ, the characters of which are given in the
following description of the only [1] species known.

UMBELLULARIA CALIFORNICA, CALIFORNIAN LAUREL

Umbellularia californica, Nuttall, *Sylva*, i. 87 (1842); Sargent, *Silva N. Amer.* vii. 21, t. 306
  (1895), and *Trees N. Amer.* 334 (1905): Jepson, *Flora W. Mid. Calif.* 191 (1901), and *Silva
  Calif.* 242, plates 10 and 76 (1910); Chesnut, in *Contrib. U.S. Nat. Herb.* vii. 349 (1902);
  Eastwood, *Trees of California*, 53 (1905); Power and Lees, in *Trans. Chem. Soc. London*,
  1904, p. 629.
Tetranthera (?) *californica*, Hooker and Arnott, *Bot. Voy. Beechey*, 159 (1833).
Oreodaphne californica, Nees ab Esenbeck, *Syst. Laurin.* 463 (1836); Hooker, *Bot. Mag.* t. 5320
  (1862).
Drimophyllum pauciflorum, Nuttall, *Sylva*, i. 85, t. 22 (1842).
Laurus regalis, Standish and Noble, *Pract. Hints on Planting*, 160 (1852).

An evergreen tree, variable in habit, occasionally attaining 120 ft. in height and
10 to 15 ft. in girth; but oftener a smaller tree or large bush, rarely reduced to a
prostrate shrub. Young branchlets green, glabrous. Leaves persistent two to six
years, coriaceous, very aromatic, alternate, simple, lanceolate or narrowly elliptical,
averaging 3 to 4 in. long and 1 to 1½ in. broad, cuneate or rounded at the base,
tapering to an acute or rounded apex, entire in margin; main veins pinnate, eight to
twelve pairs, curved, looping before they reach the margin, connected by reticulate
veinlets; dark green and shining above, duller and paler beneath, with a minute
scattered pubescence when young, ultimately glabrous; petiole ⅕ to ⅓ in.

Flowers minute in stalked simple umbels, which are solitary and axillary or
crowded near the apex of the branchlet; each umbel with four to nine flowers on
slender pedicels arising in the axils of deciduous bracts; calyx with six pale yellow
obovate lobes; petals absent; perfect stamens nine, in three series, the inner three
with a stalked orange gland on each side of the base of the filament, and alternating
with three scale-like staminodia; anthers four-celled, four-valved, the three inner

---

[1] *Umbellularia parvifolia*, Hemsley, *Biol. Centr. Amer.* iii. 77 (1882), a native of Mexico, belongs to another genus,
and is correctly named *Litsea parvifolia*, Mez, in *Jahrb. Königl. Bot. Gart. Berlin*, v. 481 (1889).

extrorse, the outer two series introrse ; ovary sessile, with a solitary ovule, and one style, crowned by a capitate stigma.

Fruit, an ovoid berry, about 1 in. long, surrounded at the base by the enlarged calyx, greenish or purplish, containing a large seed, with a hard woody outer shell, and a thin papery inner coat, destitute of albumen ; embryo with two large thick cotyledons.

The Californian Laurel is a native of California and south-western Oregon, occurring mainly in the coast ranges from the Umpqua river southward. It is less abundant on the high western slopes of the Sierra Nevada, where it ascends to 4000 ft. in the Yosemite Valley and to 5000 ft. in the Kaweah basin. It is also found on the southern slopes of the San Bernardino mountains at 2000 to 2500 ft., and on the west slope of San Jacinto below 5000 ft. ; and reaches its most southerly point in the Oriflamme Cañon near Julian.

Prof. Jepson gives a detailed account of the remarkable variation in the habit of this species. As a tree of considerable or very large size, it is found on the banks of streams, usually in rich alluvial soil. It forms tall bushy thickets on the northern walls of cañons ; and becomes a low shrub, 3 to 4 ft. high, in the chaparral of the higher parts of the coast range. A curious prostrate form grows here and there on the bluffs facing the ocean in Mendocino county.

The finest trees in California, 70 to 120 ft. in height, and 9 to 20 ft. in girth, occur in groves, unmixed with other species, on the Eel river and other streams to the northward.

I saw this tree in Oregon, where it grows on the alluvial flats of the Coquille[1] river, in mixture with *Acer macrophyllum*, and attains a large size, trees 70 to 80 ft. high and 10 to 14 ft. in girth being common. The tallest which I measured was 121 ft. high and 11 ft. in girth. The tree is most remarkable for the extraordinary density of its foliage, through which rain penetrates with difficulty and sunshine never. It sprouts freely from the stump when cut down ; and when isolated, is prone to produce abundant epicormic branches on the stem.

All parts of the tree contain a volatile oil, which is most abundant in the leaves, amounting to $7\frac{1}{2}$ per cent. This oil resembles menthol and camphor in its effects on the tongue and skin ; and when inhaled produces dizziness and headache. The dried leaves spread about drive away fleas ; and when used as a decoction are a powerful insecticide. The seeds are roasted and eaten by the Indians.

This species was discovered in 1792 by Archibald Menzies, on the shores of San Francisco Bay ; and was introduced[2] into England from Oregon by David Douglas in 1826. One of his original plants, which was living at Kew in 1862, cannot now be found.[3]   (A. H.)

We have seen no large trees in England, where it apparently only forms a bush. There is a specimen, with fruit, in the Kew herbarium, sent by Canon Ellacombe from Bitton in 1876. At Grayswood, Haslemere, a large bush

---

[1] It is reported to be abundant and of large size also on the Coos and Rogue rivers in Oregon.
[2] W. J. Hooker, *Comp. Bot. Mag.* ii. 127 (1836), where the tree is referred to as *Laurus regia*, Douglas.
[3] Cf. *Bot. Mag.* t. 5320 (1862), where the plant figured was received from Berlin.

bore flowers in May 1907. At Tortworth, a large bush bore ripe fruit on 15th October 1911, the seed of which germinated this spring in my garden at Colesborne. At Kew there is a young tree, 15 ft. high, near the Temperate House; and an older shrub on the wall near the Forestry Museum. The latter flowers sparingly in most years, but has never produced fruit.

I saw a large bushy specimen about 20 ft. high at Drove House, Thornham, Norfolk, which was bearing ripe fruit in November 1910. Major the Hon. G. Legh informed me that this tree might have been much taller, but that it had been topped by a previous tenant several years before. Considering the exposed situation close to the North Sea, it seems probable that if seedlings from the best form of this species were planted in deep shady valleys in the south-west of England, they might attain a considerable size.

The wood is hard, strong, and close-grained, taking a fine polish. It contains numerous small equal pores and many thin medullary rays. The sapwood is whitish; but the heartwood is of a rich yellow or brown colour, often with peculiar black streaks. The wood is used for making furniture, panelling, staves, shoe-lasts, etc., and is suitable for turnery. At North Bend and Myrtle Point, in south-western Oregon, there are factories where this wood is made into furniture and veneers. A figured board of this wood, which I obtained, under the name of myrtle, from the Californian State Exhibit at St. Louis in 1904, is in the collection of timbers at Cambridge University. (H. J. E.)

PRUNUS

Prunus, Linnæus, *Sp. Pl.* 473 (1753), and *Gen. Pl.* 212 (1754); Bentham et Hooker, *Gen. Pl.* i. 609 (1865).

TREES or shrubs, belonging to the order Rosaceæ. Leaves alternate, simple; usually serrate with or without glandular points to the serrations, rarely entire; petiole often with one or more glands.

Flowers perfect; solitary, or in fascicles, corymbs, or racemes; calyx-tube cup-shaped or tubular; sepals five, imbricated in the bud; petals five, white or pink; stamens ten to thirty, perigynous, inserted with the petals on the mouth of the calyx-tube; ovary one-celled, with a terminal style, and containing two pendulous ovules. Fruit, a fleshy drupe, with an indehiscent, one-seeded, smooth or rugged stone.

This genus comprises over one hundred species, mostly natives of the temperate regions of the northern hemisphere, a few being indigenous in tropical Asia and tropical America.

In the following article, only a few species, mainly of interest to foresters, are dealt with. These may be arranged as follows :—

SECTION I. *Cerasus.* Cherries.

Leaves rolled up in the bud. Flowers large, long-stalked, usually fascicled or in corymbs.

1. *Prunus Avium*, Linnæus. Europe, Asia Minor, Caucasus. See p. 1535.

Large tree. Leaves variable in shape, cuspidate-acuminate, with scattered long hairs on the midrib and nerves beneath. Branchlets glabrous. Inflorescence not leafy at the base.

2. *Prunus Cerasus*, Linnæus. Europe, Asia Minor, Caucasus. See p. 1541.

Small tree or shrub. Leaves smaller than, but similar in shape to *P. Avium*, nearly glabrous beneath. Branchlets glabrous. Inflorescence with leaves at the base.

3. *Prunus Mahaleb*, Linnæus. Europe, Asia Minor, Caucasus, Armenia, Turkestan. See p. 1542.

Small tree. Leaves broadly ovate, pubescent on each side of the midrib beneath. Branchlets densely pubescent.

SECTION II. *Padus.* Bird Cherries.

Deciduous trees or shrubs. Leaves rolled up in the bud. Flowers small in long racemes, which are leafy at the base.

4. *Prunus Padus*, Linnæus. Europe, Temperate Asia. See p. 1543.

Small tree or shrub. Leaves slightly cordate at the base ; pale beneath with axil-tufts of pubescence.

5. *Prunus serotina*, Ehrhart. North America. See p. 1546.

Large tree. Leaves tapering at the base ; pale beneath with a dense band of rusty pubescence on each side of the midrib.

SECTION III. *Laurocerasus.* Cherry Laurels.

Evergreen trees or shrubs. Leaves rolled up in the bud. Flowers small in long leafless racemes.

6. *Prunus Laurocerasus*, Linnæus. Balkan Peninsula, Asia Minor, Caucasus, North Persia. See p. 1551.

Leaves with a few remote serrations on the margin, and with one or two glands on each side of the midrib beneath.

7. *Prunus lusitanica*, Linnæus. Spain, Portugal, Azores, Madeira, and Canary Isles. See p. 1553.

Leaves regularly serrate, without glands beneath. (A. H.)

PRUNUS AVIUM, WILD CHERRY, GEAN

Prunus Avium, Linnæus, *Fl. Suec.* 165 (1755) ; Willkomm, *Forstliche Flora*, 898 (1887) ; Ascherson and Graebner, *Syn. Mitteleurop. Flora*, vi. pt. ii. 151 (1906).
Cerasus Avium, Moench, *Meth.* 672 (1794) ; Mathieu, *Flore Forestière*, 137 (1897).
Cerasus sylvestris, Loudon, *Arb. et Frut. Brit.* ii. 693 (1838).
Prunus Cerasus, var. *Avium*, Linnæus, *Sp. Pl.* 474 (1753).

A tree, attaining 80 to 100 ft. in height. Bark smooth, shining grey, often peeling off on the surface in transverse annular strips ; becoming deeply fissured and thick at the base of old trunks. Young branchlets glabrous. Leaves alternate on the long shoots, clustered at the apices of the short shoots or spurs, variable in shape and size, up to 5 in. long and 2 in. broad, ovate or obovate, elliptic or oblong, cuspidate-acuminate at the apex, cuneate or rounded at the base, biserrate or unequally serrate, the serrations tipped with a gland ; upper surface dull, glabrescent ; lower surface pale green, with scattered long hairs, mainly on the midrib and nerves ; lateral nerves, twelve to sixteen pairs, looping before reaching the margin ; petiole with a few scattered hairs, and usually with a pair of red prominent glands near its distal end.

Flowers, two to six in a cluster, appearing with the leaves, and usually situated on the short shoots, arising out of a bud, with no internal leafy scales ; calyx-tube constricted near the apex, glabrous, with five entire reflexed lobes : petals five, white, obovate-rounded, emarginate. Fruit globose, smooth, reddish, shining ; stone light brown, oval, compressed, furrowed on one edge.

This species suckers from the roots, but not so freely as *P. Cerasus* ; and when cut down, produces coppice shoots.

Varieties and Hybrid

I. In the wild and typical form of the species, the fruit is small, with little edible flesh. The cultivated cherries[1] with larger and more edible fruit derived from *P. Avium* comprise two main varieties :—

1. Var. *Juliana*. Heart or Gean Cherries. Fruit mostly black, with soft flesh.

2. Var. *duracina*. Bigarreau Cherries. Fruit mostly yellow or red, with firm flesh.

The Duke Cherries, with acid flesh, are referred to *P. Avium* as var. *regalis* by Bailey.[2]

II. The following varieties, cultivated for ornament, are peculiar in leaves, flowers, or habit.

3. Var. *decumana*, Koch, *Dendrologie*, i. 106 (1869). Leaves very large, occasionally as much as 10 to 12 in. in length, and 4 to 6 in. in breadth.

4. Var. *aspleniifolia*, Kirchner, *Arb. Musc.* 254 (1864). Leaves with deeply incised teeth.

5. Var. *salicifolia*, Dippel, *Laubholzkunde*, iii. 615 (1893). Leaves very narrow.

6. Var. *flore pleno*, Kirchner, *loc. cit.* Flowers partially or completely double.

7. Pendulous and pyramidal forms are also known.

III. The following, which is in all probability a hybrid between *P. Avium* and *P. Mahaleb* is occasionally cultivated :—

8. *Prunus Fontanesiana*, Schneider, *Laubholzkunde*, i. 617 (1906).

Prunus græca, Desfontaines, *ex* Koch, *Dendrologie*, i. 109 (1869).
Cerasus Fontanesiana, Spach, *Hist. Vég.* i. 410 (1834).

A tree, resembling *P. Avium* in habit. Young branchlets covered with dense whitish erect pubescence. Leaves about 4 in. long, 2 in. broad, similar in shape and size to those of *P. Avium*, but differing in the more crenate serrations, which are tipped with sharp minute glands, as in *P. Mahaleb*; glabrous above; pale green beneath with long pubescence on each side of the midrib; petiole pubescent, with one or two large glands. Flowers about $\frac{4}{5}$ in. across.

This tree, the origin of which is unknown, is remarkable for never producing fruit. A specimen at Kew is about 25 ft. high.

Distribution

P. Avium is widely distributed throughout nearly the whole of Europe, but appears to be rare in Spain and Italy as a wild tree; and in Russia,[3] is confined to the south-western provinces and the Crimea. It also occurs in Asia Minor and the Caucasus. It extends as far north as the province of Bergen in Norway, where Schübeler mentions a pure wood of considerable extent at Urnäs; and its remains have been found in peat-mosses in Sweden.[4]

[1] An excellent account of the cultivated cherries was given by R. Thompson in *Trans. Hort. Soc.*, Second Series, i. 248-294 (1835).

[2] *Cycl. Amer. Hort.* 1453 (1901).

[3] Köppen, *Holzgewächse Eur. Russlands*, i. 280 (1888).

[4] Willkomm, *Forstliche Flora*, 899, note (1887).

It is difficult to state in what parts of the British Isles it is truly wild; but it is certainly a native of the southern counties of England, occurring in many woods as a rare tree; while in beech woods, as in the Chiltern Hills, it is rather common, and attains a great height.

P. Avium has been found[1] in the fossil state in neolithic beds at Crossness, Essex, and at Gayfield, Edinburgh; and in interglacial deposits at West Wittering and Selsey, Sussex. (A. H.)

CULTIVATION

Though much neglected of late years by planters, the cherry is perhaps our most beautiful native tree when in full flower; and as it grows well and to a large size, on soils of only moderate quality, and suffers less than most trees from spring frosts, it should be planted for ornament as well as for timber in all suitable places.

It seems to grow best among beech, and in woods where the stem is drawn up to a good height before its crown expands, though even in exposed situations it makes a fair-sized trunk. It is indifferent to the geological nature of the soil, growing equally well on calcareous, gravelly, or sandy soils, but not on wet or heavy clays.

It is easy to raise from seed, though as a rule the fruit is eaten by birds before it ripens, and the majority of the young trees that one sees in woods are grown from suckers which spring up at a considerable distance from the trunk. I have seen a colony of young cherries, thirty paces across, all of which had sprung from the roots of one tree after felling. When it has attained a few inches in diameter it is rarely attacked by rabbits. The seed, if put in a rot-heap when ripe, will germinate the following spring, though when kept dry for long, a large proportion will lie dormant for a year before coming up. The seedlings transplant without difficulty, and may be planted out with little or no risk of loss when 3 to 4 ft. high.

For an avenue of moderate size, planted about five yards apart, I can think of no more beautiful tree than the cherry. Mr. Foljambe of Osberton told me that his father was so fond of this tree, that after he became blind he used to ask to be led out into the place where they grew at the time when they were in flower, in order that their scent might recall their beauty the better.

REMARKABLE TREES

The largest cherry tree that I have seen in England was pointed out to me by the late Prof. Fisher in Windsor Park, near the Bishopgate, and not far from the house occupied by Lady Southampton. It has probably been drawn up by beech trees round it, and measured in 1904, 93 to 95 ft. high by 9 ft. 3 in. in girth, with a trunk free from branches for nearly forty feet. It is probably past its best, for the cherry is not a long-lived tree and generally begins to decay before reaching 100 years. In Gatton Park, Surrey, on a flinty chalk hill in a wood of beech, there are

[1] Cf. C. Reid, *Origin British Flora*, 114 (1899).

some fine cherries, and the best of those now standing is about 90 ft. by 7 ft. 6 in.[1] A tree cut here contained 78 cubic feet of sound timber, some of the boards being fully 2 ft. across and quite sound.   At Walcot, Shropshire, there are some splendid vigorous young cherry trees, one of which that I measured recently being about 90 ft. high and only 4 ft. 9 in. in girth.   These may attain 100 ft. in height as they are still growing.   Plate 354 shows a wild cherry in Savernake Forest, with a burry trunk, 12 ft. 7 in. in girth at four feet, and 10 ft. 9 in. at five feet from the ground.

In Gloucestershire there was a tree on the Earl of Harrowby's estate near Campden, which, according to Loudon, measured 85 ft. by over 9 ft.   I could not find this tree when I visited Norton Court in 1906, but can well believe the correctness of this measurement after seeing the fine development of ash, oak, and chestnut at this place.   On my own land, however, though the tree grows well up to fifty or sixty years old, it does not attain these dimensions, one of 8 ft. in girth being the largest I have.

Probably one of the finest in the Chiltern Hills is a tree growing in Burke's Grove, Butler's Court, Beaconsfield, which was accurately measured in 1909 by Lord Grenfell and B. L. Majendie, Esq., R.N., as follows :—total height, 97 ft. ; height to the first branch, 67 ft. ; girth at five feet from the ground, 4 ft. 10 in.   Mr. Leslie Wood has seen a tree in a beech wood near High Wycombe, 95 ft. in height.

At Camp Wood, The Coppice, Henley, where the beautiful whitebeam grows that was figured in Vol. I. Plate 51, there is a fine tree, which measured in 1905, 74 ft. in height, by 9 ft. 10 in. in girth, at two feet from the ground, above which it divides into two stems.   In the adjoining Bolney Wood, two trees measured 79 ft. by 5 ft. 2 in., and 76 ft. by 4 ft. with stems clear of branches to forty feet, and rivalling in height the beeches amidst which they are growing.

In a wood near Riverhill, Kent, Mr. A. B. Jackson measured a tree, 70 ft. by 8½ ft. in 1908.   At Sidmouth, Devonshire, Miss Woolward measured a tree, 62 ft. by 10 ft. in 1906.   At Henham, Suffolk, there is a tree about 50 ft. high and 10½ ft. in girth, with branches which spread over an area eighty-two paces round.

On Ashampstead Common, Berks, Dr. Watney showed me a very fine tree which was in 1901 about 75 ft. by 8 ft., and was surrounded by quite a grove of suckers from the roots.

At Russells, near Watford, there is, in a thick plantation near the house, a tree about 90 ft. by 9 ft. 8 in., of which the bark, standing up in high ridges, makes the girth seem larger than it really is.   The bole of this tree is clean for about 20 ft.

At Appleby Hall, the seat of Lady St. Oswald, in north Lincolnshire, there is a tree in the shrubbery which divides into three large stems at about seven feet and measures about 65 ft. by 11 ft. 9 in.

At Alnwick Castle, Northumberland, a very fine cherry growing in a wood in 1907 measured 80 ft. by 6 ft. 10 in. with a bole about 50 ft. high, but judging from the number of dead branches it is near its end.

[1] Mr. J. S. Elliott of Cranleigh informs me that he bought twenty cherry trees out of this wood containing 807 cubic ft. ; the best ten of these averaged 57 cubic ft. each.

At Dynevor Castle there are many fine cherry trees in the slopes round the ruins, and I was told by the late Lord Dynevor that one was blown down in 1889 which at three feet from the ground was no less than 12 ft. 9 in. in girth. At Duffryn, near Cardiff, Mr. Coomber[1] measured in 1896 a tree 82 ft. high by 8 ft. 3 in. at five feet from the ground, with a bole 35 ft. high.

In Scotland the cherry is always called gean. The largest that I have seen is on the flat near the Tay at Murthly Castle, a tree 65 ft. by 11 ft. 5 in., with a bole of 9 ft. The main limb of this tree has decayed, and, in 1908, had a mountain ash about 15 ft. high growing out of the decayed stump. At Duns Castle, Berwickshire, there is a remarkably large tree which I am informed by Mr. J. Ferguson measures 42 ft. by 13 ft. 6 in. and still bears fruit, though not so much as it did 60 or 70 years ago, when the fruit of these trees was more valued in Scotland than it is now. At Gribton, near Dumfries, the seat of H. Lamont, Esq., Henry measured a tree, 56 ft. high by no less than 12 ft. 8 in. in girth, whose branches spread to a diameter of 70 ft.

At Gordon Castle there is a very fine old tree near the house 50 to 60 ft. high, which girths at two feet from the ground, where there is a large swelling, 14 ft. 2 in., and 10 ft. 5 in. at five feet. At Ardkinglas I have seen a fine old cherry by the garden wall which had a very large burr on the trunk.

At Mauldslie Castle, Lanarkshire, Mr. Renwick in 1899 measured a tree 52 ft. high and 13 ft. 2 in. in girth. In 1903, it was reported[2] to be in a state of decay, the trunk splitting, but supported by an iron band.

In Ireland, a most remarkable grove of cherry trees exists at Clonmannon, Co. Wicklow, though only ten survive out of the fifty which originally grew here on about an acre of lawn. The largest measured, in 1907, 70 ft. high by 10 ft. 10 in. in girth. Another was 55 ft. high and 12 ft. in girth at three feet from the ground, above which it divides into five great stems. Hayes[3] records at the same place in 1794, a cherry tree of the Upton-mazard kind, no less than 15 ft. in girth at five feet from the ground, being the largest cultivated cherry tree that I have ever heard of. At Mount Usher, a tree dividing into two stems near the ground, 5 ft. 10 in., and 7 ft. 2 in. in girth respectively, was 62 ft. high in 1903. The girth of the main stem at the base was 11 ft. 3 in. in 1908.

At Glenstal, Co. Limerick, a tree in a wood, clear of branches to thirty feet, was 69 ft. by 6 ft. 3 in. in 1903. At Bunratty, Co. Clare, Mr. R. A. Phillips measured a tree on 14th May 1905, when it was in full flower, 70 ft. in height and 11½ ft. in girth, with a spread of branches fifty-seven feet in diameter.

Though the cultivated varieties of cherry never, in England at least, seem to attain the size of the wild cherry, yet in some districts they become very large. In Nash's orchard at George's Green near Slough I saw, on 23rd July 1908, a tree (Plate 355) at least 60 ft. high by 9 ft. 4 in. in girth. The fruit was being gathered with the help of a ladder with 77 rungs, and the men informed me that in this orchard no less than 101 sieves (25 pounds to the sieve) had been gathered

[1] *Gard. Chron.* xx. 664 (1896). [2] *Trans. Nat. Hist. Soc. Glasgow,* vii. 83 (1903).
[3] *Treatise on Planting,* 127 (1794).

from one tree.   This was a late black cherry of moderate size known in the district as "Black bud" or "Croon's" cherry.   W. Dumbarton, a resident of the place, further told me that in an orchard belonging to Sir R. Harvey at Iver, he had gathered thirty-five years ago 165½ sieves of cherries from one tree of the Bigarreau variety, which is still alive.   This seemed to me incredible, but I was assured by Mr. Ford, steward to Sir R. Harvey, that he had heard this as a fact from natives of the place.

I saw at Golden Grove in Wales a cultivated cherry no less than 50 ft. by 9 ft. with a bole more than 20 feet high.   But as a rule the trees in orchards are cut down or decay before they arrive at anything like the size mentioned above.

I have found no account of cherry trees of great size in Europe, except one by the Rev. R. Walsh,[1] who wrote as follows :—"The second variety is an amber-coloured transparent cherry, of a delicious flavour.   It grows in the woods in the interior of Asia Minor, particularly on the banks of the Sakari, the ancient Sangarius.   The trees attain a gigantic size, they are ascended by perpendicular ladders, suspended from the lowest branches.   I measured the trunk of that from which the seeds I send were taken ; the circumference was 5 ft., and the height where the first branches issued 40 ft. ; from the summit of the highest branches was from 90 to 100 ft., and this immense tree was loaded with fruit."

TIMBER

The wood of this tree, though now little valued in the trade, is one of the best native woods for inside work, being easier to season and less liable to warp than most timbers.   It has a fine even grain which takes a good surface and polishes well. Its colour is pale pinkish when fresh, but when oiled or waxed it becomes with age a dark pinkish or brown, and is highly ornamental when cut so as to show the fine medullary rays.

I have used it with very good effect for panelling a small room at Rapsgate Park, near Colesborne.   I am informed by Mr. A. C. Forbes that the pews in the church at Gibside, Northumberland, which were made in 1812 of cherry wood, have not warped or shrunk in the least, the joints being as good as when made.   Though the sapwood in some places is worm-eaten, the heartwood is almost free from this defect.

As far as my experience goes the wood is best when the trees are felled at about 4 to 5 feet in girth, older trees being often more or less decayed at the heart. It is sold standing at 6d. a foot or less, and may be bought in the form of board at a very reasonable price, and used for furniture and chair-making, as well as for all work where toughness and strength are not specially required.   It may be made richer in colour by soaking it in lime water, and when treated in this way is very similar to pale-coloured mahogany.   (H. J. E.)

[1] *Trans. Hort. Soc.* vi. 44 (1826).

PRUNUS CERASUS, Dwarf Cherry

Prunus Cerasus, Linnæus, *Sp. Pl.* 474 (1753); Willkomm, *Forstliche Flora*, 897 (1887); Schneider, *Laubholzkunde*, i. 614 (1906); Ascherson and Graebner, *Syn. Mitteleurop. Flora*, vi. pt. ii. 147 (1906).
Prunus acida,[1] Koch, *Dendrologie*, i. 112 (1869).
Cerasus vulgaris, Miller, *Gard. Dict.* ed. 8, No. 1 (1768); Loudon, *Arb. et Frut. Brit.* ii. 693 (1838).
Cerasus acida, Gaertner, *Fl. Wettar.* ii. 185 (1800); Mathieu, *Flore Forestière*, 139 (1897).

A shrub or rarely a small tree, sending up numerous suckers from the root, and distinguishable from *P. Avium* as follows:—Leaves smaller, almost but not quite glabrous on the shining under surface; glands often absent on the petiole. Flowers, two to five in a cluster, generally on the long shoots, and not on short spurs, as in *P. Avium*; arising out of a bud, the inner scales of which are accrescent and leafy; calyx-tube glabrous, scarcely constricted at the apex, with obtuse deflexed crenate lobes. Fruit globose, shining red, smooth; stone smooth, brown.

A considerable number of the orchard cherries have been derived from this species. These constitute two well-marked groups; those with colourless juice, like the Kentish cherry, and those with coloured juice, like the Morellos. Var. *acida* (*Prunus acida*, Koch) is a shrubby form, with small leaves, and dark red sour fruit, with an ovoid stone.

The following varieties[2] are cultivated for ornament, being peculiar in leaves, flowers, or habit.

1. Var. *semperflorens*, Loudon, *Arb. et Frut. Brit.* ii. 701 (1838).

   Prunus semperflorens, Ehrhart, *Beit.* vii. 132 (1792).
   Cerasus semperflorens, De Candolle, *Fl. Franc.* iv. 481 (1805).

Flowering and fruiting throughout the whole summer. This, which is usually grafted high, is known as "All Saints' Cherry." Its origin is uncertain. Koch states[3] that it comes true from seed. There is a good specimen at Kew; and it is cultivated by Messrs. Veitch.

2. Var. *persiciflora*, Koch, *Dendrologie*, i. 111 (1869). Flowers double, rose-coloured.

3. Var. *Rhexii*, Kirchner, *Arb. Musc.* 252 (1864). Flowers double, white. This is figured as var. *ranunculiflora*, in *Flore des Serres*, xvii. t. 1805 (1867-1868).

4. Var. *cucullata*, Kirchner, *loc. cit.* Leaves puckered with swellings.

5. Var. *aucubæfolia*, Dippel, *Laubholzkunde*, iii. 613 (1893). Leaves spotted with yellow.

6. Var. *globosa*, Späth, *Cat.* 1887-1888, p. 101. A low globose shrub, with small leaves.

P. Cerasus is usually considered to be a true native of south-eastern Europe, Asia Minor, and the Caucasus; but it has escaped from cultivation in many districts

[1] The name *P. acida* has been used by various authors for different forms of the cherry. *P. acida*, Ehrhart, *Beit.* vii. 130 (1792), was applied to the cultivated sour cherries with colourless sap.
[2] Carrière, *Prod. et Fix. Vars.* 37 (1865), mentions several curious varieties which I have not seen.
[3] *Dendrologie*, i. 113 (1869).

of central Europe, and is now found apparently wild in various localities in England, Scotland, and Ireland.   There are many records [1] of *P. Cerasus* in the county floras ; but in most cases its occurrence is so rare and local, as to suggest that it is possibly only a recent escape [2] from some neighbouring orchard, from which the fruit has been carried by birds.   It is said, however, to be well established in some woods on the Pennine range in Cumberland, and in others in Surrey and Kent.   Mr. R. A. Phillips tells me that it is plentiful in old hedgerows in many parts of Ireland ; and he has seen it in a wild glen near Lehinch in County Clare.   (A. H.)

PRUNUS MAHALEB, St. Lucie Cherry

Prunus Mahaleb, Linnæus, *Sp. Pl.* 472 (1753); Ascherson and Graebner, *Syn. Mitteleurop. Flora*, vi., pt. ii. 156 (1906).
Prunus odorata, Lamarck, *Fl. Franc.* iii. 108 (1778).
Cerasus Mahaleb, Miller, *Gard. Dict.* ed. 8, No. 4 (1768); Loudon, *Arb. et Frut. Brit.* ii. 707 (1838); Mathieu, *Flore Forestière*, 140 (1897).
Padus Mahaleb, Borkhausen, *Handb. Forstbot.* ii. 1434 (1803).

A deciduous shrub or small tree, occasionally attaining 40 ft. in height.   Young branchlets pubescent with dense erect short hairs.   Leaves broadly ovate, $1\frac{1}{2}$ to 2 in. long, 1 to $1\frac{1}{2}$ in. broad ; shortly acuminate at the apex, rounded at the base ; dark shining green and glabrous above ; lighter green beneath, with dense long pubescence on each side of the midrib ; crenately serrate, each serration with a minute sharp gland ; petiole slender, glabrous, with one or two large glands.

Flowers, appearing with the leaves, in short simple corymbs, each with two small leaves at the base, and four to eight flowers above ; axis, slender pedicels, and wide-mouthed calyx glabrous ; sepals ovate, obtuse, entire.   Fruit ovoid, $\frac{1}{3}$ in. long, blackish when ripe.

I.   The following varieties are known in the wild state :—

1. Var. *transilvanica*, Schur, *Enum. Pl. Trans.* 180 (1866).   Flowers small, numerous in the corymb ; sepals reflexed.   This occurs in Transylvania.

2. Var. *Cupaniana*, Fiori and Paoletti, *Fl. Anal. Ital.* i. 561 (1896) (*Prunus Cupaniana*, Gussone, *Flor. Sic. Syn.* i. 553 (1842)).   A low shrub, with small coriaceous leaves ; flowers few in the corymb.   This occurs in mountain woods in Sicily ; and a similar, if not identical form, has been found in Dalmatia.

II.   Several varieties have arisen in cultivation :—

3. Var. *chrysocarpa*, Nicholson, in *Kew Hand-List Trees*, i. 143 (1894).   Fruit yellow.

4. Var. *globosa*, Dieck, *ex* Dippel, *Laubholzkunde*, iii. 621 (1893).   A globose compact bush.

---

[1] Mr. W. B. Crump has sent me specimens from Elland Park Wood, near Halifax, taken from small trees and bushes which resemble *P. Cerasus* in habit and are apparently wild.   The flowers, however, have a calyx with the constricted tube and entire lobes of *P. Avium* ; and in all probability these trees are hybrids between the two species.

[2] Mr. M. R. Pryor sent me a specimen from a shrub about 14 ft. high, growing in a wood at Weston Park, Stevenage. This shrub, the only one we know of apparently wild in Herts, has not increased appreciably in size during the last twenty years.

5. Var. *monstrosa*, Kirchner, *Arb. Musc.* 258 (1864).   Branches and branchlets very short and thick.

6. Var. *variegata*, Nicholson, *loc. cit.*   Leaves variegated with white.

7. Var. *albomarginata*, Dippel, *loc. cit.*   Leaves with a white edge.

P. Mahaleb is widely distributed in central and southern Europe, the Crimea, Asia Minor, the Caucasus, Armenia, and Turkestan.   In France it grows mainly on rocky ground on the limestone formation, and occurs as far north as the departments of Seine Inférieure, where it is rare, and Pas-de-Calais ; but in the west is not known wild north of La Rochelle.   In Germany, it occurs in Alsace, Baden, and Bavaria ; and, further east, is met with in lower Austria, Hungary, and Transylvania ; but is more common southwards in the Tyrol, Carniola, Dalmatia, the Balkan States, and Greece.   It also occurs in northern Spain, and in a few localities in Switzerland.

The scented kernels are sold in the bazaars of North-western India, and the tree is cultivated in Baluchistan.[1]   The young branches are utilised for pipe-stems ; and the tree is cultivated in Austria for this purpose.[2]   (A. H.)

This can hardly be considered as a timber tree, as it rarely, if ever, exceeds 40 ft. high, but it produces, under favourable conditions, a trunk of considerable thickness, which produces a wood known in France under the name of *Bois de St. Lucie*,[3] formerly much sought after by turners and cabinetmakers.   It was introduced into England in 1764.   Its principal use in England is as a stock on which to graft cultivated cherries, but it is also sometimes planted as underwood for game covert ; and Loudon says, that when grafted on the wild cherry stock it makes a much larger tree than on its own roots.   Whether this had been the cause of the large size attained by a tree at Devonshurst, Chiswick,[4] or not, I cannot say, but I measured this shortly before it was cut down to clear the site, and found it 30 ft. high by 8 ft. 5 in. in girth, the spread of the foliage being 55 ft. in diameter.   In the public gardens at Bury St. Edmunds there is a large old tree about 40 ft. by 5 ft. 9 in.

A tree in the Botanic Garden of Trinity College, Dublin, measured 40 ft. by 6 ft. in 1908.   (H. J. E.)

PRUNUS PADUS, Bird Cherry

Prunus Padus, Linnæus, *Sp. Pl.* 473 (1753); Willkomm, *Forstl. Flora*, 901 (1887); Ascherson and
    Graebner, *Syn. Mitteleurop. Flora*, vi. pt. ii. 159 (1906).
Prunus racemosa, Lamarck, *Fl. Franc.* iii. 107 (1778).
Cerasus Padus, De Candolle, *Fl. Franc.* iv. 480 (1805); Loudon, *Arb. et Frut. Brit.* ii. 709 (1838);
    Mathieu, *Flore Forestière*, 141 (1897).
Padus vulgaris, Borkhausen, *Handb. Forstbot.* ii. 1426 (1803).
Padus racemosa, Schneider, *Laubholzkunde*, i. 639 (1906).

A deciduous shrub or small tree, attaining about 30 ft. in height.   Young branchlets glabrous.   Leaves, disagreeable in odour, averaging 4 in. long and 2 in.

---

[1] Hooker, *Fl. Brit. India*, ii. 312 (1878).   [2] Mathieu, *Flore Forestière*, 141 (1897).

[3] Named after the monastery of Sainte-Lucie in the Vosges.

[4] Devonshurst was a house, with pleasure grounds, built on the site of part of the Horticultural Society's Garden at Chiswick, where a considerable number of interesting trees remained till 1904.   Cf. *Gard. Chron.* iv. 726 (1888).

broad, obovate-oblong or elliptic, acuminate at the apex, slightly cordate at the base; glabrous above, paler beneath with brown axil-tufts of pubescence; lateral nerves, 10 to 14 pairs, dividing and looping before reaching the margin, which is finely serrate, with close serrations, each ending in a sharp cartilaginous point; petiole glabrous, slender, with 1 to 4 glands near the insertion of the blade.

Flowers, appearing after the leaves, in racemes about 5 in. long, which have usually one or two small leaves near the base; axis, slender pedicels, and calyx glabrous; sepals glandular-fimbriate; petals white. Fruit globose, about $\frac{1}{3}$ in. in diameter, without a persistent calyx, shining, blackish; stone irregularly pitted.

VARIETIES

I. Besides the typical form described above, which has finely serrate leaves and pendulous racemes, and is prevalent in the plains of central Europe, the following geographical varieties are known :—

1. Var. *borealis*, Schneider, *Laubholzkunde*, i. 640 (1906).

   Var. *petræa*, Fiek, *Fl. Schles.* 119 (1881).
   Prunus borealis, Salisbury, *Prod.* 356 (1796); Schübeler, *Pflanzen-Welt Norw.* 369 (1873).
   Prunus petræa, Tausch, in *Flora*, xxi. 719 (1831).

Leaves with coarser serrations. Racemes erect. This is prevalent in the mountains of Silesia, Bohemia, and Transylvania, and in the Alps; and is also the wild form, which occurs in Scandinavia and the British Isles.

2. Var. *pubescens*, Regel, *Fl. Ussur*, No. 149, *ex* Maximowicz, in *Mel. Biol.* xi. 706 (1883). Young branchlets, under surface of the leaves, and racemes, more or less covered with reddish brown pubescence. This occurs, but mixed with the typical form, in Manchuria, Saghalien, and north China.

3. Var. *cornuta*, Henry.

   Prunus cornuta, Steudel, *Nomencl.* ii. 403 (1841).
   Padus cornuta, Carrière, in *Rev. Hort.* 1869, p. 275, fig. 64.
   Cerasus cornuta, Royle, *Illust. Bot. Himal.* 207 (1839).

Branchlets glabrous. Leaves rounded at the base, bluish green beneath with reddish brown axil-tufts. Racemes long and, like the pedicels, pubescent. Fruit large, $\frac{1}{3}$ to $\frac{1}{2}$ inch in diameter, with a smooth stone. This occurs in the Himalayas, at 6000 to 12,000 feet, and is considered by Hooker[1] to be a form of *P. Padus*; but by Koehne and Schneider to be a distinct species. It is represented at Kew by a tree 15 feet high.

II. The following have appeared in cultivation[2] :—

4. Var. *leucocarpa*, Koch, *Dendrologie*, i. 120 (1869). Fruit white to yellowish.

5. Var. *commutata*, Dippel, *Laubholzkunde*, iii. 647 (1893). Flowers and leaves appearing very early in the season. Said by Schneider to be from eastern Asia, and to be commonly cultivated under the name of *P. Grayana*.[3]

---

[1] *Fl. Brit. India*, ii. 315 (1878).
[2] *P. Laucheana*, Bolle, in Lauche, *Deut. Dend.* 652 (1882) is said to be a hybrid between *P. Padus* and *P. Virginiana*; and, according to Schneider, is sometimes known in cultivation as *P. Padus*, var. *rotundifolia*.
[3] Cf. *Garden and Forest*, i. 295 (1888). *P. Grayana*, Maximowicz, in *Bull. Acad. St. Petersb.* xxix. 107 (1883), is a distinct Japanese species.

6. Var. *bracteosa*, Seringe, in De Candolle, *Prod.* ii. 539 (1825). Leaves at the base of the raceme very large.

7. Var. *aucubæfolia*, Kirchner, *Arb. Musc.* 259 (1864). Leaves spotted with yellow.

8. Var. *aurea*, Nicholson, *Kew Hand-List Trees*, i. 229 (1902). Leaves yellow

9. Forms, pyramidal and pendulous in habit, are also in cultivation; and a variety with double flowers is mentioned by Schneider, and is in cultivation at Kew.

Distribution

P. Padus is the most widely distributed of all the species of Prunus, occurring throughout nearly all Europe, in Siberia, Manchuria, Japan, north China, the Caucasus, Persia, and the Himalayas.

In Europe, it is more common in the north, as in Scandinavia, Denmark, and northern Russia; but is also widely spread in central Europe, and extends in the south to the mountains of northern Portugal, the Sierra Nevada in Spain, Pyrenees, Apennines, and the mountainous regions of the northern Balkan States. In Russia, it extends as far north as the Kola peninsula; but is rare in the southern provinces, and does not occur on the mountains of the Crimea. Bode[1] saw a tree in Courland, 40 feet in height and 4 feet in girth.

In the British Isles, it is most common in the Highlands; but occurs in many woods in northern England, ascending to 1500 feet in Yorkshire. In Radnorshire and other parts of Wales it is in waste places more often a shrub than a tree. It has been found[2] in the fossil state in neolithic beds at Northampton, Hornsea and Sand le Meer in Yorkshire, and at Hailes near Edinburgh, and in interglacial deposits at Selsey, Sussex, and at Airdrie, Lanarkshire. In Ireland, it is widely distributed,[3] occurring in old woods and in river glens; but is absent from the southern uplands and rare on the central plain.   (A. H.)

Dr. Walker[4] states that at Drumlanrig there were in 1773 two trees about 40 feet high, one being no less than 8 feet in girth. In 1834, however, they had quite disappeared. We have seen none approaching these in size.

Schübeler says that in Norway where the tree is called " Hegg," from which the Scotch name hagberry is no doubt derived, he saw a tree at Mollenhof near Drammen, which was 58 ft. high, by 5 ft. 5 in. in girth; and in Sweden he mentions one which at 1 foot from the ground was 9 ft. in girth.

Though now seldom planted in England, the bird cherry is very ornamental when in flower, easily raised from seed, and flourishes on poor dry soil. Loudon strongly recommends the variety *bracteosa*, on account of its large pendulous racemes of flowers and fruit.

I have not heard of any use being made of its hard yellowish wood in England, but in France it was formerly used by country cabinetmakers.   (H. J. E.)

---

[1] Köppen, *Holzgewächse Europ. Russlands*, i. 300 (1888).   [2] Cf. C. Reid, *Origin British Flora*, 114 (1899).

[3] Praeger, in *Proc. Roy. Irish Acad.* vii. 95 (1901).   [4] *Essays on Nat. Hist.* 74 (1812).

PRUNUS SEROTINA, American Black Cherry

Prunus serotina, Ehrhart, *Beit.* iii. 20 (1788); Bentley and Trimen, *Medicinal Plants*, ii. t. 97
    (1880); Sargent, *Silva N. Amer.* iv. 45, t. 159 (1902), and *Trees N. Amer.* 524 (1905); Von
    Schwerin, in *Mitt. Deut. Dend. Ges.* 1906, p. 1.
Prunus virginiana, Miller, *Gard. Dict.* ed. 8, No. 3 (1768) (not Linnæus).
Cerasus serotina, Loiseleur, *Nouv. Duham.* v. 3 (1812); Loudon,[1] *Arb. et Frut. Brit.* ii. 712 (1838).
Cerasus virginiana, Michaux, *Fl. Bor. Amer.* i. 285 (1803) (not Loiseleur); Loudon,[1] *Arb. et Frut.
    Brit.* ii. 710 (1838).
Padus serotina, Agardh, *Theor. Syst.* t. 14, f. 8 (1858); Schneider, *Laubholzkunde*, i. 643 (1906).

A deciduous tree, attaining in America 100 ft. in height and 15 ft. in girth.
Bark, $\frac{1}{2}$ to $\frac{3}{4}$ in. thick, broken on the surface into small irregular scaly plates.
Young branchlets glabrous. Leaves slightly coriaceous, about 3 to 4 in. long, and $1\frac{1}{4}$
to 2 in. broad, obovate-oblong or elliptic, acuminate at the apex, tapering at the base;
shining and glabrous above; lower surface light green, with a dense band of rusty
pubescence on each side of the midrib, elsewhere glabrous; regularly and sharply
glandular-serrate; with one or two glands at the base of the leaf, or on the summit
of the glabrous petiole.

Flowers in racemes, terminating short leafy branchlets; axis and slender pedicels
glabrous; calyx cup-shaped, with short ovate sepals, which persist on the ripe fruit;
petals obovate, white. Fruit globose, slightly lobed, $\frac{1}{3}$ to $\frac{1}{2}$ in. in diameter, red
before ripening, almost black when ripe; flesh dark purple, juicy; stone obovate,
compressed, smooth, broadly ridged on the ventral suture.

This species is often confused with *P. virginiana*, Linnæus, a North American
shrub. The leaves of the latter are oval, cuspidate-acuminate, usually glabrous
beneath, with long pointed serrations; and the inner bark of the branchlets has a
strong disagreeable odour, that of *P. serotina* being aromatic and agreeable. In
the former species the calyx is deciduous; in the latter it is persistent on the fruit.

Varieties

I. This species is very variable in the wild state: and several geographical forms
have been distinguished, and ranked by American botanists as either varieties or
distinct species:[2]—

1. Var. *neomontana*, Small, *Fl. S.E. United States*, 574 (1903). Leaves
coriaceous, very large, coarsely toothed, whitish beneath; sepals pubescent.
Occurs on the higher summits of the Alleghany Mountains.

2. *Prunus Cuthbertii*, Small, in *Bull. Torrey Bot. Club*, xxviii. 290 (1901).
Branchlets, axis of the inflorescence, and pedicels pubescent. Leaves coriaceous.
A shrubby form, growing in rich sandy soil in Georgia.

---

[1] Loudon describes *P. serotina* under both *C. virginiana* and *C. serotina*; and seems to have been unacquainted with
the true *P. virginiana*, Linnæus.

[2] *Padus eximia*, Small, *Fl. S.E. United States*, 573 (1903), a glabrous form with leaves delicately reticulate-veined
beneath, occurring in river valleys in southern Texas, can scarcely be distinguished from the type.

3. *Prunus australis*, Beadle, in *Biltm. Bot. Studies*, i. 162 (1902). Branchlets, axis of the inflorescence, and pedicels pubescent. Leaves covered beneath with reddish brown pubescence. Fruit dark purple. Occurs only in one locality, on clay soil at Evergreen in Alabama, where it is a tree about 60 ft. high.

4. *Prunus alabamensis*, Mohr, in *Bull. Torrey Bot. Club*, 1899, p. 118. Branchlets, axis of the inflorescence, and pedicels pubescent. Leaves slightly pubescent beneath. Fruit purple. A tree about 30 ft. high, occurring in the mountains of Alabama and Georgia.

5. *Prunus Capuli*, Cavanilles, *ex* Sprengel, *Syst.* ii. 477 (1825); Hemsley, in *Biol. Cent. Amer. Bot.* i. 367 (1879); Bolle, in *Mitt. Deut. Dend. Ges.* 1898, p. 56.

> *Cerasus Capollin*, De Candolle, *Prod.* ii. 539 (1825); Loudon, *Arb. et Frut. Brit.* ii. 713 (1838).
> *Cerasus Capuli*, Lavallée, *Arb. Segrez.* 115, t. 34 (1885).

Leaves lanceolate, long acuminate at the apex; underneath without bands of pubescence along the midrib. Inflorescence long and slender, fruit larger than in *P. serotina*. This occurs in the mountains of New Mexico, Arizona, Mexico, and Guatemala. It is said by Loudon to have been introduced in 1820; and in 1838 a vigorous tree, trained against a wall in the Horticultural Society's Garden, retained its leaves nearly all the winter. It is cultivated in France,[1] where, according to Lavallée, it was introduced in 1867, and endured the severe winter of 1879-1880; and in Algeria attains about 30 ft. in height.

6. *Prunus salicifolia*, Humboldt, Bonpland, and Kunth, *Nov. Gen. et Spec.* vi. 190, t. 563 (1825). An evergreen tree, occurring in Colombia, Ecuador, Peru, and Bolivia. It differs little in botanical characters from *P. Capuli*. Not introduced.

II. A number of varieties have appeared in cultivation in Europe :—

7. Var. *variegata*, Zabel, *Laubholz-Benennung*, 244 (1903). Leaves variegated with white.

8. Var. *pendula*, Dippel, *Laubholzkunde*, iii. 645 (1893). Branches pendulous.

9. Var. *salicifolia*,[2] Nicholson, *Kew Hand-List Trees*, i. 144 (1894).

> Var. *phelloides*, Schwerin, in *Mitt. Deut. Dend. Ges.* 1906, p. 3.

Leaves lanceolate, long acuminate at the apex.

This variety, which has been confused with *P. Capuli*, is represented at Kew by a tree, about 30 ft. high, which flowers and fruits at the same season as the type, and is equally hardy. It may possibly have come from the United States, where narrow-leaved forms are said to occur.

10. Var. *aspleniifolia*, Kirchner, *Arb. Musc.* 260 (1864). Leaves irregularly dentate.

11. Var. *cartilaginea*, Kirchner, *Arb. Musc.* 260 (1864). Leaves very shining on both surfaces.   (A. H.)

[1] Cf. Hamelin, in *Rev. Hort.* lvi. 111 (1884), and Carrière, in *Rev. Hort.* lxiii. 62, 196, figs. 19, 20 (1891). Sargent states that plants of reputed *P. Capuli*, from France, proved hardy in the Arnold Arboretum; and he doubts their Mexican origin. I saw a tree about 40 ft. high under this name at Segrez in 1907, and raised seedlings from its fruit, which are alive at Colesborne, but do not grow as vigorously as the northern form.—H. J. E.

[2] This variety must not be confused with *Prunus salicifolia*, H.B.K., mentioned above.

Distribution

The typical form of *P. serotina* is widely distributed in North America, occurring in Canada from Nova Scotia westwards to the northern shores of Lake Superior; and in the United States southwards to Florida, and westwards to Dakota, eastern Nebraska, Kansas, Indian Territory, and eastern Texas. Further west in southern Arizona and New Mexico, it is replaced by *P. Capuli*. Sargent states that it was once very abundant in the Alleghany Mountains, reaching its largest size from West Virginia to Georgia and Alabama. In the United States it usually grows in rich moist soil, but sometimes occurs on low sandy soil and on rocky cliffs by the sea-shore in New England. Pinchot[1] states that it grows fairly well in dry situations; but it is only in moist well-drained rich soils of mild climates that the maximum development is reached, as in the southern Alleghanies, where trees 90 ft. high and 4 ft. in diameter are not uncommon. In plantations in America it grows rapidly in youth; but is looked upon as a short-lived tree. Owing to the great value of its timber it has now become scarce in all accessible regions, and large trees are hardly to be found, the largest of which I have any record being a tree measured by Dr. Schneck in Wabash Co., Illinois, which was 135 ft. high by $10\frac{1}{2}$ ft. in girth.[2]

I saw no such trees, however, in the Wabash valley; and in Canada in 1904, near Ottawa, where I found it scattered in the forest, it is a comparatively small tree. In southern Ontario, however, there are still, according to Macoun, many fine trees standing which are largely used for furniture making.

In Massachusetts, Emerson says[3] that it rarely exceeds 40 to 50 ft. in height; but on the Ohio river, Michaux measured trees from 80 to 100 ft. high, with trunks 12 to 16 ft. in girth and clear of branches to 25 or 30 ft. He recommended its culture in the Rhine valley, which, he says, has most resemblance to its native regions.

Cultivation

It is difficult to say when this species was introduced, as it was formerly confused with *P. virginiana*, which Loudon says was introduced in 1724. It has never become a common tree, and was hardly known to nurserymen until recently, when it has been planted largely as a forest tree in some places in Europe.

The only place where I have seen this tree fairly at home in England is in Sherwood Forest, where several trees have been planted on Lord Manvers' property. The first of these I found quite unexpectedly myself, and recognised it by its shining foliage; the others were pointed out to me by Mr. Foljambe, who said he had known them as American cherries for many years, though I could obtain no information as to how or when they were introduced. One of these trees was about 50 ft. by $4\frac{1}{2}$ ft., and bore a few ripe fruits in October 1905, from which I raised a plant the

[1] *U.S. Forest Circular*, No. 94 (1907).   [2] Ridgway, in *Proc. U.S. Nat. Mus.* xvii. 411 (1894).
[3] *Trees and Shrubs of Massachusetts*, ii. 516 (1875).

following spring. Another not far from the Buck Gate was 46 ft. by 6½ ft., and had a large limb broken off, from which I got a small board which shows nice colour and well-marked medullary rays.

There was a large but decayed tree at Chiswick House near Kew, which in 1904 was about 50 ft. by 7 ft., probably the same as the tree mentioned by Loudon in 1838 as being 25 ft. high, eight years after planting.

At Arley Castle a tree, probably planted about 1820, was, in 1904, 53 ft. high and 7½ ft. in girth near the ground, below the point where it gives off a large limb.

Judging from these, and from what I know of its native habitat, this species might be tried with a fair chance of becoming a small timber tree, in rich sandy soil and sheltered woods in the warmest parts of England; but the tree has such a strong tendency to become bushy, that unless carefully pruned and closely crowded it will form a large shrub rather than a tree.

Seedlings raised at Colesborne from seed which I collected near Ottawa grow slowly and seem to want more heat than they get here, but my soil probably contains more lime than this tree likes.

It is rare in Scotland, but Renwick[1] measured in 1907 a tree at Auchendrane, Ayrshire, 42 ft. high with a short bole 5 ft. 8 in. in girth, and dividing into two stems at three feet from the ground. It was planted in 1818. Walker[2] mentions a tree at Hopetoun House, which was planted in 1747, and cut down in 1788, when it was 3 ft. 10 in. in girth. It yielded a plank, a foot broad, of red wood, which was finely veined and took a good polish, equalling mahogany in appearance.

In an article on this species by Graf von Schwerin[3] a tree is mentioned as growing in the Palace garden at Rastede, Oldenburg, which at seventy-six years old is 15 metres high and 2.35 metres in girth at one metre from the ground. The photograph of this tree shows it to be a very well-shaped one with a fine head 12 metres in diameter; and a coloured plate of the leaves, flowers, and fruit is given. It is said to succeed best in dry sandy ground, and to be well worth cultivating as a forest tree on account of its beauty and the value of its wood.

This species has been tried as a forest tree in Germany. Schwappach lately reports[4] that it thrives well on fresh loamy sand, attaining about 50 ft. in height and 10 in. in diameter at twenty years old. It grows remarkably fast in youth, and is very suitable for filling up gaps in broad-leaved woods or in pine plantations. It succeeds best when mixed singly with beech, as it then forms a clean stem. When planted in groups, it is apt to become very branching. It has been tried[5] in Belgium for planting along roads, and has been successful between Curange and Zolder in Campine.

Mayr however considered that it will only be a success as a timber tree in those localities which have a warm summer, and rich light soil, for though it exists as far north as Canada, it only attains large size where the summers are longer and hotter than in any part of England, and where the soil is unusually fertile.

[1] *Trans. Nat. Hist. Soc. Glasgow*, viii. 234 (1907).   [2] *Essays on Nat. Hist.* 81 (1812).

[3] *Mitt. Deut. Dend. Ges.* 1906, p. 1.

[4] In *Zeitschrift Forst- und Jagdwesen*, xliii. 610 (1911).   [5] *Bull. Soc. Cent. Forest. Belg.* xvii. 180 (1910).

Timber, Bark

Next to black walnut, this was considered in the United States as the finest native hardwood in general use, but has now become so scarce, that it is hardly procurable for export and has been generally superseded as a furniture wood by imported mahogany. It is pale red, and when figured is extremely handsome, but such specimens are rare. A large board of it, which was given me from the State Exhibit of St. Louis at the Exhibition in 1904, was cut from a tree which grew in Cape Girardeau, County Missouri, and produced 2000 ft. of good lumber, of which more than a quarter was 2 ft. wide and over. From this a handsome table top has been made, the legs of which were cut from a tree of the same species grown at Arley Castle, of which a plank was kindly given me by Mr. R. Woodward.

Defebaugh[1] quotes the reminiscences of E. N. Mead of Buffalo, an early lumber-man in M'Kean County, Pennsylvania, as to the abundance and size of the cherry which existed there sixty years or more ago, very little of which is now left, as follows :—
" The operation with which I was connected, was a small one, only about 300 acres, but it was considered the best cherry grove in the county. We turned out a little over 3 million ft., or an average of 10,000 ft. per acre. We cut nothing under 12 in. at top, and not so small as 12 in. unless very smooth and straight. It would, I suppose, be impossible to find any stock to-day which would approach this in quality. It was pronounced the finest ever sent to the Albany market. We cut it all with a circular saw. With a modern bandsaw we could have produced lots of stock 36 inches wide. I will relate one circumstance that occurred. Two of my log-cutters sawed down most of their trees. In this case they cut entirely through the tree, driving wedges behind the saw, and the tree stood on the stump 24 hours, until a breeze toppled it over. It was over 3 ft. in diameter at the butt, made four 16 ft. logs, and stood straight as a gun barrel."

The bark[2] of _P. serotina_ has long been used medicinally, and is recognised both by the United States and British Pharmacopeias.

The fruit is small and black, like that of the Portugal laurel, and was used in America to flavour brandy, the flavour being superior to that of the common cherry.   (H. J. E.)

---

[1] _Lumber Industry of America_, ii. 618 (Chicago, 1907).

[2] Cf. Power and Moore, in _Trans. Chem. Soc._ xcv. 243-261 (1909), who have analysed the bark, which yields prussic acid and many other constituents. The same authorities, in _Trans. Chem. Soc._ xcvii. 1009 (1910), give an account of the chemical properties of the leaves. The barks of other species of Prunus are frequently used as adulterants. Cf. E. M. Holmes, in _Pharm. Journ._ 1909, p. 192.

PRUNUS LAUROCERASUS, COMMON LAUREL

Prunus Laurocerasus, Linnæus, *Sp. Pl.* 474 (1753); Bentley and Trimen, *Medicinal Plants*, ii. t. 98 (1880).

Padus Laurocerasus, Miller, *Gard. Dict.* ed. 8, No. 4 (1768).

Cerasus Laurocerasus, Loiseleur, *Nouv. Duhamel*, v. 6 (1812); Loudon, *Arb. et Frut. Brit.* ii. 716 (1838); Boissier, *Fl. Orient.* ii. 650 (1872).

Laurocerasus officinalis, Roemer, *Fam. Nat. Syn.* iii. 91 (1847); Schneider, *Laubholzkunde*, i. 646 (1906).

An evergreen large shrub, occasionally arborescent. Young branchlets glabrous. Leaves coriaceous, persistent two years, 5 to 6 in. long, 2 to 3 in. broad, obovate-oblong or narrowly elliptic, acuminate at the apex, tapering at the base, with a few remote minute serrations; glabrous; shining above; duller and lighter green beneath, and marked on each side of the midrib near the base with one or two circular glands; lateral nerves pinnate, about 8 to 10 pairs, dividing and looping before reaching the margin; petiole short, stout, without glands.

Flowers, in erect leafless racemes, about 4 in. long, arising in an axil of a leaf on the preceding year's shoot; axis and pedicels glabrous; calyx-tube wide at the mouth; sepals minute, triangular, often with peculiar teeth; petals small, wrinkled; ovary superior, glabrous, green, with a short glabrous style, and a capitate stigma. Fruit ellipsoid, ½ in. long, plum-coloured when ripe, depressed at the base; flesh scanty; stone ovoid, pointed at the apex, smooth, with a prominent ridge on one side.

VARIETIES

I. The common laurel varies in the wild state, the typical large-leaved form being common at low altitudes in the Caucasus; while at high elevations, between 6000 and 7000 feet on limestone formation, there occurs a form with shorter racemes and smaller leaves, var. *brachystachius*, Albow, *Fl. Colchica*, 68 (1895). Another form with longer narrower leaves, var. *laurifolius*, Albow, is met with in the valleys of Guria. The following appears to be the European variety of the species:

1. Var. *schipkaensis*, Späth, *ex* Dippel, *Laubholzkunde*, iii. 649 (1893). Leaves, lanceolate or narrowly elliptic, about 3 in. long, and 1 in. broad, entire in margin; glands on the back inconspicuous or absent. Racemes, 1½ in. long; calyx green; sepals broad, triangular, each with two minute reddish glands on the margin; petals white, orbicular, not wrinkled.

This is a small shrub, about 3 ft. high, which has been found wild on the Balkan range, near the Shipka Pass, and in other localities. It was introduced into cultivation by Späth, and has proved much hardier[1] than the type in Switzerland and Germany, and thrives in the United States as far north as central New York.[2] It is in cultivation at Kew and Cambridge.

---

[1] *Mitt. Deut. Dend. Ges.* 1898, p. 96.   [2] Rehder, in Bailey, *Cyc. Amer. Hort.* 1455 (1901).

Var. *serbica*, Pančic, from Servia, is similar, but not exactly identical, having a more upright growth, with obovate wrinkled leaves. It is not in cultivation in England; but has proved very hardy[1] in Germany, where it bore − 20° Cent. at Bergsdorf.

II. A large number of varieties[2] have appeared in cultivation, of which the more important are :—

2. Var. *angustifolia*, Nicholson, *Kew Hand-List Trees*, i. 145 (1894). Leaves long and narrow, scarcely an inch in breadth.

3. Var. *rotundifolia*, Nicholson, *loc. cit.* Leaves rounded at the apex. This is said[3] to be more suitable for making hedges than the type, and succeeds better than it in towns.

4. Var. *parvifolia*, Nicholson, *loc. cit.* Leaves, $1\frac{1}{2}$ in. long, $\frac{1}{2}$ in. broad, with a few coarse serrations. A low shrub, in cultivation at Kew.

5. Var. *camelliæfolia*, Nicholson, *loc. cit.* Leaves bent back, and twisted on their base. A curious form, not common in cultivation. The best specimen that we have seen is at Poles Park, Herts.

6. Var. *variegata*, Nicholson, *loc. cit.* Leaves blotched with white.

DISTRIBUTION AND CULTIVATION

The common laurel is a native of the Balkan Peninsula, Asia Minor, the Caucasus, and North Persia. It is most common in the Caucasus, where, however, it is not known in Georgia or Talysch, being confined to the west, in Abchasia, Mingrelia, and Imeritia, where the typical broad-leaved form occurs at all elevations between sea-level and 4000 ft., being replaced higher up by peculiar shrubby forms.[4] Boissier records it in Asia Minor from near Trebizond, and at the base of Mount Olympus in Bithynia. In Europe, it is recorded by Adamovic[5] for south-eastern Servia, the Balkan range between Rumelia and Bulgaria, Thrace, and Laconia in Greece.[6]   (A. H.)

The laurel was introduced to Vienna in 1576 by Ugnad from Constantinople, at the same time as the horse chestnut; and soon spread over Europe. According to Evelyn, it was first brought into England, in 1614, by the Countess of Arundel, at Wardour Castle, where, Loudon says, there were in his time a great number of very old laurels. Parkinson in 1629 says that he had seen it in fruit at Highgate; and it became very common at an early period in English gardens, and in many parts of Ireland, where old houses are often surrounded by a dense thicket of laurels, which grow in that country, as in all the moister parts of England, with great luxuriance. It is now by no means so popular as formerly, but on account of the facility with which it can be reproduced from cuttings and layers, and its persistence in coming up from the stool when cut to the ground, it is likely to remain one of our commonest garden shrubs.

[1] *Mitt. Deut. Dend. Ges.* 1897, p. 68.
[2] Cf. *Gard. Chron.* v. 620, figs. 105, 106 (1889).
[3] Cf. *Gard. Chron.* viii. 572 (1890).
[4] Radde, *Pflanzenverb. Kaukasus.* 178, 347 (1899).
[5] *Veg.-verhalt. Balkanländer*, 132, 464, 489 (1909).
[6] Cf. Halacsy, *Consp. Fl. Græcæ*, i. 498 (1901).

By far the largest on record in England, is mentioned by Mr. E. C. Batten,[1] who quotes Dr. Prior for the fact that a common laurel at Fyne Court, Somersetshire, grew to the height of 72 ft. before it was cut. At Powis Castle I have seen a bush over 40 ft. high with five stems over 4 ft. in girth. It was cultivated at Mill Hill by Collinson about 1750, and I have seen at that place what is supposed to be part of the original plant.

At Shelton Abbey, Wicklow, there was a magnificent specimen in Loudon's time, which was 45 ft. in height and 6 ft. in girth, at 90 years old.

The laurel is not hardy in most parts of Germany, and at Karlsruhe suffers much in severe winters; but at Baden-Baden and Mainau, where the humidity of the air is greater, it succeeds fairly well.[2] In the United States, it is hardy as far north as Washington, D.C.[3]

The leaves are used in medicine; and when distilled with water, yield bitter almond oil and prussic acid.[4] The fresh leaves are sometimes used for flavouring sweetmeats, custards, creams, etc.; but should be used with caution, as, on account of their poisonous qualities, they may produce injurious or even fatal effects.

The wood is, so far as I know, of no value except for firewood.

(H. J. E.)

PRUNUS LUSITANICA, PORTUGAL LAUREL

Prunus lusitanica, Linnæus, *Sp. Pl.* 473 (1753).
Padus lusitanica, Miller, *Gard. Dict.* ed. 8, No. 5 (1768).
Padus eglandulosa, Moench, *Meth.* 672 (1794).
Cerasus lusitanica, Loiseleur, *Nouv. Duham.* v. 5 (1812); Loudon, *Arb. et Frut. Brit.* ii. 714 (1838).
Laurocerasus lusitanica, Roemer, *Fam. Nat. Syn.* iii. 92 (1847).

An evergreen tree, attaining occasionally 50 or 60 feet in height, and 6 feet in girth; often shrubby. Young branchlets glabrous. Leaves persistent two years, coriaceous, but thinner in texture than those of *P. Laurocerasus*, ovate-oblong, 3 to 4 in. long, 1½ to 2 in. broad, acuminate at the apex, rounded at the base; glabrous; above shining, beneath lighter green and without glands; lateral nerves 8 to 10 pairs, dividing and looping before reaching the margin; regularly serrate, the serrations ending in glandular points; petiole about ¾ in. long, usually without glands.

Flowers in racemes, about 4 in. long, arising in the axils of the leaves on the preceding year's shoot; axis and pedicels glabrous; calyx wide at the mouth; sepals irregular, dentate; petals white. Fruit ovoid, with scanty flesh, about ⅜ in. long.

[1] *Trans. Eng. Arb. Soc.* ii. 221 (1895).   [2] *Mitt. Deut. Dend. Ges.* 1908, p. 150.
[3] Rehder, in Bailey, *Cycl. Amer. Hort.* 1455 (1901).
[4] Cf. Flückiger and Hanbury, *Pharmacographia*, 255 (1879).

VARIETIES

I. In addition to the typical form, which occurs in Spain and Portugal, the following geographical varieties are known :—

1. Var. *Hixa*, De Candolle, *Prod.* ii. 540 (1825); Lowe, *Flora Madeira*, i. 236 (1836). Leaves narrower, more oblong, about 5 in. long and 1½ in. broad. Racemes, 6 to 8 in. long, with flowers less crowded than in the type.

Indigenous in the Madeira and Canary Islands. Lowe states that in 1836, it was nearly extinct in Madeira; but a few trees remained, 40 to 60 ft. in height, and occasionally 6 ft. in girth. According to Webb and Berthelot,[1] it grows on the north-east of Teneriffe in woods, at about 2000 feet, attaining in the wood of Las Mercedes 30 ft. in height. Webb, however, informed Loudon that on Teneriffe, Grand Canary, and Palma, it occasionally reached a height of 60 or 70 ft.

2. Var. *azorica*, Nicholson, in *Kew Hand-List Trees*, i. 147 (1894). Leaves more coriaceous, and more coarsely serrate than in the type. Racemes short, densely flowered. Indigenous in St. Miguel in the Azores, where it was collected by Hunt in 1845, and later by Godman.[2] This variety, which is in cultivation at Kew, is shrubby.

II. Several varieties have arisen in cultivation as seedlings :—

3. Var. *myrtifolia*, Nicholson, *loc. cit.* A shrub of compact habit, with small leaves, 1½ to 2 in. long.

4. Var. *variegata*, Nicholson, *loc. cit.* Leaves variegated with white. This is said by Koch, *Dendrologie*, i. 125 (1869), to be cultivated in France.

DISTRIBUTION AND CULTIVATION

P. lusitanica is a native of Spain and Portugal, and of the Azores, Canary, and Madeira Islands. In Portugal, where it is called *azareiro*,[3] it is apparently not a common tree, as the only localities mentioned by Willkomm,[4] where it is indigenous, are the Serras of Bussaco, Estrella, and Gerez.[5] It is abundant in the woods of Bussaco, where Elwes[6] saw it in 1909 attaining a height of 50 to 60 feet and 5 to 6 ft. in girth.[7] On the Serra de Gerez, it grows up to 3000 ft. elevation as a scattered tree in woods composed mainly of *Quercus pedunculata*, Arbutus, holly, and sycamore. Webb informed Loudon that it formed here a small tree 20 ft. high; but growing with it was a taller tree, 60 to 70 ft. high, which he supposed to be var. *Hixa*, but which was undoubtedly the same.[8] In Spain this species is rare and

---

[1] *Hist. Nat. Isles Canar.* iii., *Phyt.* ii. 19, t. 38 (1836). The leaves figured scarcely differ from those of the Portuguese tree; but the flowers are in longer racemes.

[2] Watson, in Godman, *Nat. Hist. Azores*, 158 (1870). 　　　　[3] Broteiro, *Fl. Lusit.* ii. 252 (1804).

[4] *Pflanzenverb. Iber. Halbinsel*, 112, 312, 318, 321 (1896). 　　[5] It was collected also at Cintra by Welwitsch.

[6] In the forest of Bussaco the Portugal laurel attained a height of 50 to 60 ft., with trunks 5 to 6 ft. in girth, but it did not form an important element in the forest, either here or in the Serra de Gerez, where it was smaller. I did not see in the Serra de Gerez any tree so large as the ones mentioned by Webb.—H. J. E.

[7] Cf. J. de Vilmorin, in *Bull. Soc. Dend. France*, 1907, p. 49.

[8] Webb, *Iter. Hisp.* 48 (1838), mentions only one form of *P. lusitanica*, as occurring in the Serra de Gerez.

local, occurring according to Willkomm on the north slope of Montseny, near Barcelona, and in a few woods in Navarre.   (A. H.)

The Portugal laurel is stated by Aiton to have been introduced into the Oxford Botanic Garden in 1648; and Loudon says that this tree survived until 1828, when it was cut down, the trunk being nearly 2 ft. in diameter; but Collinson[1] states that it was first brought to England in 1719 by Fairchild, the famous nurseryman at Hoxton.

Though the Portugal laurel is hardy in the greater part of England, it is liable to be severely injured, and sometimes killed to the ground in very severe winters, and grows best in the south and west of England, especially near the sea. It ripens fruit in most seasons, and is easy to raise from seed; and though an ornamental tree of some merit, is not so popular now as it formerly was.

The oldest living tree that I know of is one at Mill Hill, where it was probably planted by Peter Collinson about 160 years ago, and is now decaying. It had five stems over 4 ft. in girth. At Fyne Court, Somersetshire, Mr. Batten,[2] quoting Dr. Prior, says that Portugal laurels of enormous size were formerly seen with their boughs bending to the ground and rooting, and an upright trunk grew from the arch thus formed. Henry saw at Belvoir Castle, in 1907, a well-shaped tree, 40 ft. in height, and 6 ft. in girth; and another at Leonardslee in 1910, 45 ft. by 4 ft. 3 in.

In Scotland it is hardy as far north as Banffshire, and grows to a very large size at Gordon Castle. At Moncrieffe, Hunter[3] mentions a specimen which in 1883 covered an area 186 ft. round, though cut back on one side, but I did not see this when I visited the place in 1907. At Biel, the seat of Mrs. Hamilton Ogilvy in East Lothian, I saw a very fine tree in 1911, with a clean trunk, 4 ft. 8 in. in girth. Bean[4] saw a specimen at Ochtertyre in 1907, which was 30 ft. high, with a spread of foliage 50 ft. in diameter; and there are two trees at Raith, with short stems, nearly 2 ft. in diameter.

In Ireland we have not noted any remarkable for their size, but, as a rule, it grows luxuriantly.

The Portugal laurel, though rarely planted in Germany, where it is supposed not to be hardy, has borne at Karlsruhe − 16° C. of frost in winter; and produces flowers and fruit every year at Mainau.[5]

The wood, as shown by a fine specimen in the Earl of Yarborough's exhibit at Lincoln in 1907, resembles that of the cherry in colour, and shows well-marked medullary rays which make it very ornamental. It seems well fitted for small cabinet work.   (H. J. E.)

---

[1] Dillwyn, *Hortus Collinson.* 11 (1843).   [2] *Trans. Eng. Arb. Soc.* ii. p. 221 (1895).
[3] *Woods, Forests, and Estates of Perthshire*, 136 (1883).   [4] *Gard. Chron.* xli. 168 (1907).
[5] *Mitt. Deut. Dend. Ges.* 1907, p. 258.

PYRUS (*continued*)

In our Article in Vol. I. pp. 141-170, the genus Pyrus has been defined, and some of the sections have been fully described. In this concluding part, an account will be given of the apples, pears, and mountain ashes, which constitute three sections of the genus.

SECTION PYROPHORUM

This section of the genus Pyrus comprises the true pears, which are deciduous trees or shrubs, with branchlets of two kinds, long shoots and short shoots, the latter spur-like and often ending in thorny points. Leaves simple, stalked, scattered on the long shoots, clustered on the spurs; in the bud rolled inwards towards the midrib. Flowers perfect, in corymbs on the short shoots; sepals five, usually persistent on the apex of the fruit, occasionally deciduous; ovary usually five-celled, rarely two- or three-celled; styles five, rarely two or three; fruit turbinate or sub-globose, with granular flesh.

The true pears comprise about sixteen species, natives of Europe, northern Africa, and extra-tropical Asia. The following synopsis gives a brief account of the wild species which are in cultivation :—

I. *Leaves deeply cut into small segments.*

   1. *Pyrus heterophylla*, Regel and Schmalhausen, in *Act. Hort. Petropol.* v. pt. ii. 581 (1878).

A small thorny tree, with glabrous branchlets. Leaves very remarkable, about 2 in. long, glabrous, deeply and pinnately cut to near the midrib into about five variously lobed and serrate narrow segments. Fruit depressed-globose, about 1 in. in diameter.

This occurs in mountain valleys at high elevations in Turkestan. It is represented at Kew by a straggling bush about 3 ft. high.

II. *Leaves sharply serrate, the serrations ending in long slender points.*

   2. *Pyrus sinensis*, Lindley, in *Trans. Hort. Soc.* vi. 397. (1826), and *Bot. Reg.* t. 1248 (1829).

Pyrus ussuriensis,[1] Maximowicz, *Prim. Fl. Amur.* 102 (1859).
Pyrus Simonii,[2] Carrière, in *Rev. Hort.* 1872, p. 28.

---

[1] This is the Manchurian pear, which is said by Ascherson and Graebner, *op. cit.* 60, to be the earliest of all pears to flower in the Berlin Botanic Garden, where it was raised from seed sent by Maximowicz.

[2] Sent to the Jardin des Plantes at Paris by Simon in 1861. This is considered by Bretschneider, *Hist. Europ. Bot. Disc. China*, 830 (1898), to be a cultivated variety, the *Pai-li* of north China, which bears delicious apple-shaped fruit of a pale yellow colour.

A tree without thorns, attaining 60 ft. in height. Branchlets and buds glabrous. Leaves, 3 to 4 in. long, ovate, acuminate, glabrous, glandular on the midrib above, distinct from all the other species in the fine serrations with long slender points. Fruit globose, about 1 in. in diameter.

A native of China, Manchuria, Korea, and Japan. Introduced in 1820. It is the *Sha-li* or "sand pear" of the Chinese. Many cultivated varieties are known, some of which have proved useful in the United States, where this species is an excellent ornamental tree, very vigorous in growth.[1] Trees at Kew about 25 ft. high are very thriving.[2]

III. *Leaves sharply serrate, the serrations without long points.*

3. *Pyrus betulæfolia*, Bunge, *Enum. Pl. Chin. Bor.* 27 (1834).

A small tree without thorns. Branchlets and buds grey tomentose. Leaves, $1\frac{1}{2}$ to $2\frac{1}{2}$ in. long, ovate to ovate-rhombic, acuminate; dark green and shining above with glands on the midrib, lighter green beneath, both surfaces retaining in summer traces of the tomentum with which they were covered in spring; petiole long, tomentose. Flowers with a two-celled ovary and two styles. Fruit depressed-globose, about $\frac{2}{5}$ in. in diameter.

A native of north China, where it is called *tu-li.* Introduced first in 1863 by Simon[3] into the Jardin des Plantes at Paris; and subsequently by Bretschneider,[4] who sent seeds to Kew and the Arnold Arboretum[5] in 1882. A tree at Kew of this origin, about 30 ft. high with pendent branches, is very thriving.

4. *Pyrus syriaca*, Boissier, *Diag. Nov. Pl. Orient.* x. 1 (1849).

A shrub or small tree, usually thorny. Branchlets and buds glabrous. Leaves, $1\frac{1}{2}$ to 2 in. long, lanceolate or obovate; acute, rounded, or mucronate at the apex; variable at the base, often very tapering, and decurrent on the petiole; glabrous; sharply and finely serrate. Fruit turbinate, $1\frac{1}{2}$ in. in diameter, with a thickened stalk.

A native of Cyprus, Syria, Asia Minor, and Kurdistan. A small tree at Kew,[6] about 15 ft. high, was obtained from Decaisne in 1874.

IV. *Leaves crenate in margin.*

* *Leaves coriaceous in texture.*

5. *Pyrus Korshinskyi*, Litwinow, in *Trav. Mus. Bot. Acad. Imp. Sc. St. Petersb.* i. 17 (1902).

A tree, height not stated. Branchlets and buds grey tomentose. Leaves coriaceous, about 3 in. long and $\frac{1}{2}$ to $1\frac{1}{2}$ in. broad, lanceolate or ovate-oblong, more or less grey tomentose on both surfaces; with coarse crenate or bi-crenate

---

[1] Cf. Bailey, *Cyc. Am. Hort.* 1470, 1471 (1901).

[2] A tree at Kew labelled *P. sinensis*, with coarse sharp-pointed serrations to the leaves, is perhaps *P. Balansæ*, Decaisne, *op. cit.* t. 6 (1871), raised from seed brought from Laristan in Persia. It is intermediate in foliage between *P. sinensis* and *P. communis*.

[3] Cf. Carrière, *Rev. Hort.* 1879, pp. 318, 319, where the plant is figured from specimens cultivated at Paris.

[4] *Hist. Europ. Bot. Disc. China*, 1053 (1898).

[5] Cf. *Garden and Forest*, vii. 224 (1894), where a figure is given.

[6] This species has been confused with *P. glabra*, Boissier, *Diag. Nov. Pl. Orient.* vi. 53 (1845), which is not in cultivation. The latter, a native of Persia, differs from all the species, in having glabrous entire lanceolate leaves.

serrations, tipped with peculiar minute sharp incurved glands; petiole long, grey tomentose.   Fruit sub-globose, $\frac{7}{8}$ in. in diameter, on stout peduncles, crowned by the persistent calyx.

This very distinct species[1] occurs in the mountains of Bokhara, Fergana, and Turkestan, at 4000 to 7000 ft. elevation.   It is represented at Kew by a tree[2] about 20 ft. high, obtained from Dieck in 1891, and by a shrub obtained from Späth in 1900.

** *Leaves not coriaceous.*

(a) *Midrib glandular above (in some of the leaves at least).*

6. *Pyrus Pashia*, Buchanan-Hamilton, *ex* Don, *Prod. Fl. Nepal.* 236 (1825).

A variable species in the Himalayas, of which the following form is in cultivation :—

Var. *Kumaoni*, Stapf, in *Bot. Mag.* t. 8256 (1906).

A tree without thorns, attaining about 50 ft. in height.   Branchlets and buds glabrous.   Leaves narrowly ovate, 2 to $3\frac{1}{2}$ in. long, $1\frac{1}{4}$ to 2 in. broad, glabrous, subcordate at the broad rounded base, contracted into a long acuminate apex, frequently glandular on the midrib above; petiole long, glabrous.   Fruit globose, 1 in. in diameter, from the apex of which the sepals fall off early.

This is represented at Kew by a grafted tree, about 25 ft. high, which has been growing in the collection of Rosaceæ for many years.[3]   The flowers are handsome, at first suffused with pink, ultimately becoming white.

(b) *Midrib without glands.*

7. *Pyrus communis*, Linnæus.   See p. 1560.

V. *Leaves entire in margin, or partly indistinctly crenulate.*

* *Midrib glandular above.*

8. *Pyrus elæagrifolia*, Pallas, in *Nov. Act. Petrop.* vii. 355, t. 10 (1793).

A spiny tree or shrub.   Branchlets and buds grey tomentose.   Leaves about 2 in. long and $\frac{3}{4}$ in. broad, obovate or obovate-lanceolate, usually tapering in the basal half, acute or rounded and mucronate at the apex, entire in margin; upper surface with scattered tomentum throughout, and peculiar dark glands on the midrib; lower surface densely covered with grey tomentum, obscuring the venation; stalks short, tomentose.   Fruit pyriform, 1 in. long, with the upper part of the peduncle much thickened.

A native of the Crimea, Caucasus, and Asia Minor.   Introduced in 1800.

There is a good specimen at Arley Castle, which Elwes found to be 44 ft. by 4 ft. in 1911.   It is forked at two feet from the ground, and bears fruit regularly in some quantity.   There is a tree of this species at Glasnevin, which has long been labelled *P. sinaica*.[4]   Elwes measured it, in 1908, as 40 ft. by 5 ft. 7 in.,

---

[1] *P. bucharica*, Litwinow, *op. cit.* i. 18 (1902), described from a sterile branch and said occasionally to have lobed and pinnatifid leaves, appears to be the same species.

[2] Both the specimens at Kew were obtained under the name *P. heterophylla*, which is a totally distinct species. Cf. p. 1556.

[3] The history of this tree is unknown.   *P. Pashia* is said by Loudon, *Arb. et Frut. Brit.* ii. 891 (1838), to have been introduced in 1825 ; but we have seen no specimens except the tree at Kew.

[4] Cf. p. 1559, note 1.

with a bole 10 ft. long ; and it appears to be one of the earliest trees planted in this garden, which was founded in 1798. A tree at Beauport, Sussex, 32 ft. by 5 ft., was bearing fruit in November 1911.

** *Midrib without glands.*

(a) *Under surface of the leaves with only traces of tomentum in summer.*

9. *Pyrus amygdaliformis*, Villar, *Cat. Meth. Jardin Strasbourg*, 323 (1807).

A spiny shrub or small tree. Branchlets and buds slightly tomentose. Leaves, 1 to 2 in. long, coriaceous; variable in shape, lanceolate to narrowly elliptical, acuminate or rounded with a mucro at the apex; covered in spring on both surfaces with grey tomentum, which disappears in greater part during summer, usually entire in margin; petiole short, slightly tomentose. Fruit sub-globose, $\frac{3}{4}$ in. in diameter, with the stalk scarcely thickened.

A native[1] of southern Europe, occurring in France on dry arid soil in the region of the olive, and spread through Spain, Italy, Istria, Dalmatia, Balkan States, Greece, and Asia Minor. In March 1910 Elwes saw a fine tree in flower on the Plan d'Aups near St. Baume, in the department of Var, at about 3000 ft. altitude. It was about 25 ft. by 6 ft.

Introduced in 1810, and occasionally seen in botanic gardens; it attains at Cambridge 30 ft. in height.

(b) *Leaves densely tomentose beneath in summer.*

10. *Pyrus nivalis*, Jacquin, *Fl. Austr.* ii. 4, t. 107 (1774).

A tree without thorns. Young branchlets and buds more or less tomentose. Leaves $2\frac{1}{2}$ to 3 in. long, $1\frac{1}{4}$ to $1\frac{1}{2}$ in. broad, elliptic to obovate-oblong, usually cuneate at the base, abruptly contracted into an acuminate apex; covered at first with a grey tomentum, persisting in summer on the under surface, and to a less extent on the shining green upper surface; margin entire or irregularly and minutely crenulate towards the apex; petiole tomentose, about 1 in. long. Fruit pyriform, about $1\frac{1}{2}$ in. in diameter.

This tree, which is known in Austria as the *schnee birn* or "snow pear," is doubtfully wild in Croatia, Hungary, and Transylvania. Schneider regards it as a cultivated form of *P. elæagrifolia*; but Focke[2] with more probability considers it to be a cross between *P. amygdaliformis* and *P. communis* which has escaped from cultivation. It appears to be closely allied to, if not identical with *P. salvifolia*,[3] De Candolle, *Prod.* ii. 634 (1825), which is naturalised in woods and hedges in central France, and often cultivated for making perry.

P. nivalis was introduced in 1826, and is represented at Kew by a tree about 25 ft. high.

[1] *P. persica*, Persoon, *Syn. Pl.* ii. 40 (1807) is probably a hybrid, arising from *P. amygdaliformis*, as there are no grounds for supposing it to be a native of Persia or Mount Sinai, as was formerly supposed. It has obovate-oblong or elliptic leaves, 2 to $2\frac{1}{2}$ in. long, 1 in. broad, entire in margin, mucronate at the rounded apex, dark shining green and slightly tomentose above, pale and with scattered tomentum beneath; buds glabrous. This is represented at Kew by a tree 20 ft. high obtained in 1875 from Decaisne, as *P. sinaica*, Thouin, in *Mém. Mus. Hist. Nat.* i. 170 (1815); and by an older tree, 30 ft. high, similarly labelled, at Cambridge. Decaisne, *Jardin Fruitier*, t. 15 (1871), however, gives a figure of *P. sinaica*, which does not agree in foliage with the tree which he sent to Kew.

[2] *Ex* Ascherson and Graebner, *op. cit.* 65 (1906).   [3] Cf. Mathieu, *Fl. Forest.* 170 (1897).

11. *Pyrus salicifolia*, Pallas, *Itin.* iii. 734 (1736).

A tree, often spiny, about 30 ft. high. Branchlets grey tomentose. Buds with brown ciliate scales, usually glabrous on the surface. Leaves, 2 to 3 in. long, averaging $\frac{1}{2}$ in. broad, linear-lanceolate, very tapering at both ends, and often prolonged at the apex into a sharp point; entire in margin; covered slightly on the upper surface and densely on the lower surface with a white silky appressed tomentum; petioles short, tomentose. Fruit turbinate, about $\frac{3}{4}$ inch in diameter, with a persistent calyx.

A native of the Crimea, Caucasus, and Armenia. Introduced in 1780, and often planted as an ornamental tree, on account of its whitish foliage, which at a distance resembles that of *Salix alba*.

P. canescens,[1] Spach, *Hist. Veg.* ii. 129 (1834), judging from a tree at Kew, about 30 ft. high, obtained from Decaisne in 1875, is possibly a hybrid of *P. salicifolia*. It has lanceolate or narrowly elliptic leaves, about $2\frac{1}{2}$ in. long, acute or mucronate at the apex, minutely crenulate, often twisted, and resembling in tomentum those of *P. salicifolia*. The buds are also like those of the latter species.   (A. H.)

PYRUS COMMUNIS, COMMON PEAR

Pyrus communis, Linnæus, *Sp. Pl.* 459 (1753); Loudon, *Arb. et Frut. Brit.* ii. 880 (1838); Willkomm, *Forstliche Flora*, 843 (1887); Mathieu, *Flore Forestière*, 167 (1897); Schneider, *Laubholzkunde*, i. 661 (1906); Ascherson and Graebner, *Syn. Mitteleurop. Flora*, vi. pt. 2, p. 60 (1906).

A tree or shrub, with numerous short shoots or spurs, which often end in thorny points. Bark smooth at first, ultimately broken on the surface into small scales. Young branchlets glabrous. Leaves, scattered on the long shoots, clustered on the short shoots, variable in size and shape, usually ovate or oval, rounded or subcordate at the base, acute or shortly acuminate at the apex; minutely crenate in margin except occasionally near the base; slightly tomentose when young, nearly quite glabrous in summer, dark green and shining above, paler beneath; petiole slender, nearly as long as or even exceeding the blade in length. In winter the buds are ovoid, pointed, shining brown, with a few glabrous ciliate scales; lateral buds, nearly as large as the terminal bud, either appressed or slightly diverging from the twig; leaf-scars crescentic, three-dotted.

Flowers, six to twelve in a leafy corymb, the axis of which, together with the pedicels, and external surface of the calyx-tube and sepals, is more or less covered with greyish tomentum; inner surface of the sepals with a dense rusty tomentum; petals white, with a short claw; styles five, free, almost as long as the fifteen to twenty stamens. Fruit turbinate, narrowing gradually towards the thickened stalk, crowned by the persistent calyx.

---

[1] *P. canescens*, Decaisne, *Jardin Fruitier*, t. 19 (1871) does not appear to agree with Spach's description.

Varieties and Hybrid

1. Var. *Pyraster*, Linnæus, *Sp. Pl.* 479 (1753), is similar to the type in foliage and thorns; but has globose fruit. It is of uncertain origin, and is seldom found in woods, being probably in most cases an escape from an adjoining orchard.

2. Var. *sativa*, De Candolle, *Prod.* ii. 634 (1825). This name is applied to the cultivated varieties[1] of the pear, which are usually large trees without thorns. They also differ from the wild type, in having larger foliage, and larger and more edible fruit. Many forms of cultivated pears are probably, however, of hybrid origin; and can scarcely be assigned to var. *sativa*.

The following, which have been described as three distinct species, are probably geographical races of *P. communis*.

3. Var. *cordata*, J. D. Hooker, *Student's Flora*, 131 (1878).

> Var. *azarolifera*, Durieu de Massoneuve, in *Bull. Soc. Bot. France*, v. 726 (1858).
> Var. *Briggsii*, Boswell-Syme, in *Journ. Bot.* ix. 182 (1871).
> *Pyrus cordata*, Desvaux, *Obs. Pl. Anjou*, 152 (1818); Decaisne, *Jardin Fruitier*, i. 330 (1871); Masters, in *Journ. Bot.* xiv. 225, t. 180 (1876); Hy, in *Bull. Herb. Boissier*, 1895, App. 1, p. 9.
> Fliche, in *Bull. Soc. Bot. France*, xlvii. 108 (1900).

A spiny shrub, said by Fliche to propagate itself freely by root-suckers. Leaves smaller than in the type, sub-orbicular to ovate, about 1 in. in width, sub-cordate at the base. Flowers smaller than in the type. Fruit, globose on the French shrub, slightly turbinate on the English plant, very small, not exceeding ½ in. in diameter; calyx persistent.

This remarkable pear is wild in the west of France, in Brittany, Anjou, and the Landes. The English form, which differs only in the shape of the fruit, was first found in a hedge between Thornbury and Wood Common in Devon, and later in two hedgerows in Cornwall[2]; but is now known to be undoubtedly wild[3] in the valley of the Wye, at Symonds Yat, on Dorward, and in Dixton parish. There is a specimen at Kew about 12 ft. high, which was obtained from Veitch in 1898.

A similar plant, *P. Boissieriana*, Buhse, *Aufzähl. Pfl.* 87 (1860), found on Mt. Elburz, in Persia, was identified by Boissier[4] with *P. cordata*; and, as no similar pear has been found in the vast region intervening between western France and northern Persia, this disjointed distribution has given rise to much speculation.[5]

4. Var. *longipes*, Henry (var. *nova*).

> *Pyrus longipes*, Cosson and Durieu, in *Bull. Soc. Bot. France*, ii. 310 (1855).

A small tree, with a few spines. Leaves, about 2 in. long and 1 in. broad, ovate, acuminate, sub-cordate, glabrous, finely and crenately serrate, on long slender petioles. This, which differs little in foliage from *P. cordata*, has also

---

[1] These are described by Decaisne, *Jardin Fruitier du Muséum* (1871-1872).

[2] Davey, *Flora of Cornwall*, 183 (1909).

[3] Riddelsdell, in *Journ. Bot.* xlix. 170 (1911). A small tree about 10 ft. high, which produces flowers very sparingly, on the cliff at Pen Moel, Chepstow, is probably wild. [4] *Fl. Orient.* ii. 653 (1872).

[5] Dr. Phené associates *P. cordata* with the island of Avalon (which means "apples"), now Glastonbury, where King Arthur is said to have been buried. Cf. *Gard. Chron.* iv. 684 (1875).

small globose fruit, about ½ in. in diameter, from the summit of which the calyx ultimately falls off completely.   It is a native of the borders of mountain streams in Algeria.   A tree at Kew, about 25 ft. high, was obtained from Decaisne in 1875.

5. Var. *Mariana*, Willkomm, in *Linnæa*, xxv. 25 (1852).

Pyrus Bourgæana, Decaisne, *Jardin Fruitier*, i. t. 2 (1871).

A small tree.   Leaves ovate, about an inch in length, rounded at the base, on very long slender petioles.   Fruit globose, about ½ in. in diameter, with a persistent calyx.   This is a little-known variety, which occurs in the Sierra Morena in Spain.

6. *Pyrus auricularis*, Knoop, *Pomol.* ii. 38 (1763).

Pyrus irregularis, Muenchhausen, *Hausvater*, v. 246 (1770).

Pyrus Pollveria, Linnæus, *Mant.* ii. 244 (1771).

Pyrus Bollwyleriana, De Candolle, *Fl. France, Suppl.* v. 530 (1815); Loudon, *Arb. et Frut. Brit.* ii. 890 (1838).

A tree, attaining about 50 ft. in height.   Branchlets and buds tomentose. Leaves, 3 to 4 in. long, 1 to 2 in. broad, elliptic, rounded, and unequal at the base, shortly acuminate at the apex; margin irregularly, coarsely, and sharply serrate or biserrate; upper surface shining green, glabrescent, with glands on the midrib; lower surface more or less covered with greyish tomentum; petiole ¾ to 1½ in. long, grey tomentose.   Flowers five to twenty, in tomentose corymbs: sepals tomentose on both surfaces; styles two to five, united and tomentose at the base. Fruit pyriform, 1 in. in diameter, reddish yellow, sweet to the taste.

This tree, which is a hybrid between *P. communis* and *P. Aria*, was first noticed at Bollweiler in Alsace, about 1650, when it was described by Bauhin.[1] It apparently does not come true from seed; and one of the seedlings, which is nearer to *P. communis*, has been named var. *bulbiformis*, Tartar.[2]   *P. auricularis* was introduced in 1786, and is occasionally seen in botanic gardens.   A large tree of it was cut down in Kew Gardens some years ago, of which a board a foot wide is now in the Cambridge Forestry Museum.   There is a fine specimen at Arley Castle, which measured 59 ft. by 4 ft. 5 in. in 1904.   Elwes saw an old tree, 43 ft. high, with a short bole, near the house at Beauport, Sussex, in 1911.   It bears fruit very sparingly, scarcely any being produced in some seasons.

DISTRIBUTION

There is no agreement amongst botanists as to the distribution of the common pear, which has long been in cultivation in Europe.   In many places, where it is now apparently wild, it is probably only an escape from orchards.   It occurs either wild or naturalised throughout the greater part of Europe, northern Africa, Asia Minor, the Caucasus, and north Persia.   According to Willkomm, it is not met with in the greater part of Scandinavia, in Finland, northern Russia, Esthonia, Livland, and the provinces east of the Volga; but it is wild in the ash forests of the Ukraine, and in the mountains of the Crimea.   In Germany, it is rare and only naturalised

[1] *Hist. Plant.* i. 59 (1650).   [2] *Wien. Obst. u. Gartenz.*, 1878, p. 26, fig. 8.

in the north-western plain; and elsewhere is seen in hedges, copses, and the edges of the broad-leaved forests. In the Swiss Alps and the Jura,[1] it ascends to 2700 ft. and in the Tyrol to 4800 ft. In France, it is found as a scattered tree in most of the broad-leaved forests of the plains and low hills; but is absent in the region of the olive. In England, it occurs as a rare tree in woods, thickets, and hedgerows from Yorkshire southwards, but it is a doubtful native, except in the form var. *cordata*. *P. communis* has been found[2] in the fossil state in neolithic peat at Crossness, Essex, determined from wood examined by the late Professor Marshall Ward.

(A. H.)

REMARKABLE TREES

It is not our intention to deal here with the cultivated varieties of pear, which are very numerous; and with regard to those used for making perry, we refer our readers to the *Herefordshire Pomona* as the most comprehensive recent work on the subject.

The largest pear tree which I have ever seen or heard of stands alone on the north side of a hill on Church Farm in the parish of Lassington, about two miles from Gloucester, in a grass field of rather strong land on the Old Red Sandstone. Whether it is, as I believe a wild pear, or not, it is on its own roots, and bears small fruit which ripen earlier than any of the perry pears of the district. It is described in Witchell's *Fauna and Flora of Gloucestershire*, 264 (1892), as being 18 feet in girth, but I measured it, in January 1909, as 16½ ft. in girth and about 50 ft. high. The trunk is about 15 ft. high, and, though hollow at the base, with a large limb broken off, seems healthy, and the branches are full of young wood. By a rough estimate it must contain at least 200 cubic feet in the trunk, and another 100 feet or more in the larger limbs, and is the oldest-looking pear tree that I ever saw (Plate 356).

Another very large pear tree grows at Hardwicke Court near Gloucester, and measures 13½ ft. in girth at a foot from the ground, dividing above this point into three main trunks.

One of the most remarkable trees in Great Britain is the pear at Holme Lacy Rectory. This is described in Littlebury's *Herefordshire Gazetteer* as having been a very old tree in 1776, when it yielded 15 to 16 hogsheads of perry in one year. At that time it covered half an acre of ground. It was described by a correspondent of Loudon in 1836 as at this time much smaller, but still healthy and vigorous.[3] Its peculiar feature is the way in which its branches after extending laterally to a considerable distance, fall to the ground and take root, giving rise to a new tree which again extends in the same way, so that it is a remarkable instance of layering. It is now impossible to say where the original trunk first stood, as it is divided into three parts, of which the principal trunk measures 59 ft. by 8 ft. 8 in., and spreads a long way into a shrubbery; another in the meadow outside the

---

[1] Lord Ducie has a thriving tree at Tortworth raised from seed sent from the Jura by Mr. G. H. Wollaston.

[2] Cf. C. Reid, *Origin British Flora*, 119 (1899).

[3] Edwin Lees figures in *Gard. Chron.* ix. 268, fig. 45 (1878), a part of this tree, which he said measured 80 feet from the base of the principal fallen trunk to the end of the branches; but I cannot recognise the part of the tree which he sketched.

garden has several stems, which spread a long way ; and the fallen stems seem sound. Perhaps the most striking part of this tree, which has probably been separately planted, is now in a meadow 200 yards off on the banks of the Wye. This has eight different stems whose various branches measured, in July 1908, 126 paces in circumference. The older trunks have very rugged bark, and twist from left to right. The late Rev. A. Ley informed me that the fruit was small ; and the local blacksmith, who makes perry from it which I found to be very fair, said that it most resembled a variety locally called " Taunton Squash."

The cultivated pear seems to attain its greatest size on rather heavy Red Sandstone soils in the neighbourhood of Worcester. One of the largest I have seen is in an orchard at Eardiston, 15½ miles from Worcester, on the road to Ludlow, and 3½ miles from Newnham Bridge Station. This pear tree is mentioned by Rider Haggard in *Rural England*, i. 340 (1902), as being 17 ft. in girth, but when I visited it in March 1907, I made it 62 ft. by 13½ ft. It grows on a steep bank sheltered from the north, on which side it is hollow at the butt. (Plate 357.)

There is an orchard of very fine old pears, many of which are now decaying and others gone, since they were described in *Herefordshire Pomona* (vol. i. p. 20) as " an orchard of Barland pears, perhaps unequalled in the world." They grow on Monkland farm between Worcester and Malvern, and, according to tradition, were planted by the Monks of Malvern, in which case they must be 300 years old. Mr. E. Lees, in *Botany of the Malvern Hills*, 62 (1843), writes of them :— " There are more than seventy lofty trees, and in a 'hit,' as it is called, the produce has amounted to 200 hogsheads. The orchard in question occupies five or six acres, and the price of perry varies from 6d. to 1s. 6d. per gallon. Supposing the average price to be £3 per hogshead, the perry produced would be worth £600, but a ' hit ' must not be expected every year, and the trees are now becoming very old."

There is another orchard not far off on Lower Woodfield Farm which Mr. Slater, forester to Earl Beauchamp, showed me in 1908, and which are also called Barland pears.[1] The best of these are about 60 ft. high, and two which I measured were 11 ft. and 8½ ft. in girth.

At Forthampton Vicarage near Tewkesbury there is another magnificent orchard of so-called " Hufcap " pears,[2] which strongly resemble the Barland in bark and habit, and like them are all grafted at about 6 ft. from the ground. These are said to have been planted in the reign of Charles II., and are growing on a strong red marl. They are in three lines, and though several of them are partially decayed,

---

[1] The Barland pear is figured by Hogg and Bull, *Herefordshire Pomona*, vol. i. plate xviii. (1876-1885), and is said to have originally grown in a field called Bare Lands, near Ledbury. Evelyn says of it in his *Pomona* :—" The horse pear and Bareland pear are reputed of the best as bearing almost their weight of spriteful and vinous liquor. They will grow in common fields of gravelly and stony ground to that largeness, as only one tree has been usually known to make three or four hogsheads." The fruit he describes as " of such insufferable taste that hungry swine will not smell to it, or if hunger tempt them to taste, at first crush they shake it out of their mouth." The authors of *Pomona*, however, say that Barland perry does not bottle well. It curdles in the bottles, and in Herefordshire is usually drunk as soon as made, when it is considered very wholesome, and singularly beneficial in nephritic complaints.

[2] The Black Hufcap pear is figured by Knight, *Pomona Herefordiensis*, plate xxiv. (1811), and is said to have been known from the seventeenth century and to be best of all the varieties. The fruit is very harsh and austere, but becomes very sweet during the process of grinding. Its perry possesses much strength and richness, and has the credit of intoxicating more rapidly than that made from any other pear. The Yellow Hufcap is a very favourite pear near Ledbury, earlier than the Black Hufcap, and bears freely, though usually in great abundance every second year. Its perry is excellent.

flower profusely, and in some years produce abundance of fruit. The largest sound tree that I measured here was over 50 ft. high and 10 ft. in girth.

In Scotland the pear rarely attains a great size; and I have seen none which were very noteworthy; but in the *Old and Remarkable Trees of Scotland* (p. 244), the Rev. R. Bremner describes a Chaumontel pear tree in the minister's garden at Banff, which was reported to be one hundred and fifty years old, and was 42 ft. by 9 ft. and 150 ft. round the branches.

Walker[1] measured in 1799 a tree at Restalrig, near Edinburgh, which was 12 ft. in girth at 2½ ft. from the ground, above which point it began to branch. This was a sort of early pear, called the "Golden Knap," which he considered to be the largest and most durable of any of the kinds of pear; and recommended it to be raised from seeds and planted out without being grafted, as these wildings, as they are called, form the most vigorous and largest trees, and should be chosen as stocks for grafting. The fruit, though often scarcely edible, is useful in making perry.

Hunter[2] mentions an orchard pear tree at Gourdie Hill near Perth, which was no less than 65 ft. by 10 ft., but we do not know if either of these is still living.

TIMBER

It seems strange that a wood, having such valuable qualities as the pear, should be practically neglected in this country. It is so hard, fine-grained, and compact that it is one of the very best woods for cogs, wood-screws, and tool-handles. It takes dye so well that when stained black it is difficult to distinguish from ebony, and for carving it is one of the finest hardwoods known.[3] At Windsor Castle there is a very beautiful panel carved in high relief from pear wood.

Like many hardwoods it is slow and difficult to season, and unless thoroughly dry is liable to warp and crack badly. When trees of sufficient size can be procured, it is better, in order to avoid warping, to cut the boards on the quarter; and as it takes a fine polish and has a rich pinkish brown colour, it might be used for chair and cabinet-making with good effect. But the greater part of the old pear trees which are grown in the Severn valley are only used for firewood, and are said to burn with a very hot and slow flame. (H. J. E.)

[1] *Essays on Nat. Hist.* 84 (1812). [2] *Woods and Forests of Perthshire*, 503 (1883).
[3] Drawing squares and curves are often made of pear wood.

SECTION MALUS

This section of the genus Pyrus comprises the apples, which are deciduous trees or shrubs, resembling the true pears in the occurrence of long and short shoots and in the arrangement of the leaves. Leaves simple, often lobed, stalked. Flowers in cymes, terminal on the short shoots; sepals 5, acuminate, either persistent and erect on the fruit or deciduous; ovary usually 5-celled, rarely 3-celled; styles united at the base. Fruit with homogenous flesh, hollowed out or rounded at the base; with papery carpels joined at the apex, free in the middle.

There are about ten species of apples, distributed in Europe, extra-tropical Asia, and North America. The following synopsis gives a brief account of the wild species, and of some of the more important hybrids, which are met with in cultivation.

I. EUMALUS, Zabel, *Laubholz-Benennung*, 185 (1903).

Leaves rolled inwards in the bud; on adult trees, serrate, without lobes or irregular teeth.

* *Calyx persistent on the fruit.*

† *Fruit depressed at both the base and apex.*

1. *Pyrus Malus*, Linnæus. See p. 1570.

†† *Fruit not depressed at the apex.*

2. *Pyrus Ringo*, Koch, *Dendrologie*, i. 213 (1869); Wenzig, in *Linnæa*, xxxviii. 37 (1874); Stapf. in *Bot. Mag.* t. 8265 (1909).

Branchlets more or less tomentose. Leaves elliptic-ovate, 2 to 3 in. long, shortly acuminate, more or less tomentose beneath, finely and sharply serrate. Fruit ovoid, about 1¼ in. long and 1 in. in diameter, yellow, hollowed at the base, crowned by the persistent sepals, which are united together at their base.

A small tree introduced by Siebold from Japan in 1856. It is supposed to be a hybrid between *P. Malus* and *P. spectabilis*. At Kew, it is very ornamental, when covered with its beautiful fruits; and is represented by two trees, one with densely tomentose leaves, the other with slightly tomentose leaves. A pyramidal variety, var. *fastigiata bifera*, Dieck, is also known, which is said to flower usually twice in the season.

3. *Pyrus spectabilis*, Aiton, *Hort. Kew.* ii. 175 (1780).

Branchlets slightly pubescent. Leaves oblong-elliptic, 2 to 3 in. long, 1 to 1¼ in. broad, acute at the apex, pubescent on the midrib beneath and on the petiole; margin with minute close serrations, which are callous and mostly incurved at their tips. Fruit globose, yellow, about 1 in. in diameter, not hollowed out at the base, crowned by the persistent sepals, which are united together at their base.

A native of north China and Japan. Introduced in 1780, and often cultivated on account of its large pink flowers, about 2 in. across, which appear early in the season. Forms with double flowers occur, one of which with very large flowers is known as var. *Riversii*.

4. *Pyrus prunifolia*, Willdenow, *Phytog.* i. 8 (1794); J. D. Hooker, in *Bot. Mag.* t. 6158 (1875).

Young branchlets tomentose. Leaves usually elliptic, about 3 in. long and 2 in. broad (occasionally a few are sub-orbicular and smaller), contracted above into a cuspidate acuminate apex, tomentose on the midrib and nerves beneath and on the petiole, conspicuously glandular on the midrib above; margin with irregular serrations, which are crenate or bluntly triangular. Fruit globose or ovoid, about an inch in diameter, yellow on one side, bright red on the other, depressed at the base, crowned by the persistent sepals, which are united together at their base.

P. prunifolia has never been found in the wild state, and is of uncertain origin. It was reputed, when introduced in 1753, to have come from Siberia. It closely resembles *P. baccata*, differing mainly in the persistent calyx of the fruit; and is supposed by Bailey[1] to be a hybrid between that species and *P. Malus*. It is cultivated at Kew; and at Bayfordbury there is a fine specimen, 35 ft. by 5 ft. 7 in., with the stem forked a few feet from the ground.

5. *Pyrus Scheideckeri*, Späth, in *Gartenflora*, liii. 417, t. 1529 (1904).

Young branchlets slightly pubescent. Leaves ovate, acuminate, about 3 in. long, with scattered pubescence on the under surface and hairy petioles; margin coarsely and sharply serrate and biserrate. Fruit, $\frac{3}{4}$ in. in diameter, globose, usually crowned by the persistent calyx, occasionally some of the sepals being deciduous.

This originated in Scheidecker's nursery at Munich, as a seedling of *P. floribunda*, but is evidently a hybrid, the other parent being probably *P. prunifolia*. It is a small tree, producing large flowers, white tinged with pink, in great abundance; and is in cultivation at Kew.

** *Calyx not persistent on the fruit.*

6. *Pyrus baccata*, Linnæus, *Mant.* 75 (1767).

Branchlets glabrous. Leaves ovate-oblong, $2\frac{1}{2}$ to $3\frac{1}{2}$ in. long, $1\frac{1}{4}$ to 2 in. wide, contracted above into a cuspidate acuminate apex; glabrous beneath, except for slight pubescence on the midrib; glandular on the midrib above; margin with shallow usually crenate serrations. Fruit globose, $\frac{3}{4}$ in. in diameter, hollowed at the base and apex, the apex being marked with a circular brown scar, no trace of the calyx remaining.

This species,[2] which is usually known as the Siberian crab, is widely distributed in eastern Siberia, Manchuria, central and northern China, and throughout the Himalayas at 6000 to 10,000 ft. altitude. It was introduced in 1784.

[1] *Cycl. Amer. Hort.* 1472 (1901).
[2] Figured by Sir J. D. Hooker in *Bot. Mag.* t. 6112 (1874), who states that, in the western parts of the Himalayas, this species becomes more pubescent in all its parts.

It is one of the most beautiful and generally cultivated of its genus, both for its flowers, and its fruit,[1] which makes a delicious jelly. It grows on good soil to a considerable size. Elwes saw at Patshull, Staffordshire, the seat of the Earl of Dartmouth, a tree, measuring 40 ft. by 7 ft., with a burry trunk.

7. *Pyrus sikkimensis*, J. D. Hooker, *Flora Brit. India*, ii. 373 (1878), and in *Bot. Mag.* t. 7430 (1895).

Branchlets tomentose. Leaves ovate to ovate-oblong, 2 to 3 in. long, ending in a long caudate-acuminate apex; tomentose beneath; margin with fine close sharp-pointed serrations. Fruit, $\frac{2}{3}$ in. in diameter, turbinate, not depressed at the base, marked at the apex with a brown circular depressed scar, no trace of the calyx remaining.

A small tree, native of Sikkim and Bhutan between 7000 and 10,000 ft. altitude. It is in cultivation at Kew, where a tree about 25 ft. high has spiny branches on the trunk.

8. *Pyrus Halleana*, Sargent, in *Garden and Forest*, i. 152 (1888).

Branchlets slightly pubescent. Leaves coriaceous, lanceolate to ovate, 2 to 3 in. long, usually tapering gradually to an acuminate apex; quite glabrous on both surfaces and on the petiole, glandular on the midrib above; margin undulate, finely serrate. Fruit globose, $\frac{1}{4}$ in. in diameter, not hollowed at the base, marked at the apex by a depressed circular scar, no trace of the calyx remaining; and containing very large seeds.

This small tree was introduced in 1863 into the United States from Japan, where, however, it is not known in the wild state. It bears beautiful pink flowers, which are usually double. The original tree in Mr. Parkman's garden in Jamaica Plain, Massachusetts, was 18 ft. high in 1888. There is a small specimen at Kew.

II. SORBOMALUS, Zabel, *Laubholz-Benennung*, 189 (1903). Leaves folded in the bud; on adult trees irregular in outline, some being lobed or dentate.

* *Calyx persistent on the fruit.*

† *Leaves covered beneath with dense grey tomentum.*

9. *Pyrus ioensis*, Bailey, in *Amer. Gard.* xii. 473 (1889).

Branchlets densely grey tomentose. Leaves ovate, acute, about 3 in. long, often with two or four lateral lobes, crenately serrate, covered beneath with a dense grey tomentum; petioles tomentose. Fruit sub-globose, $1\frac{1}{4}$ in. in diameter, depressed at the base, on a stout tomentose stalk.

This is the common crab-apple of the Mississippi basin in the United States. It was introduced at Kew in 1906. Bechtel's crab, a form with large pink double flowers, is often cultivated in the United States. *P. Soulardi*,[2] Bailey, which is wild here and there in the Mississippi valley, is supposed to be a natural hybrid between *P. Malus* and *P. ioensis*.

---

[1] The fruit which I bought at several railway stations in Siberia in May 1912 was juicy and well-flavoured.—H. J. E.

[2] Britton and Shafer, *N. Amer. Trees*, 434 (1908), say that this hybrid is cultivated in the north central States for its fruits, which are highly prized for cider and jellies, and used as a substitute for the quince where that fruit will not thrive.

†† *Leaves not covered beneath with dense grey tomentum.*

10. *Pyrus coronaria*, Linnæus, *Sp. Pl.* 480 (1753).

Young branchlets slightly tomentose. Leaves variable in shape, ovate or triangular, about 3 in. long, acute at the apex; margin irregular with large serrated teeth; lower surface green, pubescent on the nerves. Fruit 1 to 1½ in. in diameter, globose, depressed at the base, fragrant, covered with a waxy exudation.

A native of the eastern parts of Canada and the United States. Introduced in 1824, and said by Loudon [1] to have become naturalised near White Knights and Godalming, where it attained 30 ft. in height.

11. *Pyrus angustifolia*, Solander, in Aiton, *Hort. Kew.* ii. 176 (1789).

Branchlets glabrous. Leaves oblong-lanceolate, 1½ to 3 in. long, acute or rounded and apiculate at the apex, tapering at the base; margin with irregular teeth and crenate serrations; under surface glabrous except for slight pubescence on the midrib. Fruit sub-globose, 1 in. in diameter, very fragrant.

A native of the United States from Pennsylvania to Florida and Louisiana. Introduced in 1750.

** *Calyx not persistent on the fruit.*

12. *Pyrus rivularis*,[2] Douglas, *ex* Hooker, *Fl. Bor. Amer.* i. 203 (1833).

Branchlets slightly tomentose. Leaves ovate, about 3 in. long, acute or acuminate, slightly tomentose beneath; margin sharply serrate, with often two to four small lateral lobes or large teeth. Fruit ellipsoid, ½ to ¾ in. long.

A tree, 40 ft. high, occurring in western North America from Alaska to California. Introduced in 1836.

13. *Pyrus Toringo*, Siebold, in *Ann. Mus. Lugd. Bot.* iii. 41 (1856).

Malus Toringo, Siebold, *Cat. Rais.* 4 (1856); Carrière, in *Rev. Hort.* 1870-1871, p. 451, coloured plate.

Branchlets tomentose. Leaves on the short shoots, ovate, 1½ to 2 in. long, more or less tomentose beneath, sharply serrate; on the long shoots, 2 to 2½ in. long, trilobed, with a large ovate acuminate irregularly serrated terminal lobe, and two smaller triangular lateral lobes. Fruit globose, about ¼ in. in diameter.

A native of Japan,[3] introduced by Siebold in 1856.

14. *Pyrus floribunda*, Nicholson, in *Kew Hand-List Trees*, 181 (1894) (not Lindley).

Malus floribunda, Siebold, *Cat. Rais.* 5 (1859), *ex* Van Houtte, *Flore des Serres*, xv. t. 1585 (1865); Carrière, in *Rev. Hort.* 1870-1871, p. 591, coloured plate.

Branchlets slightly pubescent. Leaves ovate to elliptic, 2 to 2¾ in. long, 1 to 1½ in. broad; petiole and under surface pubescent, the pubescence dense on

---

[1] *Arb. et Frut. Brit.* ii. 908 (1838).

[2] Britton and Shafer, *N. Amer. Trees*, 435 (1908), adopt for this species the name *Pyrus diversifolia*, Bongard, in *Mem. Acad. Petersb.* ii. 133 (1833), which is strictly applicable to a pubescent variety.

[3] *Malus Sargenti*, Rehder, in Sargent, *Trees and Shrubs*, i. 71 (1903), is a very tomentose variety of *P. Toringo*, which was found by Sargent in a brackish marsh near Moronan in Japan.

the midrib and scattered elsewhere ; margin sharply serrate, some of the leaves having one or two large teeth on each side. Fruit globose, $\frac{1}{4}$ to $\frac{1}{3}$ in. in diameter, with a depressed circular scar at the apex, no trace of the calyx remaining.

This was introduced in 1856 by Siebold from Japan, where it is not known in the wild state ; and is supposed by Schneider to be a cross between *P. baccata* and *P. Toringo.* It is a shrub, producing an abundance of beautiful pink flowers, which appear with the leaves. (A. H.)

PYRUS MALUS, COMMON APPLE

Pyrus Malus, Linnæus, *Sp. Pl.* 479 (1753); Loudon, *Arb. et Frut. Brit.* ii. 891 (1838); Wilkomm, *Forstl. Flora,* 847 (1887); Ascherson and Graebner, *Syn. Mitteleurop. Flora,* vi. pt. 2, p. 74 (1906).

A tree or large shrub with scaly fissured bark. Leaves scattered on the long shoots, clustered on the short spurs, usually oval, rounded at the base, shortly acuminate at the apex, crenate in margin. Flowers, 5 or 6 in an umbellate cyme, at the apex of a short shoot; sepals 5, triangular, acuminate; petals 5, pink externally, white internally; stamens about 20, with white filaments and yellow anthers; styles 5, united at the base. Fruit sub-globose, about an inch in diameter, depressed both at the base and apex, crowned by the calyx-segments, which are not united at the base.

The wild apple, described above, is often considered to constitute two species, which, as intermediate forms are common, are best treated as two varieties :—

1. Var. *sylvestris,* Linnæus, *Sp. Pl.* 479 (1753).

> *Malus sylvestris,* Miller, *Gard. Dict.,* ed. 8, No. 1 (1768).
> *Malus acerba,* Mérat, *Pl. Env. Paris,* 187 (1812); Mathieu, *Fl. Forest,* 171 (1897).
> *Pyrus acerba,* De Candolle, *Prod.* ii. 635 (1825).

Young branchlets slightly pubescent, soon becoming glabrous. Leaves glabrous above, shining and scattered pubescent beneath ; petiole slightly pubescent. Axis of inflorescence and pedicels slightly pubescent; calyx-tube glabrous ; sepals glabrous externally, pubescent internally. This is generally supposed to be the variety indigenous in western and central Europe.

2. Var. *pumila,* Henry.

> Var. *mitis,* Wallroth, *Sched. Crit.* 215 (1822).
> *Pyrus pumila,* Koch, *Dendrologie,* i. 203 (1869).
> *Malus pumila,* Miller, *Gard. Dict.,* ed. 8, No. 3 (1768); Schneider, *Laubholzkunde,* i. 715 (1906).
> *Malus paradisiaca,* Medicus, *Gesch. Bot.* 78 (1793).

Young branchlets tomentose. Leaves ovate or oval, often cuneate at the base, crenately or sharply serrate, dull and more or less tomentose beneath. Axis of the inflorescence, pedicels, calyx, and both surfaces of the sepals, tomentose.

This variety, though often found naturalised in western Europe, is considered by Ascherson and Graebner and by Schneider to be only truly wild in south-eastern Europe, southern Russia, Siberia, Turkestan, Caucasus, and Asia Minor. It is the source of most of the cultivated apples, though some of these have arisen from var. *sylvestris*, and others from crosses of the two forms.

It is impossible here to enter into any account of the apples cultivated for their fruits ; but the following peculiar varieties may be briefly noticed.

3. Var. *astracanica*, Loudon, *Arb. et Frut. Brit.* ii. 893 (1838).

> *Malus astracanica*, Dumont de Courset, *Bot.* v. 426 (1811).
> *Pyrus astracanica*, De Candolle, *Prod.* ii. 635 (1825).

Leaves large, coarsely serrate and in part bi-serrate, tomentose beneath. Flowers and fruits on long stalks. On account of the length of the peduncles, this is considered by Schneider and by Ascherson and Graebner, to be a hybrid between *P. Malus*, var. *pumila*, and *P. baccata* ; but this is very doubtful. A tree cultivated at Kew under this name agrees with a herbarium specimen, which was collected by Schrenk in Songaria. Loudon states that this variety is "a native of about Astrachan, on the testimony of gardeners," and mentions two kinds, one with red fruit, the other with wax-coloured fruit.

4. Var. *Niedzwetzkyana*, Ascherson and Graebner, *op. cit.* 78 (1906).

> *Pyrus Niedzwetzkyana*, Hemsley, in *Bot. Mag.* t. 7975 (1904).

Adult leaves tinged with red on the midrib and nerves. Flowers deep pink. Fruit with a crimson skin and pale purple flesh. Nearly all the other parts of the tree, as the wood and bark, are also coloured red.

This remarkable variety was obtained by Dieck,[1] who introduced it in 1891, from Kashgar and the plateau of Talgar, near Vernoie, in south-west Siberia. Goeze [2] states that a similar tree is cultivated in the Caucasus. This variety is hardy at Kew, where it flowers and produces fruit.

5. Var. *apetala*, Ascherson and Graebner, *op. cit.* 78 (1906).

> *Pyrus apetala*, Muenchhausen, *Hausv.* v. 247 (1770).
> *Pyrus dioica*, Moench, *Verz. Weissenst.* 87 (1785) : Loudon, *Arb. et Frut. Brit.* ii. 892 (1838).

Flowers with two rows of sepals and 10 to 15 styles ; without petals or stamens. Fruit seedless.

The origin of this is unknown. It was said by Loudon to have been in continental gardens in his time ; but was not introduced in 1838. There is a specimen at Kew, but I have seen no flowers or fruit.[3]

6. Schneider mentions var. *pendula*, a tree with pendulous branches and branchlets ; and var. *aucubæfolia*, in which the leaves are spotted with yellow.

DISTRIBUTION

The common apple is widely distributed throughout nearly all Europe, Asia Minor, the Caucasus, north Persia, south-western Siberia, Turkestan, and the

---

[1] *Neuh. Offer. Nat. Arb. Zöschen*, 1891, p. 16, where the name is given as *Malus Medwietzkyana*, which was afterwards changed by Dieck to the spelling given above.

[2] *Gard. Chron.* ix. 461 (1891). Cf. also Graebner, in *Mitt. Deut. Dend. Ges.* 1911, p. 254, who mentions a fine specimen at Karlsruhe. Young plants vary much in the colour of the leaves.

[3] Bailey, *Cycl. Amer. Hort.* 1473 (1901), states that it is figured and described in *American Gardening*, x. 244, 279, and xi. 6 (figs.), 624.

north-western Himalayas. In many places it is undoubtedly an escape from cultivation; and its exact distribution as a wild tree cannot be defined with accuracy.

It is apparently indigenous in Europe as far north as lat. 63° 49′ in Norway, and lat. 61° in Sweden. In Russia its northern limit passes through Åbo in Finland, Lake Ladoga, Tver, Jaroslav, and Kazan; and it is said to be common in the forests of the plain in the southern provinces and in the mountains of the Crimea. It reaches in the Caucasus an elevation of 5000 ft., becoming a tree 30 to 40 ft. in height; and extends from there into the Persian province of Ghilan. It appears to be undoubtedly wild in the mountains of Turkestan, where it has been collected by various Russian travellers. Aitchison found the apple only as a cultivated tree in Afghanistan; but, according to Hooker, it is apparently wild in the north-western Himalayas, where it ascends to 9000 ft., and in western Tibet, where it reaches an altitude of 11,400 ft.

In central Europe, according to Ascherson and Graebner, the glabrous variety is scattered throughout the forests, being more common in those composed of conifers and broad-leaved species, and is usually seen in glades and on the margin of woods. In the north-western German plain, the apple is often met with remote from orchards, and is considered to be truly wild. It ascends in the Alps to about 5500 ft. elevation.

In France, Mathieu also is of opinion that the indigenous apple is the glabrous variety, occurring as a scattered tree in the forests of the plains and low hills, except in the Mediterranean region. It ascends in the Jura to about 3000 ft. altitude.

In Britain, the apple[1] is found apparently wild in copses and hedges from the Forth and Clyde southwards; and is undoubtedly wild in many parts of Ireland.

In the eastern United States, it has escaped from cultivation, and is common in woods, thickets, and road-sides, especially in southern New York, New Jersey, and Pennsylvania.[2]   (A. H.)

We must refer our readers to Loudon and to the *Herefordshire Pomona* for a full account of the cultivated apples, which we have no space to describe in this work. So far as I know, no variety of apple approaches the pear in size or age. Loudon stated that near Hereford some attained 40 ft. in height. The largest cultivated tree that I have seen is a tree in Lady Jenkins's garden at Botley Hill, Hants, adjoining the house where Cobbett once lived. This is of the variety called "Hambledon Deux ans," and in 1906 measured 47 ft. by 7 ft. and appeared sound. I was told that it had borne as much as forty bushels of fruit in one year, and often as much as twelve or fifteen bushels. Some of the last year's fruit was still edible on 21st June though shrivelled and partly rotten.

The wild crab tree seems to attain as great a size as the cultivated apples, though rarely seen under conditions where it has a chance to show its best growth. In an album of sketches by Jukes of trees at Studley Royal, which I have seen

[1] Leighton, *Flora of Shropshire*, 527 (1841), states that the glabrous and pubescent varieties are both equally wild and common. Bromfield, *Flora Vectensis*, 165 (1856), states that the pubescent form is extremely common and truly wild over most parts of the Isle of Wight, in woods, thickets, copses, hedgerows, and rough bushy places.
[2] Britton and Brown, *Illust. Flora N. United States*, ii. 236 (1897).

in the library there, is an excellent drawing of a crab which grew in the valley known as Mackershaw Trough, and in 1837 was 45 ft. high by 3 feet in girth at three feet from the ground. But this tree was no longer alive when I visited the place in 1905. In the hedgerows of the Cotswold hills it sometimes attains 30 ft. by 5 ft. Near the water tower at Barnsley Park, Cirencester, there is a tree, which in 1911 was about 45 ft. in height by 10 ft. in girth at 3 ft. from the ground, above which it divides into three stems.

Sir R. Christison[1] measured in 1876, at Kelloe, near Duns, Berwickshire, a perfectly healthy crab tree, 50 ft. high and 8 ft. in girth. It produced flowers abundantly and fruit in considerable quantity.

The crab tree comes up from seed pretty freely, but cannot be recommended for cultivation, as both its fruit and flowers are inferior to many of the exotic apples. It is not particular about soil, but seems to grow most freely on soils containing lime.

Though inferior to that of the pear tree, which it resembles in colour, hardness, and size, yet the wood of both wild and cultivated apples has some value for turnery, and was used for cog wheels and country furniture until driven out by wholesale manufacturers who use foreign wood only. I have used it for flooring blocks, for which its hardness and colour makes it very suitable, but the majority of the apple trees are usually decayed at heart before they cease to bear, and when worn out are used, like the pears, for firewood. (H. J. E.)

[1] *Trans. Bot. Soc. Edinburgh*, xii. 186 (1876).

SECTION AUCUPARIA

Small trees or shrubs, defined in Vol. I. p. 142, as constituting one group of the section Sorbus[1] of the genus Pyrus. They are characterised as follows :—Leaves deciduous, alternate, unequally pinnate, with serrate leaflets and foliaceous stipules. Flowers perfect, in terminal compound corymbose cymes ; calyx urn-shaped, with five persistent lobes ; petals five, suborbicular, white ; stamens about twenty ; ovary usually three-celled, and surmounted by three styles, occasionally two- to four-celled, with two to four styles ; ovules two in each cell. Fruit a small sub-globose pome, with acid flesh and papery carpels, which are free at the apex ; seeds two, or one by abortion, in each cell.

About twenty species of the section Aucuparia are known, widely distributed over the extra-tropical regions of the northern hemisphere. Of these *P. Aucuparia*, which is a native tree, will be described in detail. About six exotic species have been introduced.

I. Buds very glutinous, showing no white tomentum at the tip.

* *Stipules early deciduous.*

1. *Pyrus americana*, De Candolle, *Prod.* ii. 637 (1825).

Leaflets, thirteen to seventeen, 2 to 2½ in. long, ¾ in. broad ; under surface pale with scattered pubescence. Fruit, ¼ in. in diameter, said by Sargent[2] to be bright red, but purplish or bronze-coloured in England. A small tree, widely spread in North America. Introduced in 1782, and said by Loudon to be more tender than the native species ; but it appears to thrive at Kew and at Tortworth.

Var. *decora*, Sargent, *Silva N. Amer.* xiv. 101 (1902).

This differs from the type in bearing large scarlet fruit, ½ in. in diameter, It is apparently the tree often known in cultivation as *P. sambucifolia*.[3]

2. *Pyrus commixta*,[4] Ascherson and Graebner, *Syn. Mitteleurop. Flora*, vi. pt. 2, p. 90 (1906).

Leaflets, nine to thirteen, 2 in. long, ⅝ in. broad, tapering to a long caudate-acuminate apex, glabrous beneath. Fruit red, ¼ in. in diameter. A small tree, native of Japan.[5] A specimen at Kew, obtained from Späth in 1900, is about 15 ft. high and very thriving.

---

[1] *Sorbus* is regarded by many botanists as a distinct genus ; but there is no agreement amongst the various authorities as to its limits. Koehne, *Deutsche Dendrologie*, 246 (1893), includes in Sorbus only the mountain ashes. Schneider, *Laubholzkunde*, i. 667 (1906), gives it a much wider scope. Ascherson and Graebner, *Syn. Mitteleurop. Flora*, vi. pt. 2, p. 85 (1906), takes another view, which agrees practically with the arrangement given in our Vol. I. pp. 141-142.

[2] *Silva N. Amer.* iv. 79, tt. 171, 172 (1892). It is described by Sargent, *Trees N. Amer.* 356 (1905), as *Sorbus americana*, Marshall, *Arb. Amer.* 145 (1785).

[3] The true *P. sambucifolia*, Chamisso and Schlechtendal, in *Linnæa*, ii. 36 (1827), is a native of eastern Siberia, Saghalien, and Yezo ; and has not yet been introduced.

[4] This was first described as *Sorbus Aucuparia*, var. *japonica*, Maximowicz, in *Mél. Biol.* ix. 160 (1873). It is *Sorbus japonica*, Koehne, in *Gartenflora*, l. 408 (1901) (not Siebold) ; and is also named *Sorbus commixta*, Hedlund, in *Kgl. Svensk. Vet. Akad. Handl.* xxxv. 38 (1901).

[5] Var. *rufo-ferruginea*, Shirai, *ex* Schneider, *Laubholzkunde*, i. 678 (1906), has rusty red pubescence on the rachis and under surface of the leaflets. This was collected by Elwes at Chuzenji at 4000 ft. altitude, and is possibly a distinct species. It has not been introduced.

Stipules persistent till the time of fruiting.

3. *Pyrus discolor*, Maximowicz, *Prim. Fl. Amur.* 103, note (1859).

Leaflets fifteen to seventeen, remotely placed on the rachis, $1\frac{1}{2}$ to 2 in. long, ending in a long acuminate point, pale and glabrous beneath. Fruit pale pink. Readily distinguishable by its palmately cleft or lobed persistent stipules.

A small tree, wild in the neighbourhood of Peking. *Sorbus pekinensis*, Koehne,[1] in *Gartenflora*, L., 406 (1901), which was described from plants raised in Germany, is identified with this species by Schneider.[2] Small trees at Kew, obtained from Späth and Lemoine, agree with a native specimen in the British Museum, except that the buds are pubescent at the tip and not completely viscid.

II. Buds more or less covered with dense white tomentum.

Stipules early deciduous.

4. *Pyrus tianshanica*, Franchet, in *Ann. Sc. Nat.* xvi. 267 (1883).

Branchlets glabrous. Leaflets eleven to thirteen, 1 to $1\frac{1}{2}$ in. long, $\frac{2}{8}$ in. broad, green and glabrous on both surfaces. Fruit $\frac{2}{8}$ in. in diameter, bright red.

A shrub, about 10 ft. high, occurring in Turkestan, Afghanistan, and western Kashmir. Introduced[3] into Kew in 1896, and described by Sir J. D. Hooker, in *Bot. Mag.* t. 7755 (1901). This does not seem to thrive in England, as the specimens which I have seen are stunted in growth and bear small leaflets.[4]

5. *Pyrus Aucuparia*, Gaertner. See p. 1576.

Branchlets pubescent. Leaflets eleven to fifteen, about 2 in. long, pale and pubescent beneath.

**Stipules persistent till the time of fruiting.*

6. *Pyrus pohuashanensis*, Hance, in *Journ. Bot.* xiii. 132 (1875).

Leaflets thirteen to fifteen, $2\frac{1}{2}$ to 3 in. long, $\frac{3}{4}$ in. broad, pale and pubescent beneath. Fruit $\frac{3}{8}$ in. in diameter, orange-coloured. The persistent stipules are obovate and shortly toothed.

A small tree, discovered in 1874 on the Po-hua mountain, west of Peking, by Bretschneider, who sent seed to the Arnold Arboretum. The trees which were raised produced flowers and fruit[5] for several years previous to 1893; and a seedling sent in that year to Kew is now about 20 ft. high and very thriving, producing abundance of handsome fruit.

III. Buds tipped at the apex with reddish hairs; scales glabrous, ciliate in margin.

7. *Pyrus Vilmorini*, Ascherson and Graebner, *Syn. Mitteleurop. Flora*, vi. pt. 2, p. 90 (1906).

---

[1] Cf. also Koehne, in *Mitt. D. Dend. Ges.* 1906, p. 56.   [2] *Laubholzkunde*, i. 669 (1906).

[3] It appears to have been introduced into the St. Petersburg Botanic Garden in 1889. Cf. *Gard. Chron.* xxv. 389 (1899).

[4] Schneider, *Laubholzkunde*, i. 668 (1906), describes this species under the name *Sorbus thianshanica*, Ruprecht, in *Mem. Acad. St. Petersburg*, xiv. 46 (1869), and states that the leaflets are 2 in. long and over $\frac{1}{2}$ in. broad.

[5] Cf. Bretschneider, *Hist. Europ. Bot. Disc.* 1054 (1898). The trees in the Arnold Arboretum were considered for many years to be *P. discolor*, Maximowicz; but Rehder, in *Mitt. Deut. Dend. Ges.* 1901, p. 117, showed that they were the same as Hance's species, and named them *Sorbus pohuashanensis*.

Leaflets, on a winged rachis, nineteen to twenty-nine, very small, scarcely exceeding ½ in. in length, serrate only near the apex. Fruit ⅓ in. in diameter bright red.

This pretty shrub,[1] which is a native of Yunnan in China, was introduced at Kew from Les Barres[2] in 1905. (A. H.)

PYRUS AUCUPARIA, Mountain Ash, Rowan

Pyrus Aucuparia, Gaertner, *De Fruct.* ii. 45, t. 87 (1791); Loudon, *Arb. et Frut. Brit.* ii. 916 (1838); Ascherson and Graebner, *Syn. Mitteleurop. Flora*, vi. pt. 2, p. 86 (1906).

Sorbus Aucuparia, Linnæus, *Sp. Pl.* 477 (1753); Willkomm, *Forstliche Flora*, 862 (1887); Mathieu, *Flore Forestière*, 181 (1897); Schneider, *Laubholzkunde*, i. 672 (1906).

Aucuparia silvestris, Medicus, *Gesch. Bot.* 86 (1793).

A tree, occasionally attaining 50 ft. in height. Bark thin, smooth, greyish, becoming thicker, darker in colour, and fissured at the base of old trunks. Young branchlets more or less tomentose at first, ultimately glabrescent. Leaves, about 6 in. long, unequally pinnate, with a grooved rachis, tufted with long hairs and glandular at the insertion of the leaflets, elsewhere slightly pubescent; leaflets eleven to fifteen, opposite, subsessile, about 2 in. long, lanceolate-oblong, unequal at the base, acute at the apex, sharply serrate except near the base; upper surface dull green, glabrous; lower surface pale, more or less pubescent.

Flowers in large corymbose cymes, the axis and branches of which are more or less tomentose; calyx and pedicels pubescent; petals white, equal in length or shorter than the stamens; styles usually three, tomentose at the base. Fruit spherical or ellipsoid, ⅖ in. in diameter, smooth, usually red.

The terminal buds[3] are large, ovoid-conic, and covered in greater part with dense white tomentum; the lateral buds are smaller, slightly flattened, and appressed. The leaf-scars, visible in winter, are crescentic, five-dotted, and situated on prominent pulvini.

VARIETIES

I. This species varies in the wild state, as regards the amount of pubescence; and two distinct varieties are recognised :—

1. Var. *glabrata*, Wimmer and Grabowski, *Fl. Schles.* ii. 1, p. 21 (1821).

Buds, branchlets, and leaves glabrous; leaflets smaller than in the type. This is the common form at high elevations in the mountains of central and south-eastern Europe.

[1] Cf. Hutchinson, in *Bot. Mag.* t. 8241 (1909), where it is figured under the name *Sorbus Vilmorini*, Schneider, in *Bull. Herb. Boissier*, vi. 317 (1906).

[2] This was raised at Les Barres from seeds received from Père Delavay in 1889, and was described as *Cormus foliolosa*, Franchet, in Vilmorin, *Frut. Vilmorinianum*, 103 (1904); but is not, as was supposed, identical with *Pyrus foliolosa*, Wallich. The latter does not appear to be in cultivation.

[3] The stipules are described by Lubbock, in *Journ. Linn. Soc.* (*Bot.*) xxx. 492, 493 (1895).

2. Var. *lanuginosa*, Ascherson and Graebner, *op. cit.* 88 (1906).

Sorbus lanuginosa, Kitaibel, *Schult. Oest. Fl.* ii. 50 (1814).
Pyrus lanuginosa, De Candolle, *Prod.* ii. 637 (1827).

Buds, branchlets, and leaves very tomentose, the pubescence remaining on the leaflets till autumn, and on the branchlets till the second year. This is prevalent in the plains and on the low hills of south-eastern Europe.

II. Several varieties with peculiar fruit are known :—

3. Var. *dulcis*, Krätzl, in *Wiener Illust. Gartenzeit.* 1885, p. 65.

Var. *moravica*, Zengerling, *ex* Dippel, *Laubholzkunde*, iii. 367 (1893).

Fruit sweet, larger than in the type. The leaflets are remarkable, being long and narrow, with the serrations confined to near the apex.

This tree was found wild about the year 1800, in a mountain forest of the Spornhau parish in northern Moravia,[1] and was subsequently propagated by grafting. It was introduced[2] in 1885 in Sweden, where it has proved very hardy, ripening its fruit as far north as lat. 66°, where no other fruit tree can be cultivated. A tree at Kew, about 15 ft. high, was obtained from Späth in 1900.

4. Var. *rossica*, Späth, in *Mitt. Deut. Dend. Ges.*, 1896, p. 196; Koehne, in *Gartenflora*, L., 412 (1901).

Fruit sweet, leaflets as in the type. This is said to be planted in southern Russia, where at Kiev the berries, powdered with sugar and packed in little boxes, are sold and exported. This variety, which I have not seen, was introduced by Späth in 1896.

5. Var. *Fifeana*, Dippel, *Laubholzkunde*, iii. 367 (1893). Fruit yellow. The origin of this tree, which was called var. *fructu luteo* by Loudon, is unknown. There is a handsome specimen at Kew.

III. The following varieties are peculiar in habit or foliage :—

6. Var. *fastigiata*, Loudon, *loc. cit.* Branches upright. This originated in Hodgins's nursery at Dunganstown in Co. Wicklow.

7. Var. *pendula*, Kirchner, *Arb. Musc.* 293 (1864). Branches very pendulous. This when grafted six or eight feet high, makes a graceful tree. Beissner[3] describes and figures a remarkable weeping tree in the churchyard of Wiesbaden; the branches are interlaced and twisted to an extraordinary degree.

8. Var. *integerrima*, Koehne, in *Gartenflora*, L., 411 (1901), and *Mitt. D. Dend. Ges.* 1906, p. 55. Leaflets entire in margin, the upper three occasionally united together. This was found[4] in the Jena Botanic Garden, and is possibly a hybrid, though the flowers differ in no respect from those of *P. Aucuparia*.

9. Var. *asplenifolia*, Koch, *Dendrologie*, i. 189 (1869). Leaflets irregularly and deeply toothed. This is represented at Kew by a tree obtained from Dale in 1899; and appears to differ slightly from var. *laciniata*, Beissner, in *Gartenwelt*, iii.

---

[1] Cf. Willkomm, *Forstl. Flora*, 863 (1887).   [2] Cf. Hartman, in *Garden and Forest*, 1895, p. 162.
[3] In *Mitt. Deut. Dend. Ges.* 1911, pp. 246 and 247.
[4] Lange, *Danske Flora*, iii. 370 (1864), describes a similar plant, found wild in the island of Bornholm.

267 (1899). The latter is said to have been found wild in the Erz Mountains; and in a specimen at Kew, obtained from Späth in 1906, the irregular toothing is almost confined to the leaflets on the barren branches.

10. A variety with variegated leaves is mentioned by Loudon; and another with yellowish foliage is occasionally seen, which is named var. *Dirkenii* or var. *Dirkenii aurea*.

HYBRIDS

The following, which were formerly considered to be varieties of *P. Aucuparia*, are of undoubted hybrid origin.[1] They differ from this species, in having the uppermost three or five leaflets united together into one segment :—

1. *Pyrus satureiifolia*, Ascherson and Graebner, *op. cit.* 106 (1906).

Sorbus satureiifolia, Koehne, *Deut. Dendrologie*, 248 (1893).

A small tree. Leaflets nearly glabrous beneath, not decurrent on the rachis; the uppermost three united into one segment.

This appears to be identical with *Sorbus neuillyensis*, Dippel, *Laubholzkunde*, iii. 370 (1893), of which there is a tree at Kew, about 20 ft. high, obtained from Simon-Louis in 1900.

Sorbus saturejæfolia, Koch, *Dendrologie*, i. 189 (1869), is similar in the shape of the foliage, but the leaflets are described as being tomentose on both surfaces. This appears to be identical with *Sorbus subserrata*, Opiz, in *Flora*, vii. suppl. 13 (1824): and is said by Ascherson and Graebner to be inconstant as regards the union of the upper leaflets.

2. *Pyrus decurrens*, Ascherson and Graebner, *op. cit.* 106 (1906).

Pyrus lanuginosa, Hort. (not De Candolle[2]).
Sorbus decurrens, Hedlund, *op. cit.* 49 (1901).

A small tree. Leaflets tomentose beneath, more or less decurrent on the rachis; the upper three, five, or seven leaflets united into one segment.

This is represented at Kew by a tree about 25 ft. high, which is of considerable age, and apparently less vigorous in growth than *P. Aucuparia*.

DISTRIBUTION

P. Aucuparia is widely distributed throughout almost all Europe;[3] but does not occur in Portugal, southern Spain, southern Italy,[4] Dalmatia, and Greece. It is most common in northern regions, extending as far north as Iceland, the North Cape, and the Kola Peninsula; but is reduced to a low shrub beyond lat. 67° in

---

[1] *Pyrus pinnatifida*, Ehrhart, described in Vol. I. p. 163, is usually considered to be a hybrid between *P. Aucuparia* and *P. intermedia*; but Schneider, *Laubholzkunde*, i. 691 (1906), who describes it under the name *Sorbus hybrida*, Linnæus, believes it to be a true species, as it comes true from seed.   [2] Cf. var. *lanuginosa*, p. 1577.

[3] The mountain ashes resembling *P. Aucuparia* in the Caucasus, Siberia, and western Himalayas, are considered to be three distinct species by Schneider and by Hedlund.

[4] *P. præmorsa*, Gussone, *Fl. Sicul. Syn.* i. 561 (1842), is considered by Schneider to be a distinct species, inhabiting Sicily and the Madeira Islands.

Scandinavia and Russia. In Russia, it is widely spread in the forests of the plains, but does not occur south of a line passing through Orenburg, Tambof, Voronej, Kursk, and Podolia. In Germany, it grows on all soils, forming part of both the broad-leaved and coniferous forests, and often ascends to timber line, attaining an altitude of 4000 to 5000 feet. It reaches still higher elevations in the Carpathians and the mountainous districts of the Balkan states. In France it is more common in hilly and mountainous regions, often becoming a bush on elevated precipices; but occurs in a few forests in the plains of the north. In the British Isles it is probably indigenous only in mountainous and hilly districts, though it is seen as a rare tree in woods as far south as the Isle of Wight. It ascends in the Highlands to about 2600 feet. This species has been found in the fossil state,[1] in neolithic deposits at Caerwys, Flintshire. (A. H.)

CULTIVATION

Though the mountain ash in England is usually not over thirty or forty feet high, yet it has a tree-like rather than a bushy habit, and is so beautiful for its fruit and for the elegance and autumnal colours of its foliage, that it should be planted on the edges of all copses, and in hedges and pleasure grounds. No native tree surpasses it in the autumn when laden with its red berries, and though birds are so fond of these that they are mostly eaten before winter, yet where the tree is abundant, it is one of the most beautiful features of the scenery. To see its foliage at its best, one must however, go to more northern regions such as Norway, where I have seen the hillsides absolutely scarlet with it in the first week of October, mixed with the silver bark and golden leaves of the birch and aspen.

It is so easy to raise from seed, which should be sown when ripe, or treated like haws, and so indifferent as to soil and situation that it may be planted almost anywhere, but usually grows best in mountainous and northern regions.

The finest trees which I have seen are at Walcot, Shropshire, where in 1905 I measured a well-shaped specimen 56 ft. by 6½ ft. with a clean bole 20 ft. long, and at Stratton Strawless, Norfolk, where in 1907 I saw a tree 56 ft. by 5 ft., the trunk of which was covered by large burrs. Though Loudon says that the largest trees of this species are in the West Highlands, yet I have never seen or heard of any equal to those mentioned above.

In Norway where it is almost everywhere a common tree, and is called *Rogn*, the bark is the favourite winter food of the elk, and the fruits are very much liked by bears, so that in districts where these animals are found the tree becomes comparatively scarce. The largest that I have seen were rarely 30 to 40 ft., though Schübeler mentions one near Christiania 48 ft. high. He illustrates[2] a good wild tree at Akureyri, in Iceland, which shows that even in this inhospitable and treeless land the species thrives well.

No tree is more commonly seen as an epiphyte in this country than the mountain ash, whose berries are often dropped by birds on the decaying branches

---

[1] Cf. C. Reid, *Origin British Flora*, 119 (1899).   [2] *Vaextlivet i Norge*, 92, fig. 42 (1879).

or crowns of other trees, where they grow and flourish; sometimes attaining a considerable size and age, and sending their roots down to the ground. The best example I have seen of this is in Glenaffric, Inverness-shire, where an immense old alder tree about $19\frac{1}{2}$ ft. in girth had been split to the ground by a rowan (known in Gaelic as *cuerun*), which had commenced life in its head and was $4\frac{1}{2}$ ft. in girth in 1910.   Mr. Stephenson Clarke, the lessee of Glenaffric forest, told me of this tree, which he has photographed.

A good many superstitious ideas as to the value of this tree as an antidote to witchcraft are said by Loudon and others to have been formerly prevalent; but these are now dying out except in remote districts.

The wood is hard, heavy, and close-grained, of a grey or whitish colour, and so tough and strong that it was formerly used for bows[1]; but though well suited for tool-handles, hoops, and even for chair-making, it is rarely used in this country.

(H. J. E.)

---

[1] Evelyn, *Sylva*, 68 (1679), who calls it quickbeam or witchen.

MAGNOLIA

Magnolia, Linnæus, *Sp. Pl.* 535 (1735); Bentham et Hooker, *Gen. Pl.* i. 18 (1862); Nicholson, in *Gard. Chron.* xvii. 515 (1895); Rehder, in Bailey, *Cycl. Amer. Hort.* 964 (1900); Schneider, *Laubholzkunde*, i. 328 (1905).

DECIDUOUS or evergreen trees and shrubs, belonging to the order Magnoliaceæ. Leaves alternate, simple, stalked, entire, with pinnate lateral nerves, which unite and loop before reaching the margin; in most species punctate with translucent dots; petiole channelled at the base, marked with two linear scars, continuous with an annular scar around the branchlet, due to the early fall of the two connate stipules,[1] which, adnate to the petiole, formed a cylindrical sheath in the bud.

Flowers perfect, large, solitary, terminal; sepals and petals imbricate in the bud, inserted under the ovary; sepals three, often similar to the petals in size and colour, occasionally smaller and greenish; petals six, nine, or twelve, in two, three, or four rows; stamens and pistils numerous, imbricated, the stamens below the pistils on an elongated receptacle; ovary sessile, one-celled; style short, recurved, stigmatic on the inner surface; ovules, two. Fruit cone-like, composed of numerous coalesced two-seeded follicles, dehiscent on the back; seeds on long stalks, with a red or scarlet outer coat, and a minute embryo, situated at the base of the fleshy albumen.

About twenty-five species of Magnolia are known, natives of North America, China, Japan, Assam, and the Himalayas. The following key comprises the species, which are cultivated in the open air in this country :—

I. *Leaves evergreen, coriaceous.*

1. *Magnolia grandiflora*, Linnæus. United States. See p. 1583.

    Branchlets rusty tomentose. Leaves obovate-oblong, acuminate at the apex, more or less covered beneath with brownish tomentum.

2. *Magnolia Delavayi*, Franchet. China. See p. 1592.

    Branchlets glaucous, minutely pubescent. Leaves ovate-oblong or elliptic, mucronate at the apex, glaucous with scattered pubescence beneath.

II. *Leaves sub-evergreen, falling before the young leaves appear in the following spring, bluish white and pubescent beneath.*

3. *Magnolia glauca*, Linnæus. United States. See p. 1585.

    Branchlets glabrous. Leaves elliptic or oblong-lanceolate, rounded or acute at the apex.

[1] Cf. Lubbock, in *Journ. Linn. Soc. (Bot.)* xxx. 466 (1895).

4. *Magnolia Thompsoniana*, Koch.   A hybrid.   See p. 1585.

Branchlets glaucous, with a few hairs at the insertion of the leaves.   Leaves obovate-elliptic, acute at the apex.

III. *Leaves deciduous in autumn.*

* *Leaves cordate at the base.*

5. *Magnolia Fraseri*, Walter.   United States.   See p. 1590.

Branchlets glabrous.   Leaves obovate, deeply cordate and auricled at the base, pale or light green beneath.

6. *Magnolia macrophylla*, Michaux.   United States.   See p. 1589.

Branchlets densely pubescent.   Leaves obovate-oblong, cordate or sub-cordate at the broad truncate base, bluish white beneath.

** *Leaves not cordate at the base.*

(a) *Leaves large, with more than twenty pairs of lateral nerves.*

7. *Magnolia tripetala*, Linnæus.   United States.   See p. 1588.

Branchlets glabrous.   Leaves membranous, obovate-oblong, tapering towards the base and apex, pale with a scattered minute pubescence beneath.

8. *Magnolia hypoleuca*, Siebold and Zuccarini.   China and Japan.   See p. 1592.

Branchlets glabrous.   Leaves coriaceous, obovate, acute or cuspidate at the apex, bluish grey with scattered white hairs beneath.

(b) *Leaves moderate or small in size, with less than twenty pairs of lateral nerves.*

† *Leaves pale beneath.*

9, 10. *Branchlets pubescent.*

9. *Magnolia acuminata*, Linnæus.   Ontario, United States.   See p. 1586.

Large tree.   Leaves oval, acuminate at the apex, pubescent on both surfaces; lateral nerves, twelve to fifteen pairs.

10. *Magnolia parviflora*, Siebold and Zuccarini.   Japan.   See p. 1598.

Shrub.   Leaves obovate, acuminate at the apex, glabrous above, minutely pubescent beneath; lateral nerves about nine pairs.

11-13. *Branchlets glabrous.*

11. *Magnolia Campbelli*, Hooker and Thomson.   Eastern Himalayas.   See p. 1590.

Branchlets glaucous.   Leaves narrowly elliptic, acuminate, pale with a bluish tint and glabrous beneath.

12. *Magnolia Watsoni*, Hooker.   Japan (?).   See p. 1598.

Leaves obovate, acute at the apex, greyish beneath, with scattered appressed hairs.

13. *Magnolia salicifolia*, Maximowicz.   Japan.   See p. 1595.

Leaves lanceolate, tapering to an acuminate and often curved apex; pale with a bluish tint and minutely pubescent beneath.

†† *Leaves green beneath.*

14. *Branchlets glabrous.*

14. *Magnolia Kobus*, De Candolle.   Japan.   See p. 1594.

Leaves obovate, cuspidate at the apex, tapering at the base, with axil-tufts of pubescence beneath.

15-17. *Branchlets pubescent.*

15. *Magnolia denudata*, Thunberg. China. See p. 1597.

Branchlets with silvery appressed pubescence at the tip, elsewhere glabrous. Leaves obovate or oval, and broadest at the middle, acute or shortly acuminate at the apex, tapering at the base.

16. *Magnolia conspicua*, Salisbury. China. See p. 1596.

Branchlets more or less covered with appressed pubescence. Leaves obovate, cuspidate at the apex, usually rounded at the base.

17. *Magnolia stellata*, Maximowicz. Japan. See p. 1599.

Branchlets more or less covered with appressed pubescence. Leaves obovate-oblong or oblanceolate, gradually tapering to the base, variable at the apex.

(A. H.)

MAGNOLIA GRANDIFLORA

Magnolia grandiflora, Linnæus, *Syst.* ii. 1082 (1759); Loudon, *Arb. et Frut. Brit.* i. 261 (1838); Sargent, in *Bot. Gaz.* xliv. 226 (1907).

Magnolia fœtida, Sargent, in *Garden and Forest*, ii. 615 (1889), *Silva N. Amer.* i. 3, tt. 1, 2 (1890), and *Trees N. Amer.* 316 (1905).

An evergreen tree, attaining in America 60 to 80 ft. high, with a straight trunk occasionally 12 ft. in girth. Bark with thin appressed scales. Young branchlets covered with rusty red tomentum. Leaves coriaceous, persistent two years, 6 to 8 in. long, 2 to 3 in. wide, obovate-oblong or narrowly elliptic, with a short acuminate apex; upper surface dark green, shining, glabrous; lower surface covered more or less with a rusty brown tomentum, or occasionally glabrescent.

Flowers, on stout tomentose stalks, fragrant, 7 to 8 in. across; the three petaloid sepals and six (rarely nine to twelve) petals creamy white, ovate or oval, narrowed at the base. Fruit rusty tomentose.

Several varieties[1] have originated in European nurseries :—

1. Var. *exoniensis*, Loudon. Rather fastigiate in habit, with broadly elliptical leaves, rusty tomentose beneath. It begins to flower when only a few feet high. This variety appears to differ scarcely from var. *lanceolata*, Aiton,[2] *ex* Sims, *Bot. Mag.* t. 1952 (1818), though Loudon kept it distinct.

2. Var. *angustifolia*, Loudon. Leaves lanceolate, undulate in margin. Introduced from Paris in 1825.

3. Var. *Gallissoniensis*, Simon-Louis, *Cat.* 59 (1869). According to Rehder,[3] this variety has proved the hardiest in Europe; and Mouillefert[4] says that it makes the finest standard tree.

4. A variety with double flowers is known; and this peculiarity has been observed as a sport on a tree of the ordinary kind in England.[5]

M. grandiflora is a native of the south-eastern United States, extending from North Carolina along the coast to Florida, and westward through the Gulf States to

[1] Dr. J. Bedelian of Nikita, in the Crimea, writes concerning the great variability of this species in *Gard. Chron.* xlii. 390 (1907). Cf. also *ibid.* xliii. 83 (1908).   [2] *Hort. Kew.* ii. 251 (1789).

[3] In Bailey, *Cycl. Amer. Hort.* 968 (1900).

[4] *Traité des Arbres*, i. 112 (1892).   [5] Cf. *Gard. Chron.* viii. 223 (1890).

the Brazos river in Texas, ascending in the Mississippi valley to southern Arkansas and the mouth of the Yazoo river.   It is best developed in the forests of western Louisiana, where it is one of the most characteristic trees.   It usually grows in rich moist soil on the borders of river swamps and of the ponds in the pine-barrens. Plate 353, reproduced from a photograph kindly sent by Miss Cummings, shows the habit of the evergreen Magnolia in North Carolina.   (A. H.)

The date of introduction is somewhat uncertain, but it is supposed on good authority to have been grown at Exmouth by Sir John Colliton before 1737, and a tree there, of which a long account is given in *Gardener's Magazine*, xi. 70 (1835), was for many years the parent by layering of great numbers of plants, and was cut down by mistake in 1794.

This tree surpasses all others in the temperate zone except perhaps the Himalayan *M. Campbelli*, in the beauty, size, and fragrance of its flowers; but being a native of more southern climates, it only succeeds without protection in the warmest parts of Great Britain, and even then is but a poor and stunted tree compared to what it is in south-western France, Portugal, and Italy.

Though rarely planted as a standard tree we have seen it up to 20 to 30 ft. high in a few places, the best perhaps being at Powderham Castle.   It was, however, reported[1] in 1894 to be 36 ft. high by 4 ft. 8 in. in girth at East Cowes Castle, and 50 ft. by 5 ft. at Rozel Bay, Jersey.   A tree at Gunnersbury House measured 31 ft. by 2 ft. 8 in. in 1911.   There are two trees in the Azalea garden at Kew, about 23 ft. high; and a fine specimen trained against the Museum, which is about 35 ft. high.

On a wall even in cold parts of England it has survived a temperature below zero, though it only flowers after two good seasons, and then often so late that the flowers are cut off by early frosts.   Though it sets seeds in hot summers, I am not aware that they ever ripen here; and plants which I have raised from foreign seed grow slowly and want greenhouse treatment for some years.

The largest trees that I have seen in Europe of this species are in the garden of Baron Soutelinho (Mr. A. Tait) at Oporto, one of which, now nearly dead, was $11\frac{1}{2}$ ft. in girth, and another, also showing signs of decay, was 62 ft. by 11 ft.   A third with a fine stem clean of branches to 35 ft. was in perfect vigour and 60 ft. by 7 ft. in 1909.   In France the largest that I know are in the public gardens of Bordeaux, one of which, that had been transplanted when already an old tree, was 59 ft. by 6 ft. 3 in. in 1909.

On the Isola Madre in Lake Maggiore, I measured a tree in 1906 which was 70 ft. by 6 ft. 9 in. and ripened seed freely.

Sargent says that this species has the hardest, heaviest, and best wood of the American Magnolias, but it is little used, even in the United States, and hardly known in commerce.   I believe, however, that some of it is mixed with the so-called canary or white wood (*Liriodendron*), and is not easy to distinguish without careful examination.   It is a close-grained wood of pale creamy yellow, or brownish yellow colour, showing a minute silver grain, and looks as if it would take a fine polish.

(H. J. E.)

[1] *Gard. Chron.* xvi. 286, 375 (1894).

MAGNOLIA GLAUCA

Magnolia glauca, Linnæus, *Syst.* ii. 1082 (1759); Loudon, *Arb. et Frut. Brit.* i. 267 (1838); Sargent, *Silva N. Amer.* i. 5, t. 3 (1890), and *Trees N. Amer.* 317 (1905).

A tree, occasionally attaining in America 50 to 70 ft. in height and 10 ft. in girth, but usually much smaller. Branchlets slender, glabrous. Leaves remaining on the branches usually throughout the winter till the young leaves appear in spring, but in England some fall earlier; elliptical or oblong-lanceolate, about 4 inches long and 2 in. broad, rounded or acute at the apex; thin in texture; upper surface bright green, shining, glabrous; lower surface bluish white and covered more or less with fine white pubescence.

Flowers, on slender glabrous stalks, creamy white, fragrant, globose, 2 to 3 in. across; sepals membranous and shorter than the nine or twelve obovate petals. Fruit pink, 2 in. long, glabrous.

1. Var. *longifolia*, Loudon. This variety, which I have not seen, is said by Sargent to have lanceolate leaves, and to continue flowering for a period of two or three months. According to Loudon, it originated in Belgium, and was considered to be of hybrid origin.

2. The following is usually considered to be a hybrid, between *M. glauca* and *M. tripetala* :—

Magnolia Thompsoniana, Koch, *Dendrologie*, i. 369 (1869); Sargent, in *Garden and Forest*, i. 269, fig. 43 (1888).

Magnolia glauca, var. *major*, Sims, *Bot. Mag.* t. 2164 (1820).
Magnolia glauca, var. *Thompsoniana*, Loudon, *Arb. et Frut. Brit.* i. 267 (1838).
Magnolia major, Schneider, *Laubholzkunde*, i. 334 (1905).

A small tree. Young branchlets glaucous, glabrous except for a few hairs at the insertions of the leaves. Leaves obovate-elliptic, 6 to 8 in. long, $1\frac{1}{2}$ to 3 in. wide, acute at the apex; upper surface light green, glabrous except for pubescence on the midrib; lower surface whitish, with a greyer tint than *M. glauca*, covered more or less with a fine pubescence. Flowers white, fragrant, 5 to 6 in. across; sepals shorter than the petals, greenish, reflexed as in *M. tripetala* when the flower opens, and not so early deciduous as in *M. glauca*; petals nine, obovate-oblong, contracted into a narrow claw.

According to Sabine,[1] this was raised in Thompson's nursery at Mile End in 1808, in which year the Magnolias fruited freely. A tree of *M. tripetala* was growing close to the tree of *M. glauca*, from which the seed was obtained that gave rise to the new plant. The latter has much larger flowers than those of *M. glauca* and in some respects shows the influence of *M. tripetala*.

M. Thompsoniana, like some hybrids, is hardier than either of the parents, being much less liable to injury from spring frosts, and is commoner in cultivation than *M. glauca*. (A. H.)

[1] In *Trans. Hort. Soc.* iii. 205 (1823).

M. glauca is a native of the eastern parts of the United States, where it is known as Sweet or Swamp Bay, being an evergreen tree in the south, and becoming deciduous in the north.   It occurs in one or two stations in Massachusetts and Long Island, and is distributed along the coast from New Jersey to Florida, ranging inland to Franklin County in Pennsylvania ; and extending through the Gulf States to south-western Arkansas and Trinity River in Texas.   It usually grows in swamps and along the borders of the ponds in the pine barrens ; and attains its largest size in Florida.

This species was introduced by Banister, who sent it to Bishop Compton, at Fulham, in 1688.   Loudon mentions several places in England where it was cultivated in 1838, and states that it frequently ripened seed.   It is now a rare tree, seldom seen [1] except in botanic gardens.   The finest specimen that we know is at White Knights, an old tree about 30 feet high in 1905.   It is said by Sargent to grow better when grafted on stocks of *M. acuminata* than on its own roots.   (H. J. E.)

MAGNOLIA ACUMINATA, Cucumber Tree

Magnolia acuminata, Linnæus, *Syst.* ii. 1082 (1759) ; Sims, *Bot. Mag.* t. 2427 (1823) ; Loudon, *Arb. et Frut. Brit.* i. 273 (1838) ; Sargent, *Silva N. Amer.* i. 7, tt. 4, 5 (1890), and *Trees N. Amer.* 319 (1905).

A deciduous tree, attaining in America 60 to 90 ft. in height, and 10 to 12 ft. in girth.   Bark ½ in. thick, furrowed, scaly.   Young branchlets, with dense appressed whitish pubescence towards the base, elsewhere with scattered long hairs.   Leaves oval, 6 to 9 in. long, 4 to 5 in. broad, acuminate at the apex, broad and rounded or cuneate at the base ; upper surface dark green, dull, with a scattered minute pubescence on the surface, denser on the midrib ; lower surface pale, with scattered wavy white hairs ; petiole pubescent.

Flowers on pubescent stalks, campanulate, greenish yellow or glaucous green, about 2 to 3½ in. long ; sepals membranous, soon reflexed ; petals six, ovate or obovate, pointed, upright, those of the outer row much broader than those of the inner row.   Fruit glabrous, dark red, 3 in. long.

In winter the branchlets are reddish, glabrous ; marked with V-shaped six-dotted leaf-scars, the two apices of which are continuous with a linear scar encircling the stem and indicating the fall of the stipule.   Buds surrounded by a single scale, pubescent with silky hairs, the terminal bud much larger than the lateral buds.

1. Var. *cordata*, Sargent, in *Amer. Journ. Science*, xxxii. 473 (1886), *Silva N. Amer.* i. 8, t. 6 (1890), and *Trees N. Amer.* 320 (1905).

Magnolia cordata, Michaux, *Fl. Bor. Amer.* i. 328 (1803) ; Loudon, *Arb. et Frut. Brit.* i. 275 (1838).

A small tree, with leaves more greyish pubescent beneath than in the type, and rarely cordate at the base.   Flowers smaller, bright canary yellow.

---

[1] Bunbury, *Arb. Notes*, 55 (1889), states that two trees planted at Barton, Suffolk, in 1861, died in a few years.

This variety is said by Loudon to have been brought from America[1] by Lyon in 1801, the original tree in Loddiges' nursery being about 15 ft. high in 1838. No wild tree exactly similar to this variety has been discovered; but forms approaching it in the pubescence and shape of the leaves and in the small size and colour of the flowers have been found on the Blue Ridge in South Carolina, and in central Alabama. We have seen no large specimens in England. Bean[2] saw at Herren-hausen, Hanover, a tree 35 ft. high by 3 ft. 1 in. in 1908.   (A. H.)

DISTRIBUTION AND CULTIVATION

M. acuminata extends from western New York and southern Ontario, where Macoun gives the Niagara Falls as its only natural station, southward along the Alleghany Mountains, to southern Alabama, Kentucky, Tennessee, Arkansas, and Mississippi. It seems to be rare in the north, and attains its greatest size and abundance only in the rich woods of the lower valleys of east Kentucky and Tennessee.

It was introduced into cultivation in England by Bartram, who sent plants to Peter Collinson in 1736.

It is the only Magnolia which as yet has grown to be a large tree in this country, and seems to require less summer heat, and to endure more severe frost than any of the other American species; but it only becomes a fine tree in warm rich soils in the southern half of England. I am not aware that it has ever ripened seed[3] in this country; but Loudon says that seedlings are preferable to plants raised from layers, and that they were used as stocks on which to graft other species of Magnolia. Masters says[4] that it is one of the very best of trees for towns.

Loudon mentions as the largest specimen known to him, one recently cut down at Thorndon Hall, the seat of Lord Petre in Essex, which was nearly 7 ft. in girth; and there was another in the same park 37 ft. by 7 ft., which survived until 1903. He figures one at Syon 49 ft. high in 1838, which may not be the same as one now growing in the Church Walk there, and recorded by Jackson as 51 ft. 5 in. by 3 ft. 7 in. in girth.

The finest tree known to us (Plate 358) grows at West Dean Park, and when I saw it in 1906 was 60 ft. by 7 ft., with a clean trunk about 25 ft. high. There is a much taller but ill-shaped tree branching near the ground at Albury, which was about 75 ft. high in 1905. At Claremont in 1903 I saw a tree about 55 ft. by 6½ ft., which in 1883 was 40 ft. high. In 1910 Mr. Bean[5] found it to be 60 ft. high. At The Mote, Maidstone, I measured a handsome tree 68 ft. by 4 ft. 9 in. in 1911. Another standing close to it was 56 ft. by 6 ft.

At Heanton Satchville, in North Devon, in 1905 I saw a straight well-shaped pyramidal tree 57 ft. by 5 ft. 3 in. At Arley Castle there are two fine trees 65 ft.

---

[1] Sargent, in *Garden and Forest*, 1889, p. 338, states that two specimens in the Harvard Botanic Garden are known to have been fully grown trees in 1842.   [2] *Kew Bull.* 1908, p. 392.
[3] Bunbury, *Arb. Notes*, 55 (1889), says that the fruit always dropped off from the tree at Barton, before it was half ripe.
[4] *Gard. Chron.* vi. 474 (1889).   [5] *Kew Bull.* 1910, p. 164.

and 62 ft. high by over 6 ft. girth, which were planted in 1820. At Barton, Suffolk, a tree over 70 ft. high in 1904, when it appeared to be dying at the top, was planted in 1826. This tree was not injured by the severe winter of 1860-1861.

At Westonbirt a tree, 62 ft. by 5 ft. 9 in. in 1906, had been split nearly to the ground, but had been so well repaired by hoops round the trunk, that the two parts were growing together. At Chatsworth, Mr. A. B. Jackson measured in 1908 a tree 50 ft. by 4 ft. 6 in. At Essendon Place, Herts, there is a very narrow slender tree, which Henry found to be 56 ft. high by 3 ft. 4 in. in girth in 1907. At Merton Hall, Norfolk, a tree, which was raised from seed in 1862, measured 50 ft. by 4 ft. 4 in. in 1908. At Nuneham Park, Oxon, there is a tree which was 40 ft. by 4 ft. in 1907. At Fawley Court, Henley, a fine specimen was 40 ft. by 5½ ft. in the same year.

In Scotland, the finest tree we have heard of is one recorded by Hunter[1] as growing in the American garden at Dunkeld, which measured 40 ft. by 3 ft. 4 in. in 1883; but I did not see it when I visited this place in 1906. At Biel, East Lothian, I saw a healthy tree about 30 ft. by 5 ft. in 1911.

In Ireland probably the tallest tree is one at Curraghmore, which Henry found to be 60 ft. by 4 ft. 6 in. in 1907. At Narrow Water Castle, Co. Down, Captain Hall informed us in 1907, that there was a remarkable tree, 41 ft. by 5½ ft. This had not grown in height for some years past, owing to the top catching the wind, but it increased enormously below; and the branches, widely spreading and reaching the ground, now cover an area 192 ft. in circumference.

The tallest tree in Europe, if correctly measured, is probably one at Schloss Dyck, near Dusseldorf, in Germany, which was reported[2] in 1904 to be 30 metres in height and 2.12 metres in girth.

The wood of this tree resembles that of *M. grandiflora*; but, judging from Hough's specimen, is rather darker in colour, and even more like the wood of *Liriodendron*. Hough says that it is largely used for doors and wainscots, and for bowls, troughs, and wooden ware; but I have never seen this wood in England, or heard of its being imported under its own name.   (H. J. E.)

MAGNOLIA TRIPETALA, Umbrella Tree

Magnolia tripetala, Linnæus, *Syst.* ii. 1082 (1759); Loudon, *Arb. et Frut. Brit.* i. 269 (1838); Sargent, *Silva N. Amer.* i. 13, tt. 9, 10 (1890), and *Trees N. Amer.* 321 (1905).
Magnolia umbrella, Lamarck, *Encycl.* iii. 673 (1789).
Magnolia frondosa, Salisbury, *Prod.* 379 (1796).

A deciduous tree, attaining in America 30 to 40 ft. in height and 5 ft. in girth. Young branchlets glabrous. Leaves, variable in size, usually 12 to 20 in. long and 6 to 8 in. broad, obovate-oblong, tapering towards the base and apex, the latter ending in an acuminate point; upper surface green, glabrous; lower surface pale, with a scattered minute pubescence, densest on the midrib and nerves. In winter,

[1] *Woods of Perthshire*, 52 (1883). An old tree at Blair, near Dalry, Ayrshire, is said by the gardener to flower annually and to have been 45 ft. high in July 1912.   [2] *Mitt. Deut. Dend. Ges.* 1904, p. 19.

the glabrous branchlets[1] have broad oval leaf-scars, with numerous dots arranged in two or three series. The buds are glabrous, glaucous.

Flowers on slender glaucous glabrous stalks, cup-shaped, eight to nine inches across, disagreeable in odour; sepals light green, becoming reflexed; petals, six or nine, coriaceous, white, ovate, pointed, those of the outer row much longer and broader than those of the inner rows. Fruit, $2\frac{1}{2}$ to 4 in. long, glabrous, pink.

This species, which has the leaves crowded at the summits of the flowering branches, is widely distributed in the region of the Alleghany Mountains, from the valley of the Susquehanna River in Pennsylvania southwards to Kentucky, Alabama, Tennessee, north-eastern Mississippi, and Arkansas, and extending nearly to the coast in the south Atlantic States. It usually grows on the banks of mountain streams or on the edges of swamps.　(A. H.)

This species was introduced into England in 1752, and seems to have been commoner in Loudon's time than it is now. He mentions trees at Cobham (Kent), Syon, Golden Grove, Croome, and Walton House, 30 to 36 ft. high, which cannot now be found; and states that at Deepdene self-sown seeds had produced plants. We have only seen a few small trees, none of which look very thriving, at Kew; Holkham; Stanage Park, Herefordshire; and Canford, Hants.　(H. J. E.)

MAGNOLIA MACROPHYLLA

Magnolia macrophylla, Michaux, *Fl. Bor. Am.* i. 327 (1803); Loudon, *Arb. et Frut. Brit.* i. 271 (1838); Sargent, *Silva N. Amer.* i. 11 tt. 7, 8 (1890), and *Trees N. Amer.* 320 (1905).

A deciduous tree, attaining in America 30 to 50 ft. in height and 5 ft. in girth. Young branchlets covered with a dense short pubescence, retained in part in the second year. Leaves, the largest of any of the species in cultivation, 12 to 25 in. long, 6 to 10 in. broad, obovate-oblong, acute or acuminate at the apex, cordate or subcordate at the truncate base; upper surface bright green, glabrous; lower surface bluish white, with a scattered fine pubescence, which is denser on the midrib.

Flowers, on stout tomentose stalks, white, cup-shaped, fragrant, 10 to 12 in. across when expanded; sepals membranous, much narrower than the six ovate concave thick creamy white petals, which become reflexed above the middle. Fruit pubescent, pink, nearly 3 in. long.　(A. H.)

This species is a rare tree in the south-eastern United States, occurring from North Carolina and south-eastern Kentucky, southwards to Florida, Alabama, Mississippi, Louisiana, and central Arkansas. It usually grows in sheltered valleys in deep rich soil.

It was introduced into England in 1800. None of the trees at Arley, Chiswick, and White Knights, mentioned by Loudon, can now be found; but there is a fine specimen at Claremont, which, in 1910, was 40 to 45 ft. in height, and 2 ft. 11 in. in girth.[2]　(H. J. E.)

---

[1] Cf. Foerste, in *Bot. Gaz.* xx. 80, t. 8 (1895).　　[2] Bean, in *Kew Bull.* 1910, p. 163.

MAGNOLIA FRASERI

Magnolia Fraseri, Walter, *Fl. Carol.* 159 (1788); Sargent, *Silva N. Amer.* i. 15, tt. 11, 12 (1890), and *Trees N. Amer.* 322 (1905).
Magnolia auriculata, Lamarck, *Ency.* iii. 645 (1789); Loudon, *Arb. et Frut. Brit.* i. 276 (1838).
Magnolia auricularis, Salisbury, *Parad. Lond.* i. t. 43 (1806).

A tree, attaining in America 40 ft. in height and 4 ft. in girth, often dividing near the base into diverging stems. Young branches glabrous. Leaves 8 to 12 in. long, 5 to 6 in. wide, obovate, acute at the apex, deeply cordate and auricled at the base; both surfaces and petiole glabrous; pale beneath.

Flowers on stout glabrous glaucous stalks, opening after the leaves, cream-white, sweet-scented, 6 to 9 in. across; sepals quickly deciduous; petals six or nine, obovate, acuminate, membranous, contracted below the middle. Fruit glabrous, pink, 4 to 5 in. long, the ripe carpels ending in long subulate persistent tips.

1. Var. *pyramidata*, Nuttall, *Gen. N. Amer. Plants*, ii. 18 (1818).

Magnolia pyramidata, Pursh, *Fl. Amer. Sept.* ii. 382 (1814); Edwards, in *Bot. Reg.* t. 407 (1819); Sargent, *Trees and Shrubs*, iii. 101 (1903), and *Trees N. Amer.* 323 (1905).

Flowers much smaller than in the type, 3 to 4 in. across. Fruit, 2 to $2\frac{1}{2}$ in. long, the ripe carpels ending in short incurved persistent tips.

This variety occurs in rich alluvial soil in the coast region from southern Georgia through western Florida to southern Alabama. It was introduced into Loddiges' nursery by Lyon in 1818.

M. Fraseri is a native of the southern Alleghany Mountains, from south-western Virginia to northern Georgia and Alabama, eastern Tennessee, and northern Mississippi; most abundant and of its largest size on the upper waters of the Savannah River in South Carolina. The typical form is a native of mountain valleys.

M. Fraseri was introduced into England in 1786; and seems to be very uncommon. A tree at Kew, which belongs to var. *pyramidata*, is about 20 feet high; and produces small flowers freely from the middle of May to the end of June, which are almost yellow at the time of opening, becoming a rich cream colour as they fade. At Leonardslee,[1] it succeeds in a sheltered position. (A. H.)

MAGNOLIA CAMPBELLI

Magnolia Campbelli, J. D. Hooker and Thomson, in J. D. Hooker, *Illust. Him. Plants*, tt. 4, 5, (1855); J. D. Hooker, *Fl. Brit. India*, i. 41 (1872), and in *Bot. Mag.* t. 6793 (1885); King, in *Ann. Bot. Gard. Calcutta*, iii. pt. 2, p. 208, tt. 51, 52 (1891); Gamble, *Indian Timbers*, 9 (1902).

A large deciduous tree, attaining occasionally in the Himalayas 150 ft. in height, and 12 to 20 ft. in girth. Young branchlets glabrous, glaucous. Leaves

---

[1] *Gard. Chron.* xli. 223 (1907).

membranous, narrowly elliptic, 4 to 8 (rarely 12) in. long, 2 to 3 (rarely 4) in. broad, shortly acuminate at the apex, slightly pubescent when young, but glabrous on both surfaces when mature, light green above, pale beneath; petioles glabrous.

Flowers appearing before the leaves, globose, 6 to 10 in. across, delicately scented, deep rosy pink[1] externally, cream-white internally; sepals three, similar in shape and colour to the nine or twelve petals, all elliptic-oblong, and rounded at the apex. Fruit about six inches long, cylindric, with red seeds. (A. H.)

This species is a native of Sikkim and Bhutan, occurring, according to Gamble and Hooker, at elevations of 8000 to 10,000 ft.; but I have seen it in the Rangirun forest, near Darjeeling, below the road leading to Pashok, which cannot be higher than 7000 ft. Hooker[2] describes it as a large forest tree with black bark, often 80 ft. high and 12 to 20 ft. in girth; and Gamble says that, according to King, specimens 150 ft. high were common in Sikkim in 1849, but that the demand for building timber and tea-box boards has made large trees scarce. Though the tree itself is not a handsome one, it is impossible to exaggerate its beauty when seen in April standing leafless, but covered with its immense rosy flowers among the chestnuts, oaks, and other trees, in one of the most beautiful forests in the world.

Repeated attempts were made to introduce this species by seed; but on arrival the fleshy albumen was always found decayed and the minute embryo killed. Ultimately, about 1880, living plants were sent to Kew from the Calcutta Botanic Gardens.

It is perfectly hardy at Kew, and has attained a considerable size in the milder parts of this country; but apparently requires a greater degree of heat and moisture in summer, and remains usually in a bushy state. It flowered for the first time in Europe in 1885 in Mr. Crawford's garden at Lakelands, Cork, and was then figured in the *Botanical Magazine*. It has since flowered at several places, as in 1898 at Veitch's nursery[3] at Exeter, where it had been growing in the open for twelve years, and at Leonardslee[4] in 1907, when it was about 20 ft. high. At Abbotsbury,[5] where it was about 25 ft. high in 1903, it is the earliest of all the species in flower, being two or three weeks in advance of *M. stellata* and *M. conspicua*. A specimen at Belgrove,[6] Cork, produced 147 flowers in 1903; but the late Mr. Gumbleton stated that in other years either no buds were set or the flowers never opened, having been killed by frosts and cold winds. Nicholson mentions[7] a fine specimen at Fota, 25 ft. high in 1895, which produced beautiful flowers of a much richer tint than those at Lakelands. (H. J. E.)

[1] The flowers are variable in the depth of the pink hue; and are occasionally white.
[2] *Him. Journ.* i. 125 (1854), where he mentions that *Rhododendron Dalhousiæ* often grows epiphytically upon its branches.
[3] *Gard. Chron.* xxiii. 89, fig. 33 (1898). [4] *Ibid.* xlii. 3 (1907).
[5] *Ibid.* xxxiii. 174, fig. 73 (1903). This tree was, when I saw it in July 1912, about 30 ft. high and 2⅓ ft. in girth.—A. H. [6] *Ibid.* xxxiii. 172 (1903) and xlii. 33 (1907). [7] *Ibid.* xvii. 515 (1895).

MAGNOLIA DELAVAYI

Magnolia Delavayi, Franchet, *Pl. Delav.* 33, tt. 9, 10 (1889); Sprague, in *Bot. Mag.* t. 8282 (1909).

An evergreen tree, about 30 ft. high. Young branchlets glaucous, minutely pubescent. Leaves coriaceous, persistent two years, ovate-oblong or elliptic, 7 to 12 in. long, and 4 to 7 in. wide, rounded and mucronate at the apex; dull green and glabrous above; lower surface glaucous, with scattered fine pubescence; midrib beneath very prominent and, like the petiole, very stout and more or less pubescent.

Flowers creamy white, fragrant, about 7 in. across; sepals three, oblong, reflexed; petals about seven, spatulate-obovate. Fruit about 5 in. long.

This is a native of Yunnan in south-western China, where I saw it growing in rocky situations and woods at 5500 to 7000 ft. above sea-level. It was introduced by Wilson in 1900; and flowered at Kew in 1908. It is scarcely hardy at Kew,[1] though it thrives against a wall; but it will probably succeed better in the milder parts of England, Wales, and Ireland. Both the foliage and flowers are very handsome. (A. H.)

MAGNOLIA HYPOLEUCA

Magnolia hypoleuca, Siebold and Zuccarini, in *Abh. Ak. München*, iv. pt. ii. 187 (1845); Sargent, in *Garden and Forest*, i. 304, fig. 49 (1888); Matsumura, in *Journ. Coll. Sc. Imp. Univ. Tokyo*,[2] xii. 284 (1899); Shirasawa, *Icon. Ess. Forest. Japon*, i. text 70, t. 39, figs. 13-29 (1900); Skan, in *Bot. Mag.* t. 8077 (1906); Mayr, *Fremdländ. Wald- u. Parkbäume*, 481 (1906).

A deciduous tree, attaining 100 ft. in height and 10 ft. in girth in Japan. Young branchlets glabrous. Leaves coriaceous, often crowded at the ends of the branches, obovate or obovate-elliptic, about 8 to 15 in. long and 6 to 8 in. broad, acute or cuspidate at the apex: upper surface light green, glabrous; lower surface bluish grey, with scattered curved white hairs, often with dense appressed pubescence on the midrib; petiole pubescent.

Flowers, opening when the leaves are nearly fully grown, very fragrant, 6 to 8 in. across, creamy white or white; sepals coriaceous, tinged with red: petals nine, coriaceous, obovate-spatulate, rounded or cuspidate at the apex. Fruit red, 5 to 8 in. long.

M. hypoleuca is a native of China and Japan. In central and western China it is commonly cultivated around dwellings in mountainous districts at elevations between 2500 and 4500 ft. It is known to the Chinese as *hou-p'o*, its bark[3] being esteemed as a valuable drug, which is exported to all parts of China. Neither

---

[1] Cf. *Kew Bull.* 1909, p. 235. A tree in the Temperate House at Kew is about 20 ft. high.

[2] Recorded as doubtfully wild in the Liu Kiu Islands.

[3] Cf. Hanbury, *Sc. Papers*, 266 (1876), and Bretschneider, *Bot. Sinic.* iii. 472 (1895). Père David, *Journ. Trois. Voy. Chine*, ii. 360 (1875), mentions a large plantation of this tree in the province of Kiangsi.

Wilson nor myself ever found the tree in a wild state in China; but it is probably indigenous in some of the unexplored districts. It is mentioned in the earliest Chinese Herbal, which was compiled about 200 B.C.; and it is extremely unlikely to have been introduced from Japan. (A. H.)

This beautiful tree is called *Honoki* in Japan, where it is most common and attains a large size in the forests of Hokkaido. It also occurs in the mountains of Hondo, at elevations of 2000 to 5500 ft. The largest that I saw were about 100 ft. high, usually with tall clean stems. This species requires a rich moist soil and a considerable rainfall in summer; and produces valuable timber. The wood[1] is firm and uniform in texture, and of a yellowish or greenish white colour, and is largely used for drawing-boards, musical instruments, lacquer work, and many other purposes. It seems very similar to the wood of *M. grandiflora*, and is said to be little subject to warping and splitting. Its charcoal is highly prized in Japan for polishing lacquer-ware and metallic mirrors, and for finishing cutlery.

Though introduced[2] as long ago as 1865 by Thomas Hogg into the United States, where it has proved hardy in the north, it does not seem to have attracted attention in this country till recently; but on account of its large beautiful leaves, sometimes 2 ft. long, and its fragrant flowers, it is well worth cultivation in the south and west of England. Plants which I raised from seed, grew slowly at first, and for the most part died when planted out, owing, I believe, to the presence of lime in the soil. The oldest tree appears to be one at Grayswood, Haslemere, which was obtained from Yokohama in 1884, and flowered in 1905. A tree at Kew, which flowered in the same year, when it was 14 ft. high, was obtained from Japan in 1890. One at Trewidden, Cornwall, which was planted about 1893, is 30 ft. high by 2½ ft. in girth at one foot from the ground; this produces flowers abundantly; and bore ripe fruit in September 1911. Another at Enys, was about 20 ft. high in 1911. Wilson sent seeds to Messrs. Veitch, who now have the Chinese plant growing in their nursery at Coombe Wood.

In Ireland I have seen it growing well at Baronscourt, Co. Tyrone, the seat of the Duke of Abercorn, where the soil and climate seem to be less favourable than in many parts of Ireland.

This species was raised[3] in 1877 from seed in the botanic garden at Heidelberg, and flowered there when about 20 ft. high in 1898. Mayr who had a high opinion of it as a forest tree, suitable for producing good timber quickly in central Europe, introduced it in 1890 into the experimental garden at Grafrath near Munich, where a tree attained 20 ft. in height at ten years old. Schwappach[4] states that the experimental plots of this species at Eberswalde are very thriving, the trees being 42 ft. high after seventeen years' growth. The growth is rapid in youth, like the ash, which it also resembles in its requirement for space. Schwappach recommends planting it, mixed with oak and beech, on soils suitable to these species. Count Von

---

[1] Figured by Mayr, *op. cit.* t. xviii.

[2] According to *Garden and Forest*, i. 304, fig. 49 (1888), where the figure is taken from a tree in New York, which was 28 ft. high in 1898. [3] *Semaine Horticole*, 1900, p. 199.

[4] In *Zeitschrf. Forst. u. Jagdwesen*, xliii. 604 (1911), and *Mitt. Deut. Dend. Ges.* 1911, p. 12.

Schwerin, who gives a good account[1] of *M. hypoleuca*, states that the seeds, when sent from Japan, dry up and lose their germinating power, unless they are packed in their fleshy covering in charcoal dust or peat powder.　　　(H. J. E.)

MAGNOLIA KOBUS

Magnolia Kobus, De Candolle, *Syst.* i. 456 (1818); Maximowicz, in *Mél. Biol.* viii. 507 (1872); Shirasawa, *Icon. Ess. Forest. Japon*, i. text 71, t. 39, figs. 1-12 (1899); Matsumura in *Journ. Sc. Coll. Imp. Univ. Tokyo*,[2] xii. 284 (1899); Masters, in *Gard. Chron.* xxxvii. 265, *Supply. Illust.* (1905); Bean, in *Bot. Mag.* t. 8428 (1912).
Magnolia glauca, var. α, Thunberg, *Flora Jap.* 236 (1784).
Magnolia tomentosa, Thunberg, in *Trans. Linn. Soc.* ii. 336 (1794) (in part).

A deciduous small tree. Young branchlets slender, glabrous. Buds pubescent. Leaves membranous, averaging 3 to 5 in. long, and 2 to 3 in. wide, occasionally up to 6 in. long and 4 in. wide; obovate, gradually tapering to the cuneate base, cuspidate-acuminate at the apex; upper surface green, glabrescent, often pubescent on the midrib; lower surface lighter green, with pubescence on the lateral nerves, forming axil-tufts at their junctions with the midrib; margin ciliate; lateral nerves eight to ten pairs; petiole with scattered long hairs.

Flowers, appearing before the leaves, 4 in. across; sepals very small, narrow, quickly deciduous; petals six, obovate, rounded or emarginate at the apex, white, with a purple median line externally. Fruit, dark brown, 4 in. long, often curved and contorted.

1. Var. *borealis*, Sargent, *Trees and Shrubs*, ii. 57 (1908).

Magnolia Kobus, Sargent, in *Garden and Forest*, vi. 64, fig. 11 (1893), and *Forest Flora of Japan*, 9, fig. 3 (1894) (not De Candolle).

A large tree, attaining in Japan 80 ft. in height and 6 ft. in girth, with a straight trunk, covered with dark slightly fissured bark; leaves larger than in the type; flowers with pure creamy white sepals.

This variety, according to Sargent, is a native of Yezo, where it grows plentifully in the forests of the hills around Sapporo. It also occurs in northern Hondo. So far as we know, it is not in cultivation in England.

The typical form of the species is a small tree, about 20 to 30 ft. high, which is occasionally seen at considerable elevations in the Hakkone and Nikko mountains, and is recorded for Fujiyama by Hayata.[3] Shirasawa states that the wood is harder than that of *M. hypoleuca*; but owing to the rarity of trees of a large size, is little used. It is known in Japan as *Kobushi*.

M. Kobus was introduced into the United States by Thomas Hogg, and was distributed from Parsons' nurseries under the name *M. Thurberi*. Sargent states that in New England it is the hardiest, most vigorous, and fastest growing of all the Magnolias. Rehder, however, says that it flowers sparingly, and is not showy. A tree in the Arnold Arboretum flowered when fifteen years old.

[1] In *Mitt. Deut. Dend. Ges.* 1904, p. 1.　　　[2] Recorded as doubtfully wild in the Liu Kiu Islands.
[3] *Vegetation of Mt. Fuji*, 56 (1911).

M. Kobus was introduced into England by Maries in 1879; and one of the original trees[1] at Coombe Wood, which died in 1906, was about 20 ft. high in 1902. Some of the specimens of *M. stellata* at Kew were grafted on stock of *M. Kobus*, obtained from this source. This species was again introduced[2] about 1887 by Messrs. R. Veitch and Sons, Exeter, who reported that it proved fast in growth, some of the plants, transplanted two or three times, having attained 12 to 14 ft. in height in seven or eight years.

We have seen no large specimens in England, except one at Abbotsbury, which has borne flowers for the past ten years, and was about 25 ft. high in July 1912. A tree at Kew, obtained from Harvard in 1889, is only 15 ft. high. One in Victoria Park, Bath, about 13 ft. high, produced flowers very freely in 1911, but set no fruit.[3] In Mr. Thomas Irvine's garden at Newry, there is a fine specimen, planted about twenty-one years ago, which is 18 ft. high. It produces flowers in abundance every year, and is very thriving.   (A. H.)

MAGNOLIA SALICIFOLIA

Magnolia salicifolia, Maximowicz, in *Bull. Acad. St. Petersb.* xvii. 418 (1872), and *Mél. Biol.* viii. 509 (1872); Shirasawa, *Icon. Ess. Forest. Japon*, i. text 72, t. 40, figs. 18-31 (1899); Sargent, in *Garden and Forest*, vi. 65, fig. 12 (1893), and *Forest Flora of Japan*, 10, fig. 4 (1894).

Buergeria (?) *salicifolia*, Siebold and Zuccarini, in *Abh. Akad. München*, iv. pt. ii. 187 (1845).

A small deciduous tree, attaining in Japan, about 30 ft. in height and 4 ft. in girth. Bark smooth. Young branchlets glabrous, slightly glaucous. Leaves membranous, lanceolate, 3 to 4 in. long, 1 to $1\frac{1}{4}$ in. broad, gradually tapering to an acuminate and often curved apex, cuneate at the base; upper surface light green, glabrous; lower surface pale, more or less covered with a minute pubescence; punctate with numerous translucent dots; lateral nerves, about twelve pairs, yellowish, as is also the midrib; petiole glabrous.

Flowers campanulate, 3 to 4 in. across; sepals green, much smaller than the petals, spreading, early deciduous; petals six, white, oblong-spatulate, spreading, slightly reflexed at the apex, about $2\frac{1}{2}$ in. long. Fruit narrowly cylindric, 3 in. long, pale brown; seeds compressed, triangular, almost black.

This little-known species is a native of Japan, where it grows in mountain forests in Hondo, at 2000 to 5000 feet elevation. It was introduced[4] in 1892 into the Arnold Arboretum by Professor Sargent, who collected seed on Mount Hakkoda; and is represented at Kew by a small specimen obtained from Yokohama in 1906. This, as well as a shrub at Arley, flowered[5] for the first time in 1912. The flowers are handsome and fragrant.   (A. H.)

---

[1] Note in Arboretum Herbarium, Kew. It is impossible to say whether Maries's introduction was typical *M. Kobus* or var. *borealis*; but all the specimens which we have seen in cultivation belong to the former.

[2] *Gard. Chron.* xxxvii. 265, *Supply. Illust.* (1905).

[3] It was reported, in *Gard. Chron.* xxxvi. 322 (1904), to have produced fruit, when about 8 ft. high in 1904.

[4] It is mentioned in Veitch's *Catalogue of Trees and Shrubs*, 1902, p. 45.

[5] Cf. *Gard. Chron.* li. 222, 245, fig. 99 (1912).

MAGNOLIA CONSPICUA

Magnolia conspicua, Salisbury, *Parad. Lond.* t. 38 (1806); Sims, *Bot. Mag.* t. 1621 (1814); Loudon, *Arb. et Frut. Brit.* i. 278 (1838).

Magnolia precia,[1] Correa, in Ventenat, *Jard. Malm.* 24, note 2 (1803); Schneider, *Laubholzkunde*, i. 331 (1905).

Magnolia yulan, Desfontaines, *Hist. Arb.* ii. 6 (1809).

A deciduous tree, attaining in China 30 to 50 ft. in height. Young branchlets more or less covered with appressed white pubescence. Leaves obovate, or obovate-oblong, about 4 to 6 in. long and $2\frac{1}{2}$ to 4 in. wide, cuspidate-acuminate at the apex, usually rounded and unequal at the base; upper surface light green, more or less covered with a minute pubescence, dense on the midrib and nerves; lower surface lighter green, similarly pubescent with longer white hairs; minutely punctate with translucent dots; network of veins beneath wrinkled and prominent; petiole pubescent.

Flowers, appearing in spring before the leaves, campanulate, sweet-scented, about six inches across, pure white; sepals resembling the petals; petals six, fleshy, concave, oblong-obovate or spatulate. Fruit brownish, 3 to 6 in. long.

Numerous varieties[2] of this species have arisen in cultivation in Europe, which are supposed to be of hybrid origin.

1. *Magnolia Soulangiana*, Soulange-Bodin, in *Ann. Soc. Hort. Paris*, 1826, p. 90.

Magnolia conspicua, var. *Soulangiana*, Lindley, in *Bot. Reg.* t. 1164 (1828); Loudon, *Arb. et Frut. Brit.* i. 278 (1838).

Leaves similar to those of *M. conspicua*; but usually narrower in proportion to their length, longer acuminate at the apex, and tapering at the base. Flowers, later in opening, fragrant, delicate green at first, the sepals and petals becoming white inside, and purplish outside.

This was raised about 1820 at Fromont, near Paris, from the seeds of a plant of *M. conspicua*, which stood near one of *M. denudata*, in front of the château of M. Soulange-Bodin; the flowers of the former, it was supposed, being fertilised by the pollen of the latter. It resembles *M. denudata* in the later opening and colour of the flowers.

2. *Magnolia Lennei*, Topf, *ex* Van Houtte, *Flore des Serres*, xvi. tt. 1693, 1694 (1867). Leaves similar to those of *M. conspicua*, broadly obovate-oval, but with a longer acuminate apex, and tapering at the base. Flowers, appearing with the leaves, deep crimson outside, very fragrant.[3] This originated in the Salvi Garden at Vicenza in Italy; and was sent from there in 1850 to Topf, at Erfurt, who named it after Lenné, director of the Potsdam Garden. Whether it is a seedling of *M. conspicua* or of *M. Soulangiana* is unknown.

---

[1] This name was published without any description, and cannot therefore be used, although the earliest.

[2] Var. *purpurascens*, Maximowicz, in *Mél. Biol.* viii. 509 (1872), is a variety with purple flowers, cultivated in Japan.

[3] According to *Gard. Chron.* xxvi. 379 (1899), it produced, in 1893 and 1899, pink fruit with orange seed at Straffan, Kildare. It ripened fruit in 1911 at Enys in Cornwall.

Var. *rosea grandiflora* was raised from a seed of *M. Lennei*, and is one of the most beautiful varieties.

3. Numerous other hybrids are in cultivation, which mainly differ in colour and time of flowering, as var. *Alexandrina*,[1] var. *Norbertiana*, and var. *nigra*.[2]

M. conspicua is a native of China, where it has been found wild in the mountainous districts of Hupeh and Szechwan. It is cultivated extensively by the Chinese, who call it *yü-lan*; and it is mentioned in their earliest literature. It is not considered to be a native of Japan, where it was probably introduced from China by the early Buddhist monks.

It was introduced into England by Sir Joseph Banks in 1789: and is commonly cultivated for its beautiful flowers, which appear in spring the earliest of all the species, except *M. stellata* and *M. Campbelli*. The flowers are liable to be injured by frosts and cold east winds. The finest specimen is probably one[3] at Gunnersbury House, which was 31 ft. high by 2½ ft. in girth in 1911. A tree, said[4] to be over seventy years old, and about 25 ft. high, was growing at Slocock's Nursery at Woking, in 1898, and bore flowers described as pure white suffused with purple. A fine specimen at La Fantaisie, Jersey, is figured in *Gard. Chron.* xxxvi. 59, fig. 25 (1904). One trained against the wall of Hornby Grange, Northallerton, and 30 ft. high, was reported[5] to have borne 2000 flowers in 1896.   (A. H.)

MAGNOLIA DENUDATA

Magnolia denudata,[6] Lamarck, *Encycl.* iii. 675 (1789); Schneider, *Laubholzkunde*, i. 330 (1905).
Magnolia obovata, Thunberg, in *Trans. Linn. Soc.* ii. 336 (1794) (Excl. *Icon. Kaempf.* t. 43).
Magnolia purpurea, Curtis, *Bot. Mag.* t. 390 (1797); Loudon, *Arb. et Frut. Brit.* i. 282 (1838).
Magnolia discolor, Ventenat, *Jard. Malm.* t. 24 (1803).

A deciduous shrub, rarely exceeding 10 ft. in height. Young branchlets glabrous except near the tip, where they are covered with silvery appressed pubescences. Leaves membranous, about 4 to 5 in. long and 2 to 3 in. broad, obovate, or oval and broadest at the middle, acute or shortly acuminate at the apex, tapering at the base; upper surface dark green, with scattered appressed hairs, denser on the midrib and nerves; lower surface pale green, glaucescent, with similar pubescence, confined mainly to the midrib and nerves; punctate with minute translucent dots; margined with scattered cilia; lateral nerves eight to ten pairs; petiole with appressed pubescence.

Flowers, appearing before the leaves, without scent, campanulate; sepals small, ovate-lanceolate, greenish yellow, spreading, and slightly reflexed; petals six, purple outside, white inside, about 3½ in. long, broad, ovate, obtuse, slightly fleshy. Fruit brownish.

[1] Cf. Loudon, *Gard. Mag.* xix. 269 (1843), where this variety is said to flower later than *M. conspicua*, and earlier than *M. Soulangiana*.

[2] According to *Hortus Veitchii*, 370 (1906), var. *nigra* was introduced from Japan by J. Gould Veitch. Cf. Nicholson, in *The Garden*, xxv. 276, fig. 434 (1884). It has dark plum-coloured flowers.

[3] Figured in *Gard. Chron.* ix. 591, fig. 5 (1891).   [4] *Gard. Chron.* xxiii. 262 (1898).

[5] *Ibid.* xix. 494 (1896).   [6] This is the oldest name, and Lamarck's description is satisfactory.

1. Var. *gracilis*, Dippel, *Laubholzkunde*, iii. 151 (1893).

Magnolia gracilis, Salisbury, *Parad. Lond.* t. 87 (1806); Loudon, *Arb. et Frut. Brit.* i. 283 (1838).

Branchlets slender; leaves narrower; flowers smaller, dark purple.   Introduced from Japan in 1804.

M. denudata is a native of China, where it has not, however, so far as I know, been collected in the wild state.   It is mentioned in the earliest Chinese literature under the name *mu-lan*, which it still bears.   It is cultivated throughout China and Japan, having been introduced into the latter country at an early period.   It was brought to England from Japan by Thunberg in 1790; and is occasionally cultivated, though it is not so ornamental as the hybrids.   It produces its flowers just after *M. conspicua*.   (A. H.)

MAGNOLIA PARVIFLORA

Magnolia parviflora, Siebold and Zuccarini, in *Abh. Akad. München*, iv. pt. ii. 187 (1846);
Maximowicz, in *Mél. Biol.* viii. 509 (1872); J. D. Hooker, in *Bot. Mag.* t. 7411 (1895);
Shirasawa, *Icon. Ess. Forest. Japon*, ii. t. 17, figs. 1-5 (1908).

A deciduous shrub,[1] attaining about 10 ft. high in Japan.   Young branchlets covered more or less with a minute pubescence.   Leaves membranous, 4 to 5 in. long, 2½ to 3 in. wide, obovate or oval, shortly acuminate at the apex, rounded at the base; upper surface shining dark green, glabrous; lower surface bluish white, more or less covered with a scattered minute pubescence, denser and longer on the midrib and nerves; lateral nerves about nine pairs, greenish, as is also the margin; petiole pubescent.

Flowers appearing in June after the leaves, on long stalks; cup-shaped, fragrant, 3½ to 4 in. across; sepals pink, nearly as large as the petals, soon reflexed; petals six, obovate, very concave, white.

This handsome species is a native of Japan, where it occurs in mountain woods at high elevations on the main island and in Kiusiu.   It was introduced into Kew from Yokohama in 1893, and flowered in the following year.   It is said to be rather tender at Leonardslee[2]; but thrives at Westonbirt, where a small shrub was bearing flowers on 25th May 1912.   (A. H.)

MAGNOLIA WATSONI

Magnolia Watsoni, J. D. Hooker, in *Bot. Mag.* t. 7157 (1891); Masters, in *Gard. Chron.* xvi. 188,
fig. 29 (1894); Nicholson, in *Gard. Chron.* xvii. 516, fig. 72 (1895).

A deciduous shrub.   Young branchlets glabrous.   Leaves slightly coriaceous, about 5 to 6 in. long and 3 to 3½ in. broad, obovate to elliptical, acute at the apex, tapering at the base; upper surface light green, pubescent on the midrib,

---

[1] Maximowicz, *loc. cit.*, states that this is occasionally a large tree; but Shirasawa gives its height as about 10 feet.
[2] *Gard. Chron.* xli. 222 (1907).

glabrous elsewhere; lower surface greyish, thinly covered with scattered appressed hairs, dense on the midrib and nerves; lateral nerves about fifteen pairs, with the midrib and margin yellowish; petiole pubescent.

Flowers appearing at the same time as the leaves, shortly stalked, 5 to 6 in. across, highly fragrant, resembling in odour those of Calycanthus; sepals oblong, pink, ultimately deflexed; petals six to nine, obovate, concave, spreading, cream-coloured.

This species was described by Sir Joseph Hooker from a shrub, which was purchased in the Japanese Court of the Paris Exhibition in 1889, and which flowered at Kew in the following year. It is unknown in the wild state in Japan, and is possibly a hybrid between *M. hypoleuca* and *M. parviflora*. At Stevenstone, North Devon, there is a fine shrub,[1] 18 ft. high, with seven main branches, which bore about 100 flowers in June 1909. (A. H.)

MAGNOLIA STELLATA

Magnolia stellata, Maximowicz, in *Bull. Acad. Imp. Petersburg*, xvii. 418 (1872) and *Mél. Biol.* viii. 509 (1872); J. D. Hooker, in *Bot. Mag.* t. 6370 (1878); Masters, in *Gard. Chron.* vii. 618 fig. 102 (1890), and xxxix. 260, figs. 108, 109 (1906); Nicholson, in *Gard. Chron.* xvii. 516, fig. 73 (1895).

Magnolia Halleana, Parsons, in *The Garden*, xiii. 572, fig. 132 (1878).

Buergeria stellata, Siebold and Zuccarini, in *Abh. Akad. München*, iv. pt. ii. 186 (1846).

A deciduous shrub or small tree. Young branchlets more or less covered with appressed silvery long hairs. Leaves membranous, about 3 in. long and 1 in. broad, obovate-oblong or oblanceolate, gradually tapering to the base; rounded and emarginate, acute, or shortly acuminate at the apex; upper surface glabrous; lower surface green, glabrous between the nerves, which with the midrib and petiole are more or less covered with appressed pubescence or are glabrescent; lateral nerves about ten pairs; margin non-ciliate.

Flowers opening before the leaves, about 3 in. across, sweet-scented, short-stalked; sepals similar to the petals; petals nine to eighteen, narrow, linear-oblong, at first spreading, then reflexed, white, with a faint pink streak externally. Fruit, about 2 in. long; carpels cuspidate, only a few ripening.

This species is a native of Japan, where it is wild in woods in central Hondo, and is everywhere cultivated. It was introduced into cultivation in the United States by Dr. Hall[2] in 1862, and was put into commerce by S. B. Parsons of Flushing, New York, under the name *M. Halleana*. It was introduced[3] from Japan into England by Messrs. Veitch, and flowered for the first time in this country in their nursery at Coombe Wood in 1878. It is very hardy and produces flowers most profusely, at an early age, when it is hardly 2 feet high; and is now seen in many gardens. (A. H.)

---

[1] *Kew Bull.* 1909, p. 337.

[2] An account of the plants introduced by Dr. Hall from Japan, is given in *The Garden*, xiii. 572 (1878), reproduced in *Gard. Chron.* xlv. 275 (1909).

[3] Cf. *Hortus Veitchii*, 370 (1906). This form has the petals slightly suffused with pink, and has been named var. *rosea*.

HALESIA

Halesia, Ellis, in Linnæus, *Syst. Nat.* 1044 (1759), and in *Phil. Trans.* li. 931 (1761) (not Browne[1]); Bentham et Hooker, *Gen. Pl.* ii. 669 (excl. *Pterostyrax*) (1876); Perkins, in Engler, *Pflanzenreich*, iv. 241, *Styracaceæ*, 94 (1907); Schneider, *Laubholzkunde*, ii. 582 (1911).

Mohria, Britton, in *Garden and Forest*, vi. 434 (1893) (not Swartz[2]).

Mohrodendron, Britton, in *Garden and Forest*, vi. 463 (1893); Sargent, *Silva N. Amer.* vi. 19 (1894), and *Trees N. Amer.* 754 (1905).

Carlomohria, Greene, in *Erythea*, i. 236 (1893).

DECIDUOUS trees or shrubs, belonging to the order Styracaceæ. Branchlets slender, with chambered pith; one large bundle-scar in the centre of each leaf-scar. Buds all axillary, no true terminal bud being formed, two or three superposed above each leaf-scar, with a few imbricated scales. Leaves simple, alternate, stalked, penninerved, without stipules, denticulate in margin, more or less stellate-pubescent.

Flowers regular, perfect, in fascicles or short racemes, arising from the axils of the leaf-scars of the previous year's branchlet; calyx-tube obpyramidate, four-ribbed, with a short four-toothed limb; corolla epigynous, campanulate, divided into four or five shallow or deep lobes. Stamens, eight to sixteen in one series, adnate to the tube of the corolla, included; filaments nearly free or more or less connate; ovary, two- to four-celled, inferior in greater part, gradually contracted into an elongated style, which is stigmatic at the apex; ovules, four in each cell, two ascending and two pendulous. Fruit, a drupe, crowned by the calyx-tube and the thickened persistent style, dry, indehiscent; with a thick exocarp, produced into two or four wings; containing a thick and bony obovate stone, gradually narrowed at the base into an elongated stipe, one- to four-celled. Seed solitary in each cell.

Halesia comprises three species, natives of the United States, one of which, *Halesia parviflora*[3] is a small shrub, not known in cultivation in Europe.

Pterostyrax, Siebold and Zuccarini, *Fl. Jap.* 96 (1835), was included under *Halesia* by Bentham and Hooker, *Gen. Pl.* ii. 669 (1876); but is a distinct genus, comprising three species, shrubs or small trees, natives of China and Japan, with

[1] *Halesia*, Patrick Browne, *Hist. Jamaica*, 205 (1755), was applied to a West Indian tree, which is a species of *Guettarda*, a genus founded by Linnæus in 1753.

[2] *Mohria*, Swartz, *Syn. Fil.* 159 (1806), is a genus of ferns, with one species, in South Africa.

[3] *H. parviflora*, Michaux, *Fl. Bor. Amer.* ii. 40 (1803), is a little-known shrub of southern Georgia and Florida, which has never been properly described. The plant described under this name by Lindley, *Bot. Reg.* t. 952 (1825), and Loudon, *Arb. et Frut. Brit.* ii. 1190 (1838), is *Styrax americana*, Lamarck. Cf. Perkins, *Styracaceæ*, 76 (1907), and also a note by Smith, in Rees' *Cyclopædia*, under the article on *Halesia*.

foliage and stellate pubescence similar to that of the species of *Halesia*. *Pterostyrax* differs from the latter as follows :—Branchlets with solid pith. Flowers numerous in axillary and terminal panicles, arising on the current year's branchlets ; corolla five-partite ; stamens ten, exserted, unequal. Fruit small, either obovate and five-winged, or cylindric and ten-ribbed.

Of the three species, one appears to be in cultivation,[1] and deserves a brief mention :—

Pterostyrax hispida, Siebold and Zuccarini, in *Abh. Akad. München*, iv. 3, p. 132 (1846). A small tree, 20 to 30 ft. high, native of the main island of Japan and central China, which was introduced[2] about 1870. It is perfectly hardy at Kew, coming into flower late in June, and ripening occasionally its seed. It was lately figured in *Bot. Mag.* t. 8329 (1910) from a specimen in Canon Ellacombe's garden at Bitton.   (A. H.)

HALESIA DIPTERA

Halesia diptera, Ellis, in *Phil. Trans.* li. 931, t. 22 B (1761); Linnæus, *Sp. Pl.* 636 (1762);
    Loudon, *Arb. et Frut. Brit.* ii. 1191 (1838); Perkins, *Styracaceæ*, 97 (1907); Schneider,
    Laubholzkunde, ii. 583 (1911).
Halesia reticulata, Buckley, in *Proc. Acad. Sci. Philad.* 1860, p. 444.
Mohria diptera, Britton, in *Garden and Forest*, vi. 434 (1893).
Mohrodendron dipterum, Britton, in *Garden and Forest*, vi. 463 (1893); Sargent, *Silva N. Amer.* vi.
    23, t. 259 (1894), and *Trees N. Amer.* 756 (1905).

A large shrub or small tree, not exceeding 30 ft. in height. Young branchlets with a few very scattered stellate hairs. Leaves membranous, ovate, 4 to 5 in. long, 2½ to 3½ in. wide, acuminate at the apex, rounded or cuneate at the base ; margin with remote serrations tipped with long cartilaginous points ; glabrous on both surfaces, except for scattered stellate hairs on the midrib and nerves ; green beneath ; petiole about ½ in. long, with slight stellate pubescence.

Flowers on long slender pedicels, in fascicles of three or four ; calyx pubescent externally, with four short distinct triangular teeth ; corolla white, divided nearly to the base into four or five lobes ; stamens usually eight, rarely ten to sixteen with pubescent filaments, which are united together in their lower half ; ovary usually two-, occasionally three-celled, tomentose ; style tomentose. Fruit oblong, compressed, 1½ to 2 in. long, with two large opposite wings, and two or three additional slight ridges ; stone narrowly obovate, conspicuously furrowed ; seeds acuminate at the ends.

This is a native of the coast region of the United States, from South Carolina to northern Florida, and westward to Texas, ascending in the Mississippi valley to

[1] The other species, not yet introduced are :—
(1) *Pterostyrax psilophylla*, Diels, *ex* Perkins, *Styracaceæ*, 103 (1907). A shrub found by Wilson in central China, closely allied to *P. hispida*.
(2) *Pterostyrax corymbosa*, Siebold and Zuccarini, *Fl. Jap.* 96, t. 47 (1835). A shrub, native of Japan.
[2] Cf. Koch, *Dendrologie*, ii. pt. i. 198 (1872).

central Arkansas. It is usually a large shrub, growing on the borders of swamps and in other wet situations. It is hardy in gardens as far north as Philadelphia.

H. diptera is said by Loudon to have been introduced into England in 1758; and a shrub trained against a wall produced flowers and fruit many years ago in Loddiges' nursery. It appears to be extremely rare in cultivation at the present time, the only specimen which I know being one at Kew, about 15 ft. high, obtained from Meehan in 1896. This has not yet produced flowers or fruit.   (A. H.)

HALESIA CAROLINA, SNOWDROP TREE

Halesia carolina, Linnæus, *Syst. Nat.* 1044 (1759); Perkins, *Styracaceæ*, 94 (1907); Schneider, *Laubholzkunde*, ii. 583 (1911).
Halesia tetraptera, Ellis, in *Phil. Trans.* li. 932, t. 22 A (1761); Linnæus, *Sp. Pl.* 636 (1762); Loddiges, *Bot. Cat.* t. 1173 (1827); Loudon, *Arb. et Frut. Brit.* ii. 1190 (1838).
Mohria carolina, Britton, in *Garden and Forest*, vi. 434 (1893).
Mohrodendron carolinum, Britton, in *Garden and Forest*, vi. 463 (1893); Sargent, *Silva N. Amer.* vi. 21, tt. 257, 258 (1894), and *Trees N. Amer.* 755 (1905).
Carlomohria carolina, Greene, in *Erythea*, i. 246 (1893).

A tree, attaining occasionally in America 80 to 90 ft. in height and 9 ft. in girth, but usually much smaller, and often a large bush throwing up several stems from the ground. Bark divided into broad rounded scaly ridges. Young branchlets stellate-pubescent. Leaves oval, 4 to 6 in. long, 2 to 3 in. wide, thicker in texture than those of *H. diptera*, acuminate at the apex, rounded or cuneate at the base; margin minutely serrate, with very short callous points; glabrescent above; lower surface pale or whitish, more or less densely covered with stellate pubescence; petiole about ½ in. long, covered with stellate hairs.

Flowers, opening in spring just after the leaves begin to unfold, on long slender pedicels, in fascicles of three or four; calyx glabrous, four-ribbed, with four minute deltoid ciliate teeth; corolla bronzy red before opening, white when open, with four short lobes, the division between the lobes not exceeding one-third the length of the corolla; stamens, ten to sixteen, with glabrous filaments, connate only at the base; ovary four-celled, glabrous; style glabrous. Fruit, ripening late in autumn, and persisting on the tree in winter; ellipsoid, 1½ in. to 2 in. long, with four broad wings; stone obovate, obscurely ridged; seeds rounded at the narrow ends.

This species as seen in cultivation is very variable; and the following hybrids or varieties are known :—

1. *Halesia stenocarpa*, Koch, in *Wochenschr. Gärtn. u. Pflanzenk.* i. 190 (1858), and *Dendrologie*, ii. pt. i. 200 (1872).
Leaves similar to the type in shape, consistence, and pale colour beneath; but with more distinct serrations, and with sparser stellate pubescence on the lower surface. Flowers: corolla deeply divided to near the base into four lobes; filaments with scattered hairs, connate at the base or at some distance above it; ovary and style glabrous. Fruit with four narrow wings.

This is intermediate between *H. carolina* and *H. diptera*; and, as Koch points out, is probably a hybrid. It has the deeply partite corolla and hairy filaments of the latter species, but resembles the former in foliage. The wings of the fruit are four in number, as in *H. carolina*, but are much narrower than in that species.

This variety or hybrid occurs in the wild state, as shown by specimens from Carolina and Florida in the Kew Herbarium. It appears to be much commoner in cultivation[1] than the species. All the old trees in England, and one in the Botanic Garden at Berlin dating from the time of Willdenow, belong to *H. stenocarpa*.

2. Var. *mollis*, Lange, in *Bot. Tidsk.* xix. 1, p. 258, fig. 2, a-g (1894).

Leaves in shape, colour beneath, and minute serrations, similar to the type, but densely tomentose with stellate hairs on the lower surface. Corolla[2] deeply divided to near the base into four lobes; filaments pubescent. Fruit as in the type.

This, which mainly differs from the typical form in the deeply divided corolla, is of unknown origin. It is represented at Kew by a shrub about 5 ft. high, which was planted in 1887.

3. Var. *glabrescens*, Lange, in *Bot. Tidsk.* xix. 1, p. 257, fig. 1 (1894).

Leaves narrowly elliptic, up to 6 in. long and 2 in. broad, gradually tapering to a long acuminate apex, sparingly stellate-pubescent beneath. Corolla[2] with four shallow lobes. Fruit with four very narrow wings.

This, the origin of which is unknown, is represented at Kew by a dense shrub, about 12 ft. high, which suckers very freely, and produces flowers and fruit, as described above.

4. Var. *Meehani*, Sargent, in *Garden and Forest*, v. 611 (1892).

Halesia Meehani, Meehan, in *Garden and Forest*, v. 534, fig. 91 (1892).

A round bush, attaining about 12 ft. in height. Leaves thick, wrinkled, pale, and on young vigorous plants often conspicuously glandular-serrate. Flowers smaller than in the type, with a short calyx-tube, and a cup-shaped corolla not narrowed at the base; pedicels not exceeding $\frac{1}{2}$ in. in length.

This was found in Meehan's nursery at Germantown, Philadelphia, as a solitary plant in a bed of seedlings raised from the seed of a tree of *H. carolina*. It bore fruit abundantly, and one of the seedlings raised from it seemed exactly similar to *H. carolina*. According to Wyman,[3] the flowers last longer than in the other kinds of *Halesia*. It is not in cultivation, so far as I know, in England. (A. H.)

DISTRIBUTION

H. Carolina was first described by Catesby in the *Natural History of Carolina*. It ranges from the mountains of West Virginia to southern Illinois, and south-

---

[1] *Halesia tetraptera*, var. *dialypetala*, Rehder, in *Mitt. Deut. Dend. Ges.* 1907, p. 75, described from a tree in the Arnold Arboretum, is probably *H. stenocarpa*.

[2] Lange describes the corolla in var. *mollis* as divided to the middle, and in var. *glabrescens* as deeply divided to the base; but in specimens at Kew identical in other respects with these varieties, the corolla is as stated above. It is possible that the flowers may vary in these varieties, when they are raised from seed.

[3] In Bailey, *Cyc. Amer. Hort.* 710, fig. 1017 (1900).

ward to Florida, Alabama, and Mississippi. It grows on rich wooded slopes and river banks, attaining its largest size in the forests on the western slopes of the southern Alleghany Mountains, where it sends up tall straight trunks, sometimes 3 ft. in diameter and 50 to 60 ft. high.

Sargent says that the wood is light, soft, and close-grained, of a light-brown colour, but it does not seem to have any special use or value in America, and is not mentioned in Hough's *American Woods*.

This species, or the form *H. stenocarpa*, was introduced in 1756, when John Ellis raised plants from seeds sent from America by Dr. Alex. Graham. The plant figured as *H. tetraptera* by Sims, *Bot. Mag.* t. 910 (1805), shows the petals deeply divided to the base, and is *H. stenoptera*; and all the old trees that we have seen are of this form.

Though one of the most ornamental flowering deciduous trees that we have, the Snowdrop tree has never become common in cultivation, and like many old favourites has been neglected for the numerous new introductions from China and Japan. Though usually seen rather as a shrub than a tree, it has in a few places attained such large dimensions that it may rank with Arbutus, Laburnum, and Magnolia among trees of the third rank in size, but of the first in beauty. It seems to thrive best in a warm sandy loam, free from lime; to require a long and warm summer to ripen its wood properly; and to be proof against the most severe frosts when well established in the south and east of England. It ripens seeds in warm summers only, but I have not succeeded in raising plants from English-grown seeds.

Loudon records the finest trees known to him at Purser's Cross and Syon, 30 ft. high and 4 to 4½ ft. in girth, but these seem to be no longer living. The tallest that we have seen is in an outlying part of the woods at Pains Hill, where a tree, forked at the base and almost prostrate, is 48 ft. from the root to the top. The two stems were 3 ft. 6 in. and 3 ft. 2 in. in girth; and the tree was bearing fruit when I saw it in 1908.

Another very large tree grows in the grounds of Mr. Boardman, Town Close House, Norwich, and when figured by Grigor in 1841 was 29 ft. high and 4 ft. in girth, with a circumference of 33 yards. When I saw it in 1908 it was 32 ft. by 6 ft. 3 in., forking at about 4 ft. from the ground; and one of the limbs was 4½ ft. in girth. Its fruit was nearly ripe in October. There is a handsome tree at Leonardslee, about 25 ft. by 2 ft., with a clean stem 15 ft. long. There is a fine specimen in Colonel Duncombe's grounds at Waresley Park, Herts, which is about 35 ft. high by 4½ ft. in girth. Mr. Wyndham Fitzherbert reports[1] a tree in a garden at Kidderminster, 28 ft. in height and 5 ft. in girth, with a spread of branch of 48 ft. He states that some small trees planted in decomposed peat made astonishing growth, attaining 20 feet in height in twelve years. At Milford Lodge, Craven Arms, there is a fine tree about 30 ft. in height.

We have seen no specimens in Scotland or Ireland.   (H. J. E.)

[1] *Gardening Illustrated*, 19th November 1910.

MORUS

Morus, Linnæus, *Gen. Pl.* 283 (1737); Bureau, in De Candolle, *Prod.* xvii. 237 (1873); Bentham et Hooker, *Gen. Pl.* iii. 364 (1880); Bailey, *Cycl. Amer. Hort.* 1033 (1901).
Morophorum, Necker, *Elem. Bot.* iii. 255 (1790).

DECIDUOUS trees and shrubs belonging to the order Moraceæ. Leaves alternate, distichous, simple, stalked, serrate, with or without lobes; palmately three- to five-nerved at the base, pinnately nerved above; stipules lateral, lanceolate, early deciduous.

Flowers monœcious or diœcious, in solitary spikes,[1] which arise on the base of the current year's shoot in the axil of a leaf or of a deciduous scale; calyx deeply divided into four lobes; corolla absent. Staminate spikes elongated, cylindric; stamens four, inserted opposite the rounded calyx-lobes, beneath the minute rudimentary ovary; filaments ultimately exserted, uncoiling like a spring at the moment of dehiscence of the two-celled anthers, and forcibly scattering the pollen. Pistillate spikes short, oblong; ovary sessile, one-celled, with a solitary pendulous ovule; style divided nearly to the base into two stigmatic lobes. Each ovary results in an achene, enclosed in the calyx, which becomes thickened and fleshy. The whole mass of achenes, closely packed together in one spike, forms a multiple fruit, the mulberry superficially resembling a blackberry (*Rubus*), but very different in structure.

In winter, the species of Morus show the following characters :—No terminal bud is formed, the tip of the branchlet dying off in early summer, and leaving a scar at the apex of the twig. Buds all axillary, arranged in two ranks, ovoid, acuminate; scales, five to seven, closely imbricated in two ranks, the inner accrescent and falling in spring, marking the base of the shoot with ring-like scars. Leaf-scars on prominent cushions, nearly circular, concave with a slight rim, marked on the surface by an irregular group of dots. Stipule-scars linear, one on each side of the leaf-scar.

About eight species of Morus are known, occurring in North America, Central America, Western South America, Western Asia, China, Japan, Indo-China, and the high mountains of the Indian Archipelago. Three species have been long in cultivation which are distinguishable as follows :—

1. *Morus nigra*, Linnæus. Native country uncertain, probably Western Asia. See p. 1606.

---

[1] Androgynous spikes have often been observed on *M. alba* and *M. rubra*, growing in the Arnold Arboretum. Cf. *Garden and Forest*, viii. 223 (1895).

Leaves rarely lobed, deeply cordate at the base, shortly acuminate at the apex ; margin ciliate ; lower surface pubescent throughout.

2. *Morus rubra*, Linnæus.   North America.   See p. 1608.

Leaves rarely lobed, slightly cordate at the base, contracted above into a long acuminate apex ; margin slightly ciliate ; lower surface pubescent throughout.

3. *Morus alba*, Linnæus.   China, Japan.   See p. 1609.

Leaves often lobed, variable in size and shape, thinner in texture than those of the preceding species ; margin non-ciliate ; lower surface glabrous, except for pubescence on the midrib and nerves.

The following species, lately introduced, may be briefly noticed :—

4. *Morus cathayana*, Hemsley, in *Journ. Linn. Soc.* (*Bot.*) xxvi. 456 (1894).

A tree, about 20 to 30 ft. high.   Young branchlets densely pubescent.   Leaves large, 4 to 5 in. long, ovate, cordate, cuspidate-acuminate, usually without lobes, crenate in margin ; very scabrous above with minute blackish tubercles ; lower surface softly pubescent throughout.   Fruit, cylindric, 1 in. long ; styles as long as the ovary.

This was discovered by me in 1888 in the mountain forests of western Hupeh in China.   It has lately been introduced by Wilson ; and a plant at Kew, obtained from the Arnold Arboretum in 1907, is about 4 ft. high.   (A. H.)

MORUS NIGRA, BLACK MULBERRY

Morus nigra, Linnæus, *Sp. Pl.* 986 (1753) ; Loudon, *Arb. et Frut. Brit.* iii. 1343 (1838) ; Bentley and Trimen, *Medicinal Plants*, t. 229 (1880) ; Willkomm, *Forstl. Flora*, 541 (1887) ; Mathieu, *Fl. Forestière*, 292 (1897).

A tree, attaining about 50 ft. in height.   Bark thick, fissured into broad scaly plates.   Young branchlets with a scattered downy pubescence.   Leaves (Vol. IV. Plate 267, Fig. 1), subcoriaceous, 4 to 6 in. long, 3 to 5 in. wide, broadly ovate, deeply cordate at the base, acuminate at the apex ; upper surface dark green, shining, with scattered short pubescence ; lower surface pale green, covered throughout with a short downy pubescence ; margin ciliate, with coarse triangular serrations ; petiole $\frac{3}{4}$ in. long, pubescent.   Staminate spikes, $\frac{3}{4}$ to $1\frac{1}{2}$ in. long.   Pistillate spikes, $\frac{1}{2}$ in. long, on a short pubescent peduncle ; style and stigmas pubescent.   Fruit, about 1 in. long, black, very shortly stalked.

Lobed leaves[1] are rarely seen on adult trees, but are usual on root-suckers.

The native country of the black mulberry cannot be ascertained with certainty.[2] It has been cultivated in southern Europe from a very early period, but there is no evidence that it is indigenous in Italy or Greece. Boissier[3] and De Candolle[4] suppose it to be truly wild in the districts in Persia, bordering on the Caspian Sea ; but

[1] Var. *laciniata*, Loudon (*M. laciniata*, Miller, *Dict.* ed. 8, No. 2 (1768)).   This is only an individual variation, and cannot be retained as a distinct variety.

[2] Hehn, *Wanderings of Plants and Animals*, 290 (1888), states, without giving any evidence, that the mulberry is a Medo-Persian tree.

[3] *Fl. Orient.* iv. 1153 (1879).   [4] *Origin of Cultivated Plants*, 152 (1886).

Koch[1] never found it in the wild state in all his travels in the Orient; and Radde[2] states that it is only naturalised in the Caucasus.

According to Loudon, it was first introduced[3] into England in 1548, when a few trees were planted at Syon, one of which still survives. Canon Ellacombe,[4] however, believes that its introduction was much earlier, and adduces in support of this, that *morat*, a favourite drink in Anglo-Saxon times was a kind of mead, flavoured with mulberries. *Morus*, however, signified blackberry as well as mulberry; and *morat* may have been flavoured with blackberries. (A. H.)

Though the mulberry was often planted in old gardens and the remains of some still exist which may be over 300 years old, we are unable to find a tree of great size or age which is not more or less decayed at present.

The oldest, supposed to have been planted in the sixteenth century at Syon House, by the botanist Turner, was in Loudon's time 22 ft. high, and still exists; and though a wreck is one of the largest we know. Loudon speaks of one at Wardour Castle 40 ft. high, but I could find no trace of it when I was there in 1904. Among the drawings of the late E. Lees, there is one made in 1858 of a tree at the White Ladies, Worcester, partly prostrate, which seemed of unusual size, but I cannot hear whether it still exists. At Wotton House, Gloucester, a fine tree, 11 ft. in girth at 3 ft. from the ground, is said to have been planted by Queen Elizabeth. At Stoke Edith, Herefordshire, I saw in 1905, a very old tree with a prostrate trunk which had thrown up several stems. In front of the headmaster's house at Eton, Henry measured, in 1907, a tree 30 ft. by 8 ft. 3 in. There are old trees at Christ's, Jesus, and Emmanuel Colleges at Cambridge, which may have been planted in obedience to the edict of James I. in 1605, recommending the cultivation of silkworms and offering mulberry seed to all who would sow them. Christ's College purchased 300 trees in 1608. There are old trees at Dunster Castle and at Pembroke College, Oxford. One at Abington Park, Northamptonshire, is said to have been planted in 1778, by David Garrick.

In Scotland it exists on walls, but rarely fruits. In Ireland we have seen no large specimens, but Loudon mentions one in the grounds at Terenure, near Dublin, which was 25 feet high with a head 130 ft. in circumference.

The mulberry is a tree which has lost its former popularity in this country, and is now seldom planted, but as it requires more heat and sun than it usually gets in England, it is better[5] on a wall than as a standard, if grown for its fruit. It is perhaps best propagated by means of large cuttings, which soon take root, and its fruiting is improved by careful pruning. It ripens seed in hot seasons, but the seedlings are very slow in growth, some that I raised in 1901 from home-grown seeds being now only 2 to 3 ft. high, as their young wood ripens badly.

The wood of the mulberry is very like that of Robinia in texture, colour, and useful properties; but yellow when fresh, it acquires in the course of time a brownish

---

[1] *Dendrologie*, ii. pt. i. 444 (1872).   [2] *Pflanzenverb. Kaukasusländ.* 182 (1899).

[3] In *Gard. Chron.* 1868, p. 79, it is said that title-deeds, in the possession of Sir Henry Austen, showed that a mulberry tree, which was formerly growing at Shelford, was planted as a sapling in 1537.

[4] *Plant-lore of Shakespeare*, 176 (1896).   [5] Cf. Williams, in *Trans. Hort. Soc.* ii. 91 (1817.)

red tint, while that of the Robinia becomes brownish yellow. On a transverse section the pores of the outer part of each annual ring are minute and occur in groups, while in Robinia they are considerably larger in size and scarcely grouped. The sapwood is narrow, only 3 to 5 annual rings, and is white in colour. The wood is used in France for cooperage, wheelwright's work, vine props, and trenails. It takes a good polish and is often used for making furniture. The bark yields a fibre, which is sometimes used for cordage. Mulberries are used medicinally for the preparation of a syrup.[1]   (H. J. E.)

MORUS RUBRA, RED MULBERRY

Morus rubra, Linnæus, *Sp. Pl.* 986 (1753); Loudon, *Arb. et Frut. Brit.* iii. 1359 (1838); Sargent,
    Silva N. Amer. vii. 79, t. 320 (1895), and *Trees N. Amer.* 303 (1905).
Morus canadensis, Poiret, in Lamarck, *Encycl.* iv. 380 (1797).
Morus scabra, Willdenow, *Enum.* 967 (1809).
Morus tomentosa, Rafinesque, *Fl. Ludovic.* 113 (1817).
Morus riparia, Rafinesque, *New Fl.* iii. 46 (1836).
Morus reticulata, Rafinesque, *Am. Man. Mulberry Trees*, 28 (1839).

A tree, attaining in America 70 ft. in height, with a short trunk 10 to 12 ft. in girth. Bark dark brown, divided into irregular scaly plates. Young branchlets covered with a dense minute pubescence, glabrous in the second year. Leaves (Vol. IV. Plate 267, Fig. 3) 3 to 5 in. long, $2\frac{1}{2}$ to 4 in. wide, broadly ovate or orbicular, usually abruptly contracted into a long acuminate apex; broad, rounded or slightly cordate at the base; coarsely serrate; dull green with a scattered pubescence above; light green beneath and covered with a soft downy pubescence; petiole $\frac{3}{4}$ in. long, pubescent. Staminate spikes pendulous, 2 in. long, on pubescent peduncles. Pistillate spikes, 1 in. long, on short pubescent peduncles, densely flowered; style glabrous, short, with long stigmatic lobes. Fruit, about 1 in. long, at first bright red, becoming dark purple or nearly black and sweet when ripe.

The leaves on ordinary branches are usually entire or with one or two lobes; but on vigorous young shoots are often deeply three-lobed with oblique and rounded sinuses.

1. Var. *tomentosa*, Bureau, in De Candolle, *Prod.* xvii. 246 (1873).

Leaves scabrous above, pale and tomentose beneath. Louisiana, Texas. A large-fruited form of this was introduced into cultivation in America by Munson in 1889, as the Lampsas mulberry.[2]

The Red Mulberry is widely distributed in North America,[3] from southern Ontario and Massachusetts southwards to Florida, and westwards to Michigan, Nebraska, Kansas, and the Colorado river in Texas. It is most abundant and of its largest size in the basin of the lower Ohio river and on the foothills of the southern Alleghany Mountains. Sargent gives[4] a good figure of a tree in Alabama, and

---

[1] Cf. Flückiger and Hanbury, *Pharmacographia*, 544 (1879).
[2] Bailey, *Cycl. Amer. Hort.* 1035 (1901).
[3] A closely allied species, *M. tiliæfolia*, Makino, in *Tokyo Bot. Mag.* xxiii. 88 (1909) is found in Japan.
[4] In *Garden and Forest*, vii. 24, fig. 3 (1894).

states that there was one at Augusta, Georgia, 20 ft. in girth at 3 ft. from the ground; but this large size is exceptional.

This species has been tried, but with indifferent success, for feeding silkworms in America. It is occasionally planted in the southern States for its fruit, which is mainly used for feeding pigs and poultry. Bailey,[1] however, states that three of the named varieties of fruit-bearing mulberries belong to this species, as well as a yellow-leaved variety which is cultivated for ornament.

M. rubra is said to have been introduced into England, early in the seventeenth century; but is now scarcely known in cultivation, the only specimen which we have seen being a small tree at Kew, which died two or three years ago.

The timber, according to Sargent, is orange-coloured, with a thick light-coloured sapwood, soft and coarse-grained; but tough and resisting decay in contact with the soil. It is occasionally used in the United States for fencing, cooperage, and boat-building. (A. H.)

MORUS ALBA, WHITE MULBERRY

Morus alba, Linnæus, *Sp. Pl.* 986 (1753); Loudon, *Arb. et Frut. Brit.* iii. 1348 (1838); Willkomm, *Forstliche Flora*, 540 (1887); Hooker, *Fl. Brit. India*, v. 492 (1888); Mathieu, *Flore Forestière*, 290 (1897); Gamble, *Indian Timbers*, 634 (1902).

A tree, usually 30 to 40 ft. in height, but attaining, according to Mayr, much larger dimensions in Japan. Bark thick, fissured into broad scaly plates. Young branchlets with scattered minute pubescence. Leaves (Vol. IV. Plate 267, Fig. 2), thin and membranous, very variable in size and shape, ovate or oval, often with deeply indented sinuses, forming irregular lobes; base rounded, truncate, or widely cordate; apex obtuse, acute, or shortly acuminate; margin non-ciliate, irregularly serrate; upper surface shining, usually glabrous; lower surface dull green, pubescent on the midrib and nerves, elsewhere glabrous; petiole $\frac{3}{4}$ to $1\frac{1}{2}$ in. long, glabrescent. Pistillate spikes on long slender peduncles; style glabrous or papillose, divided or not divided to the base into stigmatic lobes. Fruit, variable in size, $\frac{1}{2}$ to $1\frac{1}{2}$ in. long, white or reddish.

The seedling [2] has two cotyledons, raised above the ground, dark green above, pale beneath, about $\frac{1}{3}$ in. long, tapering gradually to a short petiole, and faintly veined with a midrib and a few lateral nerves; primary leaves irregularly dentate.

No plant is more variable than *M. alba*; and possibly under this name are included two or three distinct species, natives of China and Japan, where the tree has been so long in cultivation, that it is almost impossible at the present day to distinguish the cultivated and possibly the hybrid forms, from those which are truly indigenous. For a complete account, Bureau,[3] who enumerates sixteen main varieties, may be consulted. The following are worthy of note:—

1. Var. *mongolica*, Bureau, in De Candolle, *Prod.* xvii. 241 (1873).

---

[1] *Cycl. Amer. Hort.* 1034 (1901). [2] Tubeuf, *Samen, Früchte, u. Keimlinge*, 53, 114 (1891).

[3] In De Candolle, *Prod.* xvii. 238-245 (1873). Several of these varieties are now considered to be distinct species, as *M. serrata*, Roxburgh, *M. indica*, Linnæus, and *M. lævigata*, Wallich, which are wild in the Himalayas.

Leaves large, with three to seven lobes, and a long acuminate apex ; serrations large, triangular, bristle-pointed. Fruit reddish in colour and insipid in flavour. This is common in the mountains around Peking, and was found by David near Jehol.

2. Var. *stylosa*, Bureau, in De Candolle, *Prod.* xvii. 243 (1873) ; Shirasawa, *Icon. Ess. Forest. Japon*, ii. t. 6, figs. 1-11 (1908).

> *Morus stylosa*, Seringe, *Desc. Muriers*, 225, t. 22 (1855).
> *Morus japonica*, Audibert, *ex* Seringe, *loc. cit.*

A shrub rarely exceeding 10 ft. high. Leaves small, about 3 in. long and $1\frac{1}{2}$ in. broad, very polymorphic, on the same individual simple, variously lobed or laciniate ; serrations usually ending in long points ; upper surface scabrous with short hairs ; lower surface pubescent on the midrib and veins. Spikes slender ; style twice as long as the ovary, dividing above into two pubescent stigmas. Fruit small, $\frac{1}{2}$ in. long, reddish, with few achenes.

This, which is probably a distinct species, is wild throughout the mountains of central and western China ; and also occurs in Korea and Japan. According to Seringe, it is tender in France ; and Bailey says that it suffers when young in the northern United States.

3. Var. *tatarica*, Loudon, *Arb. et Frut. Brit.* iii. 1358 (1838).

> *Morus tatarica*, Linnæus, *Sp. Pl.* 986 (1753).

A shrub or small tree, with numerous slender branches ; leaves usually without lobes, broadly elliptic, obtuse at the apex, with blunt serrations.

This has been found in Russia, growing on tracts inundated by the Volga and the Don, apparently wild, but probably only naturalised. It is much hardier than the type ; and was introduced in 1784 into England, and in 1875 into the United States, where in the western prairie States it is used as a windbreak, and for producing fence-posts and fuel.[1]

Numerous varieties have arisen in cultivation :—

4. Var. *latifolia*, Bureau, in De Candolle, *Prod.* xvii. 244 (1873).

> Var. *multicaulis*, Loudon, *Arb. et Frut. Brit.* iii. 1348 (1838).
> *Morus latifolia*, Poiret, in Lamarck, *Encyc.* iv. 381 (1797).
> *Morus multicaulis*, Perrottet, in *Ann. Soc. Linn. Paris*, 1824, p. 129.

A shrub, dividing near the base into many stems, and suckering very freely. Leaves large, usually without lobes, minutely tuberculate on the upper surface, which also shows here and there peculiar swellings. This is cultivated for feeding silkworms in southern and central China ; and was introduced into France in 1821.

5. Var. *macrophylla*, Loddiges, *ex* Loudon, *Arb. et Frut. Brit.* iii. 1349 (1838).
This differs little from the preceding, except in habit, having a single stem.

6. Var. *venosa*, Delile, in *Bull. Soc. Agric. Hérault*, 1826, p. 13.
Leaves marked beneath with white prominent veins. This is said to have originated in Europe.

[1] Cf. Pinchot, *U.S. Forest Circ.* No. 83 (1907).

7. Var. *colombassa*, Seringe, *Descrip. Muriers*, 206 (1855). Leaves distant on the branches, small, thin, lobed.

8. Var. *pyramidalis*, Seringe, *Descr. Muriers*, 212 (1855). Branches vertical, similar in habit to the Lombardy poplar.

9. Var. *pendula*, Dippel, *Laubholzkunde*, ii. 10 (1892). Branches pendulous.

10. Var. *constantinopolitana*, Loudon, *Arb. et Frut. Brit.* iii. 1358 (1838).

> *Morus constantinopolitana*, Poiret, in Lamarck, *Encycl.* iv. 381 (1797).

A tree with thick and twisted branches, resembling in habit *Robinia Pseudacacia*, var. *tortuosa*; leaves thick in texture. The origin of this peculiar form is unknown. It is well figured by Seringe, *Desc. Muriers*, 210, t. v. (1855).

The White Mulberry is undoubtedly a native of China, where it is common wild in the mountainous districts of the northern and central provinces. The cultivation of this species and the rearing of silkworms can be traced back to the remotest times of Chinese civilisation.[1] In Japan, where it is called *Kuwa*, several varieties of this species appear to occur wild; but the broad-leaved form, used for rearing silkworms, was probably introduced, with the art of sericulture, from China in the third or fourth century of our era. Mayr[2] states that *M. alba* grows wild and attains a large size in the virgin forests of central Yezo; and he measured a mulberry tree in the Bonin Isles 100 ft. in height and 10 ft. in girth.[3]

Complete details of the introduction of this species and the silk-worm industry into the Levant and Europe are given by Loudon, and need not be repeated here. The tree has been widely naturalised in many countries, as in Persia, Armenia, the Caucasus, Asia Minor, and in south-eastern Russia. It is only known in the cultivated state in Greece, Italy, and France.

The White Mulberry was introduced into England in 1596, but has apparently never succeeded in attaining a great size or age, the largest mentioned by Loudon being one at Syon, 45 ft. high and 6 ft. in girth in 1838. This tree no longer exists; but there are two young trees about 20 ft. high. Most trees which I have seen in this country are of small size and growing in botanic gardens, as at Kew, where the varieties are well represented. Elwes found a tree at Henham Hall, which was, in 1909, 20 ft. high with a head twenty paces round. Another at Beauport, Sussex, was 25 ft. by 3 ft. 3 in., and was bearing fruit in October 1911. This species is said[4] to have set fruit in the open garden at Dalkeith in July 1894.

The finest specimen that I have seen in Europe, was growing on the roadside near Zvornik in the Drina valley in Servia, and measured 65 ft. by 8 ft. in 1909.

(A. H.)

---

[1] Cf. Bretschneider, *Bot. Sinic.* ii. 328 (1892). [2] *Fremdländ. Wald- u. Parkbäume*, 485 (1906).

[3] The mulberry of the Bonin Isles, which may be a distinct species (*M. indica?*), is known in Japan as *Shima-guwa*. It produces a rich yellow wood with beautiful markings, which is highly valued, and used for high-class cabinet-making. I purchased a board of this wood at a high price in Tokio, part of which is in the Cambridge Forestry Museum.—H. J. E.

[4] *Trans. Bot. Soc. Edin.* xx. 237 (1894).

EUCALYPTUS

Eucalyptus, L'Héritier, *Sertum Anglicum*, 18, t. 20 (1788); Bentham and Mueller, *Flora Austra-liensis*, iii. 185 (1866); Bentham et Hooker, *Gen. Pl.* ii. 707 (1876); F. von Mueller, *Eucalyptographia*, decades i.-x. (1879-1884); Naudin, in *Ann. Sc. Nat.* xvi. 337-430 (1883), and *Descript. Emploi Eucalpt. Europe*, 1-72 (Antibes, 1891); Masters, in *Gard. Chron.* xxi. 148 (1884), and ix. 176 (1891); M'Clatchie, *U.S. Dept. Agric. Forestry Bulletin* No. 35, pp. 1-106 (1902); Maiden, *Revision Genus Eucalyptus*, i. 1-24 (1903); Ingham, *Agric. Exper. Station, Berkeley, California, Bulletin* No. 196, pp. 1-114 (1908).

EVERGREEN trees or shrubs, belonging to the order Myrtaceæ. Bark on young trees smooth and peeling off; on old trees variable—(*a*) remaining smooth; (*b*) persistent and rugged at the base, but smooth on the upper part of the trunk and on the branches; (*c*) persistent and fibrous; (*d*) persistent, very hard, and deeply furrowed; or (*e*) persistent and dividing into separate scales on the trunk.

Leaves on young plants beyond the seedling stage, and also on suckers, opposite, horizontal, sessile, cordate: variable on adult trees, (*a*) in some species, remaining always opposite, horizontal, sessile, and cordate; (*b*) in other species, becoming alternate and stalked, after the first six or eight leaves on the seedling; or (*c*) in most of the species, remaining opposite for several years, subsequently becoming alternate and stalked. Usually the alternate leaves are similar on both surfaces, with the stalk twisted, so that the blade is placed in a vertical or oblique plane; but in a few species the upper surface of the alternate leaves is darker in colour than the lower surface, the petiole not being twisted, and the blade remaining in the horizontal plane. Leaves nearly always glabrous,[1] odorous, with pellucid or con-cealed oil-dots; venation pinnate, the branches always converging to an intra-marginal vein, which is either close to or at some distance from the edge.

Flowers usually in pedunculate umbels, rarely reduced to a single sessile flower; peduncles in most species solitary and axillary, or occasionally lateral at the base of the current year's shoot below the leaves, or in some species, clustered in terminal panicled umbels. Calyx of two parts: (*a*) the tube persistent, adnate to the ovary, and truncate and entire (or rarely 4-toothed) after the fall of the (*b*) lid or operculum, which covers the stamens in the bud. Petals none, unless represented in a few species by a membrane under the operculum. Stamens numerous,[2] inserted close to the edge of the calyx in several rows, free or rarely united at the base into four clusters, always deciduous; all fertile or some of the outer without anthers; filaments thread-like, usually inflexed (rarely straight) in the bud; anthers dorsifixed,

---

[1] In the seedlings of a few species the leaves and branchlets are hairy. In a few species the stalk is inserted above the base of the leaf, so that the blade is peltate; but this condition only persists in the adult foliage of one species.

[2] Kerner, *Nat. Hist. Plants*, Eng. trans., ii. 107, 449, 782 (1898), points out that the walls of the capsules of the Eucalypti are remarkably thick and strong, to protect the seeds against desiccation during long periods of drought. As, however, no rain occurs when the trees of most species are in flower, the pollen is left quite unprotected. There is no corolla, and the top of the calyx falls off, so that the whole of the stamens, often a hundred or more in number, are completely exposed.

their two cells parallel and distinct or divergent and confluent at their apex, opening usually by longitudinal slits, rarely by terminal pores. Ovary, free at the apex, two to six-celled; style long; stigma convex or flat, undivided; ovules numerous in each cell, the majority remaining unfertilised. Fruit, consisting of the enlarged truncate calyx-tube, enclosing a hard woody resinous capsule, which is provided with three to five valves, which are either wholly or partially exserted or entirely enclosed. Seeds numerous, but very few fertile; minute, polygonal, winged in a few species.

In Tasmania all the species take at least twelve months for the flower-bud to reach maturity, and another year for the fruit to perfect.[1] None of the species at low altitudes have any constant flowering period, and may be found in flower in any month of the year, one tree flowering at midsummer, while another beside it will not flower till winter. The flowers are fertilised by honey-feeding birds.

The leaves of all the species, and often the young branchlets, flowers, and fruits contain numerous oil-vesicles. The leaves of many species are distilled to yield this oil, which is variable in composition, but is volatile and antiseptic and used for pharmaceutical purposes.[2] The bark of several species exudes a resin,[3] called *kino*, which contains tannin in commercial quantities; and it is on account of this resin, that the name gum trees was applied to the genus.

In many species of Eucalyptus and other genera of Myrtaceæ and Proteaceæ, which are natives of Australia, the leaf-blades, as was pointed out by R. Brown,[4] are not placed horizontally like those of European broad-leaved trees; but are, by the twisting of their stalks, set vertically. This is a provision to lessen evaporation in the dry climate of Australia, since the narrow edge and not the broad surface of the leaf is directed towards the sun. In the Australian species of Acacia, the same adaptation is effected by the non-development of the blade of the leaf, the stalk becoming expanded, simulating the blade and forming a so-called phyllode, which is also directed vertically. In the Australian forests of Eucalypti and Acacias, the linear edges of the leaves and phyllodes cast little shadow, and sunlight streams on the ground. Behr[5] describes a typical forest in South Australia, as follows: "As a rule a dense meadow sward, in most cases accompanied by a light park-like forest of gigantic Eucalypti, whose crowns, however, never meet. The smooth stems, freed from their outer layers of cortex, stand apart at definite and often regular distances."

The currently accepted opinion that some species of Eucalyptus in Australia are the tallest and largest trees in the world is based on records for *Eucalyptus amygdalina*,

---

[1] Rodway, *Wild Flowers of Tasmania*, 49 (1910), states, however, that *E. globulus* and many other species take two years from the first appearance of the bud to the fall of the operculum, and another two years to mature fruit. *E. calophylla* of West Australia flowers in a few months, and takes a year to produce ripe fruit. The seeds are retained and remain quite fertile for several years in the capsules, which open their valves often only when exposed to forest fires. Seedlings usually spring up in consequence in burnt-over lands. Most of the species also when killed by fire, rapidly regenerate by suckers from the roots.

[2] Cf. R. T. Baker and H. G. Smith, *Researches on Eucalypts* (1902). A complete set of oils of 109 species, with herbarium specimens of the timber and bark, are preserved in the Pharmaceutical Museum, Bloomsbury Square, London. Cf. *Pharmac. Journ.* 1904, p. 187.

[3] This is called Australian or Eucalyptus kino, and is different from the officinal kino, which is produced by *Pterocarpus Marsupium*, a large deciduous tree of central and southern India. Cf. Flückiger and Hanbury, *Pharmacographia*, 894 (1879).

[4] *Botany of Terra Australia*, i. 62 (1814).

[5] *Linnæa*, xx. 546 (1847). Schimper, *Plant Geography*, 495, 528, figs. 260, 261, 262 (1903) gives a general account of these forests, with pictures of the various types.

which were accepted as correct by Baron von Mueller, who states in his *Eucalyptographia* that four trees of this species were 415, 471, 410, and 420 ft. in height respectively. Dr. A. S. Ewart of the Melbourne University considers[1] that these measurements were grossly exaggerated; and instances the fact that the tree which many years previously D. Boyle had measured on the Dandenong range as 420 ft., was found by Fuller in 1889 to be only 220 ft. in height, with a girth of 48 ft. at six feet from the ground. Dr. Ewart states that although *E. amygdalina* (including *E. regnans*), is the tallest species in Australia, it rarely exceeds 300 ft. He gives as the maximum heights accurately recorded the following: a tree growing on the edge of a ridge found by Professor Kernot to be 302 ft.; others growing in thick groves, which were found by Perrin, Davidson, and Fuller to be 271, 294, 296½, 297, and 303 ft. The tallest Australian tree that has been correctly measured[2] appears to be one of *E. regnans*, on Mt. Baw-Baw, Gippsland, 91 miles from Melbourne, which Maiden[3] gives as 326 ft. in height, and 25 ft. 7 in. in girth.

About 150 species of Eucalyptus are known, the greater part of which are natives of Australia and Tasmania, only three or four species occurring in New Guinea, the Moluccas, and Timor, and one species[4] in the Bismarck Archipelago (New Britain) and the Philippine Islands.

The following notes deal shortly with some of the tender species,[5] which have seemed to succeed in the British Isles for a time; but which we do not consider worthy of a lengthened notice.

Eucalyptus alpina, Lindley, a shrub confined to the summit of Mt. William, 5000 ft. altitude, fifty miles north of Melbourne. Planted out in Arran[6] in 1884, it attained 14 ft. in height in eleven years, and seemed to be very hardy, flowering in the spring of 1888. It was, however, ultimately killed in 1894. At Kinloch Hourn, the seedlings which were planted out did not long survive.

E. amygdalina,[7] Labillardière, a large tree abundant in Tasmania, and occurring in many localities in Victoria and New South Wales. There are three trees 20 ft. high, probably of this species, at Menabilly. Seedlings raised at Kinloch Hourn in 1890 were all killed in the winter of 1894-95, except one in a sheltered valley, which was cut to the ground, and died subsequently. Another batch of seedlings raised in 1895 were killed in 1899-1900. Planted at Cromla in Arran[8] in 1895, it grew rapidly, and was 20 ft. in 1905. Mr. John Paterson, who sent a specimen branch, states that this tree was 30 ft. high in June 1911. We have a specimen from Brodick of a young tree planted out in 1909. At Abbotsbury, it was killed in 1908, when 20 ft. high, by 16° of frost. Mueller says that near Lake Maggiore in

---

[1] In *Phil. Trans. Roy. Soc.* Series B, vol. 199, p. 367 (1908). Cf. Hemsley, in *Gard. Chron.* xlvii. 69 (1910).

[2] The tallest tree in the world, that has been accurately measured, was the Redwood on Eel River, California, which was found by Sargent in 1896 to be 340 feet in height. Cf. our Vol. III. p. 692, and also p. 690, note 3.

[3] Maiden, *Forest Flora*, ii. 161-165 (1907), gives accurate measurements of several trees that Ewart does not mention.

[4] *E. Naudiniana*, Mueller. Cf. Maiden, *Revis. Gen. Eucalypt.* ii. 79 (1910). The distribution of this species is remarkable.

[5] Some of these may be hardy in Mr. Heard's garden at Rossdohan in Kerry, and also in the Scilly Isles; but the climate of these localities is exceptionally free from frost.

[6] Landsborough, in *Trans. Bot. Soc. Edin.* xx. 522 (1896). In *ibid.* xxiii. 147 (1905), Landsborough mentions a tree at Cromla in Arran, planted in 1895, which was 20 ft. high in 1905.

[7] The tree recorded in *Gard. Chron.* xxvi. 790 (1886) under this name, as 60 ft. high at Fota, is *E. pauciflora.* Cf. p. 1632. The tree at Dalkeith, mentioned under this name in *Journ. Roy. Hort. Soc.* xviii. 76 (1895) was probably incorrectly named.   [8] Landsborough, in *Trans. Bot. Soc. Edin.* xxiii. 148 (1905).

Italy, it endured a temperature of 18° Fahr., proving hardier than *E. globulus* or *E. rostrata*; and grew astonishingly fast, attaining a height of 60 ft. in nine years.

E. regnans, Mueller, closely allied to, if not a variety of, *E. amygdalina*, did not prosper in Arran[1]; and seedlings were speedily killed at Kinloch Hourn.

E. Beauchampiana, Treseder, *ex* Masters, in *Gardeners' Chronicle*, xxxviii. 3, fig. 3 (1905), and xxxix. 174 (1906). This species was raised at Truro by Messrs. Treseder and Co., who inform us that they obtained the seed direct from New South Wales. They consider it to be hardier than the other species in their nursery, having withstood 20° of frost without injury. Their largest specimen is 30 ft. high and 18 in. in diameter; but has not yet flowered. This species was identified by Maiden, to whom I sent specimens, with *E. Stuartiana*,[2] Mueller, the Apple Eucalyptus of the cool tablelands of Victoria and New South Wales. Coming from this region, it is unlikely to prove very hardy; and was killed[3] in the severe winter of 1908-1909, both at Wisley, where the temperature fell to 7.5° Fahr., and at Myddelton House, Herts, where the thermometer placed on the grass registered −1° Fahr. There is, however, a tree at Tregothnan, planted eight years, which was bearing fruit and measured 25 ft. high in February 1911. Another at Mount Usher, which was raised from seed obtained from the Sydney Botanic Garden in 1904, was 20 ft. high and bearing flowers in October 1911. There are two small trees about 15 ft. high at Menabilly, which are labelled *E. Stuartiana*; but differ from *E. Beauchampiana* in having glaucous foliage.

E. botryoides, Smith, the Bastard Mahogany of Queensland, New South Wales, and Victoria. A specimen of this planted[4] in Arran in 1896 grew well for several years, but was ultimately killed by drainage from a manure heap. According to Mr. Birkbeck, this species bears more frost in winter than *E. globulus*, but less in spring. It was, however, killed at Abbotsbury in 1908.

E. citriodora, Hooker, a native of Queensland. According to Mr. Birkbeck, this only survived a few years out of doors at Menabilly; and seedlings were soon killed at Kinloch Hourn. It attained 25 feet at Tregothnan; but was killed to the ground in a severe winter.

E. ficifolia,[5] Mueller, a native of New South Wales. This was reported[6] to have flowered in the open air at Cove House, Tiverton, Devon; but Mr. W. North-Row informs us that this was an error, as the plant is growing in the border of a cool greenhouse. At Monreith, where it was growing on a south-east wall, it was killed[7] in the severe winter of 1908-1909.

[1] Landsborough, in *Trans. Bot. Soc. Edin.* xx. 522. In *ibid.* xxiii. 149 (1905), a tree planted in 1892 is reported to have been, though cut by frost, 13 ft. high in 1905. The Rev. Dr. Landsborough informed Mr. Birkbeck that the tree at Roseneath, referred to *E. regnans* in *Gard. Chron.* xxvi. 715 (1886), was *E. viminalis*.

[2] *E. Beauchampiana* is erroneously referred by Masters to *E. cinerea*, Mueller. It is doubtful if the name *E. Stuartiana*, Mueller, can be retained for the "Apple Eucalyptus," as it was first applied by Mueller to another species. Cf. p. 1645, note 2.

[3] *Journ. Roy. Hort. Soc.* xxxvi. 374 (1910).   [4] *Trans. Bot. Soc. Edin.* xxiii. 149 (1905).

[5] Figured by J. D. Hooker in *Bot. Mag.* t. 7697 (1900), from a specimen which flowered in a cool greenhouse at Arbroath, Scotland. There are two good trees in the Temperate House at Kew.

[6] *Gard. Chron.* xlii. pp. 376, *Suppl. Illust.* and 418 (1907).

[7] *Journ. Roy. Hort. Soc.* xxxvi. 374 (1910).

E. hæmastoma, Smith, a large timber tree of New South Wales and Queensland. Seedlings planted at Kinloch Hourn and in the Isle of Arran[1] were soon killed.[2]

E. leucoxylon, Mueller, the Iron-Bark of New South Wales, Victoria, and South Australia. This was killed at Abbotsbury in the severe weather of 1908. A tree at Rossdohan, Kerry, of this species, as well as it can be identified from a barren branch, was 33 ft. by 2 ft. 3 in. in 1910.

E. obliqua, L'Héritier, the Stringy Bark, an immense tree, abundant in Tasmania and forming the great part of the hill forests, ascending, according to Sir J. D. Hooker, to 4000 feet; also common in Victoria, New South Wales, and South Australia. We have specimens, without flowers or fruit, from two trees at Menabilly, which are probably this species. The largest is reported to have been 40 ft. by 2 ft. 8 in. in January 1911; the smaller one being 26 ft. by 1 ft. 11 in. A tree at Tregothnan, planted about five years and 20 ft. high in 1911, may also be referred to this species. Seedlings did not survive at Kinloch Hourn; but Mr. T. A. Dorrien-Smith informed Mr. Birkbeck in 1894 that he had a thriving tree at Tresco, Scilly Isles, which was 30 ft. high and twenty years old. This species was killed at Abbotsbury in the severe weather of 1908. As it ascends to a high elevation in Tasmania, seeds from there should produce hardy plants; but it appears to be rare in collections, and possibly only lowland forms have been tried.

E. resinifera, Smith, a tall tree, occurring in New South Wales and Queensland, where it is known as Red Gum. This was recorded[3] as one of the plants uninjured at Ventnor, Isle of Wight, in the severe winter of 1879-1880; but it is very doubtful if the tree was correctly named; and we have not seen or heard of any specimens of this species in the open air in the British Isles.

E. rudis, Endlicher, a native of West Australia. Planted out in 1887 at Cromla in Arran,[4] it had attained 15 ft. high in 1895, when it lost all its branches and leaves. Afterwards it sprouted from the root, and was 22 ft. high in 1905. Seedlings planted out at Kinloch Hourn were speedily killed. At Bradfield, Devon, it was reported[5] to have been cut to the ground in 1885; but it afterwards sent up strong shoots. It was killed at Abbotsbury in the severe winter of 1908.

E. stellulata, Sieber, a small tree, occurring in Victoria and New South Wales. This has produced flowers and fruit at Rossdohan, Kerry; but it is unlikely to prove hardy anywhere except in the extreme west of Kerry or in the Scilly Isles. It has not been tried, so far as we know, except at Rossdohan.

The following key, based on the characters of the foliage, will serve to distinguish the species in cultivation in the open air in Great Britain and Ireland, and includes only those which have attained a considerable age, and have borne flowers and fruit :—

---

[1] Landsborough, in *Trans. Bot. Soc. Edin.* xx. 524 (1896).

[2] A tree cultivated under this name at Abbotsbury is *E. urnigera*; and another so-named at Leonardslee is apparently a form of *E. Gunnii*, which has survived 26° of frost. Sir E. G. Loder has tried *E. globulus, E. coccifera, E. urnigera, E. pulverulenta, E. amygdalina,* and *E. viminalis*: but these are always killed by 20° to 24° of frost.

[3] By Ewbank, in *Journ. Roy. Hort. Soc.* viii. 10 (1887).

[4] Landsborough, in *Trans. Bot. Soc. Edin.* xx. 520 (1896), and xxiii. 149 (1905).

[5] J. W. in *Gard. Chron.* xxvi. 754 (1886).

I. *Leaves on adult trees, opposite, sessile, cordate at the base.*

  1. *Eucalyptus cordata*, Labillardière, Tasmania. See p. 1620.

      Young branchlets quadrangular, glaucous, roughened with oil-glands. Leaves suborbicular or broadly ovate, $2\frac{1}{2}$ in. long, 2 in. broad; both surfaces glaucous and roughened with numerous raised oil-glands; crenate in margin.

  2. *Eucalyptus pulverulenta*, Sims. New South Wales, Victoria, and Queensland. See p. 1622.

      Young branchlets terete or angled, smooth. Leaves ovate, $1\frac{1}{2}$ in. long and broad; glaucous, smooth; entire or slightly undulate in margin.

II. *Leaves on adult trees, alternate, stalked.*

     * *Leaves 5 in. or more in length.*

  3. *Eucalyptus globulus*, Labillardière. Tasmania, Victoria, and New South Wales. See p. 1623.

      Young branchlets green, quadrangular. Leaves thick, lanceolate, very falcate, 6 to 9 in. long, 1 to $1\frac{1}{2}$ in. wide.

  4. *Eucalyptus pauciflora*, Sieber. Tasmania, South Australia, Victoria, and New South Wales. See p. 1631.

      Young branchlets usually more or less covered with a glaucous bloom. Leaves thick, lanceolate, about 5 in. long and $\frac{3}{4}$ in. broad, usually falcate; main lateral veins longitudinal.

  5. *Eucalyptus viminalis*, Labillardière. Tasmania, Victoria, South Australia, and New South Wales. See p. 1633.

      Young branchlets green, very slender, with four linear ridges. Leaves narrowly lanceolate, 5 to 6 in. long, $\frac{1}{2}$ to $\frac{3}{4}$ in. broad, falcate, tipped at the apex with a long slender curved or twisted filamentous point.

     ** *Leaves less than 5 in. long.*

     (a) *Leaves and branchlets glaucous.*[1]

  6. *Eucalyptus coccifera*, J. D. Hooker. Tasmania. See p. 1635.

      Young branchlets covered with a thick whitish bloom. Leaves lanceolate, $2\frac{1}{2}$ to 3 in. long, $\frac{2}{3}$ to $\frac{3}{4}$ in. wide, thick in texture, green or glaucous, tipped with a long slender curved point.

  7. *Eucalyptus Gunnii*, J. D. Hooker. High altitudes in Tasmania, Victoria, and New South Wales. See p. 1638.

      Young branchlets more or less glaucous. Leaves ovate-lanceolate, 2 to $2\frac{1}{2}$ in. long, $\frac{5}{8}$ to $\frac{7}{8}$ in. broad, thick in texture, shining green, tipped with a short point.

     (b) *Leaves and branchlets not glaucous.*

     † *Leaf margin entire.*

  8. *Eucalyptus whittingehamensis*, Nicholson. A hybrid, probably originating in Tasmania. See p. 1642.

      Young branchlets[1] usually green, terete. Leaves lanceolate, 3 to $3\frac{3}{4}$ in. long, $\frac{5}{8}$ to $\frac{3}{4}$ in. wide, usually straight, greyish green, entire in margin.

---

[1] Some of the descendants of *E. whittingehamensis* have glaucous branchlets.

9. *Eucalyptus acervula*, Miquel.   Tasmania, Victoria, New South Wales, South Australia, and Queensland.   See p. 1645.

Young branchlets green, terete.   Leaves lanceolate, or ovate-lanceolate, 4 in. long, $1\frac{1}{4}$ in. broad, usually straight, dull green, often undulate in margin.

10. *Eucalyptus vernicosa*, J. D. Hooker.   Tasmania.   See p. 1646.

Young branchlets green, angled.   Leaves alternate or sub-opposite, narrowly ovate or elliptical, $\frac{1}{2}$ to 2 in. long, $\frac{1}{2}$ to 1 in. broad, very thick and coriaceous, shining as if varnished on both surfaces.

†† *Leaf-margin distinctly crenate.*

11. *Eucalyptus Muelleri*, T. B. Moore.   Tasmania.   See p. 1647.

Young branchlets reddish.   Leaves usually straight, lanceolate, 3 to 4 in. long, $\frac{3}{4}$ to 1 in. wide, thick in texture, shining polished green on both surfaces.

12. *Eucalyptus urnigera*, J. D. Hooker.   Tasmania.   See p. 1649.

Young branchlets green, tinged with red.   Leaves ovate-lanceolate, 3 to $3\frac{1}{2}$ in. long, 1 to $1\frac{1}{2}$ in. broad, often falcate, thick in texture, dull yellowish green.   (A. H.)

CULTIVATION

If one may judge from the numerous references in horticultural literature to this genus, none has been more persistently tried in various parts of the country ; and yet when we come to record the small number of trees which have endured our climate for more than a few years, it must be acknowledged that none has proved more disappointing.   Even in those parts of the south and west where the thermometer only falls to 15° or 20° Fahr. at long intervals, not more than a few species have long endured, and possibly the absence of summer heat is the cause, quite as much as the cold and damp of winter.   Though the Eucalypti seem indifferent as regards soil, and grow for a few years with great rapidity, yet with some exceptions they are short-lived, and die off suddenly after an inclement season, or blow down, when they become tall enough to be exposed to gales.

No one in England seems to have tried to graft the more tender species on stocks of the hardier ones, though, judging from experience in other genera, it might be possible to do this successfully.[1]   On account of their leaves, flowers, bark, scent, and habit, all so unlike those of any European trees, they form an attractive feature in gardens and pleasure grounds ; and are so easy to raise from seed, that the certainty of their death after a few years will not deter gardeners from planting them.

A great many species have been tried at different times in the open in the British Isles ; but only a few have proved hardy even in the mild climate of the south and west ; and the only really large and old trees which have survived are those at Powderham, Penmere, Garron Tower, and Whittingehame.   The Whittingehame hybrid is perhaps the hardiest of all, and is the only species which has succeeded out of doors at Kew.

---

[1] In *Gard. Chron.* xxv. 145 (1899) an interesting article on this question will be found, largely taken from a paper by M. Felix Sahut in the *Annales de la Société d'Horticulture de l'Herault*, which should be consulted by any one wishing to propagate Eucalypti by budding or grafting.

Mr. Robert Birkbeck, to whom we are much indebted for information concerning the cultivation of this genus, has tried forty or fifty species at Kinloch Hourn, on the west coast of Scotland, and sums up his experience as follows :—"*E. vernicosa* is the hardiest of all the species ; *E. Gunnii*, *E. coccifera*, *E. cordata*, and *E. urnigera* may be considered quite hardy ; *E. globulus* will stand 15° of frost only ; and *E. amygdalina* about the same. Both *E. viminalis* and *E. pauciflora* are more tender than *E. globulus*. *E. alpina* is slightly hardier, but is killed by 20° of frost." Mr. Birkbeck gives a long list[1] of species which completely failed at Kinloch Hourn, and tells us that *E. alpina*, *E. amygdalina*, *E. viminalis*, and *E. globulus* were killed at Inveraray.

At no place have more species been tried in the open air than at Lady Ilchester's garden at Abbotsbury, in Dorsetshire ; and Mr. Kempshall, head gardener there, sent us a list of forty species, all of which succumbed in the cold winters of 1907 and 1908, except the following :—*E. coccifera*, *E. cordata*, *E. Gunnii*, *E. Muelleri*, *E. vernicosa*, and *E. urnigera*, the last named being in his opinion the hardiest species.

Mr. Birkbeck gives the following rules for their propagation and planting :— Raise them from seed under glass ; pot when 2 or 3 in. high ; repot often as they never do well after their roots are pot-bound. Give them some bone-meal, keep under glass till 3 ft. high, and plant out in good soil in May or June, when they are about two years old. Keep them well staked, as they are easily blown down, but keep the fastenings loose, as they increase in girth quickly.

When planting has to be done on a large scale, seedlings should be treated exactly like pines and other conifers, *i.e.* they should, when about 6 in. high, be transplanted in the nursery to promote the formation of lateral rootlets, which will render easy their establishment in the ground where they are to remain permanently. Seedlings should not be dried up when being moved, and ought to be transplanted in cool cloudy weather, and watered for some days afterwards.

With regard to economic planting of the Eucalypti in warmer parts of the world than our own islands, we cannot enter into any details concerning the selection of the proper species ; and must refer our readers to the excellent papers of Naudin,[2] which deal with their cultivation in the south of France, and to those of M'Clatchie[2] and Ingham,[2] which give a complete account of the results already obtained in California, where many species have been tried. The admirable report[3] on Cyprus by Mr. D. E. Hutchins, whose long experience in South Africa enables him to speak with authority, may also be consulted. (H. J. E.)

[1] A list of thirty-five species tried at Kinloch Hourn is given in *Trans. Bot. Soc. Edin.* xx. 525 (1896). Mr. Birkbeck wrote a further account of his experience in *Gard. Chron.* xxv. 84 (1899).

[2] See the titles of these papers at the beginning of this article. Cf. also *U.S. Forest Service Bull.* No. 87, pp. 1-47 (1911), which gives an account by Zon and Briscoe of the species which are planted in Florida, with notes on their cultivation.

[3] D. E. Hutchins, *Report on Cyprus Forestry*, pp. 64-67 (London, 1909) ; cf. also the same author, in Flint and Gilchrist, *Science in South Africa*, 395-396 (Cape Town, 1905).

EUCALYPTUS CORDATA, Tasmanian Heart-leaved Gum

Eucalyptus cordata, Labillardière, *Pl. Nov. Holl.* ii. 13, t. 152 (1806); J. D. Hooker, *Fl. Tasm.* i. 132 (1860), and *Bot. Mag.* t. 7835 (1902); Bentham and Mueller, *Fl. Austral.* iii. 224 (1866); Mueller, *Eucalyptographia*, Dec. viii. (1882); Masters in *Gard. Chron.* iii. 798, fig. 111 (1888); Maiden, in *Rep. Austral. Assoc. Advance. Science, Hobart*, 1902, p. 374; R. T. Baker, in *ibid.* p. 344; Rodway, *Tasmanian Flora*, 58 (1903); Parsons, in *Gard. Chron.* xlvii. 168, *Suppl. Illust.* (1910).

A small tree, rarely exceeding 30 ft. high in Tasmania. Bark smooth, the older bark being shed in scales. Young branchlets quadrangular, glaucous, roughened with reddish oil-glands. Leaves (Plate 365, Fig. 1) on adult trees, opposite, in decussate pairs, sessile; suborbicular or broadly ovate, averaging $2\frac{1}{2}$ in. long and 2 in. broad; cordate and clasping at the base; rounded or acute, rarely emarginate, at the apex, which is usually tipped with a short triangular point; similar in colour on both surfaces, green, or more or less covered with a whitish bloom; margin reddish, revolute, distinctly crenulate; oil-glands very numerous, unequal, pellucid, the larger ones roughening the surface as minute protuberances; lateral nerves few, slender, spreading from the midrib at an angle of 80°.

Flowers in axillary umbels of threes, usually glaucous; peduncle stout, glaucous, $\frac{1}{4}$ in. long; calyx-tube campanulate, sessile, $\frac{1}{3}$ in. long, covered with oil-glands, usually with two lateral ridges; operculum cap-shaped, rounded, with a short conical umbo; stamens all perfect, inflexed in the bud; anthers ovate, with distinct parallel cells. Fruit hemispheric, but slightly contracted at the summit, where it is about $\frac{1}{2}$ in. in diameter; glaucous and slightly roughened with oil-glands; rim narrow, slightly elevated; capsule deeply enclosed, the four valves when open scarcely reaching the level of the rim.

Seedling,[1] with a terete tuberculate stem; cotyledons slightly emarginate, three-nerved, transversely oblong; primary leaves sub-sessile, opposite, acute, followed by crenate sessile leaves.

This species was discovered in 1792 by Labillardière near Recherche Bay in Tasmania, and appears to be confined[2] to this island, where it is rare and local, being recorded also for Mount Brown, Huon Road, Campania, and the Tasman Peninsula. It is without any economic value, and has no popular name in the colony; but Rodway calls it Heart-leaved Gum. It sometimes remains shrubby, and bears flowers and fruit when only 3 ft. high; but is usually a small erect tree, occasionally attaining 30 ft. in height, according to Mueller, who adds that Mr. Coombs found a tree 50 ft. high and 18 inches in diameter on the Sandfly river.

E. cordata appears to have been early introduced, as Sir J. D. Hooker knew

---

[1] Lubbock, *Seedlings*, i. 531 (1892).

[2] It is said by Deane and Maiden, in *Proc. Linn. Soc. N.S. Wales*, 1901, p. 126, to occur at Rockley Road, near Bathurst in N.S. Wales; but the specimen, which I have seen from that locality in the Cambridge Herbarium, is *E. pulverulenta*; and R. T. Baker, in *Rep. Austr. Assoc. Advance. Sc. Hobart*, 1902, p. 344, considers that the Bathurst tree is certainly the latter species.

about sixty years ago a plant trained on a wall at Kew, of which there is a specimen in the herbarium, gathered in 1851. This tree was eventually killed by frost. This species often flowers under glass at Kew, forming a pyramidal tree,[1] about 15 ft. high, and is much used in the young state for bedding out.[2]

At Menabilly there is a fine tree, which Mr. Bennett reported to be 50 ft. high by 2½ ft. at three feet from the ground in 1909. Colonel Trefusis tells us that at Porthgwidden,[3] in the same county, it produces flowers and is 28 ft. high and 1 ft. 2 in. in girth. A. B. Jackson saw a tree at Heligan 30 ft. high in 1909. It is also growing at Redruth and at Pencarrow in Cornwall, and at Exeter, where it has frequently flowered ; but is ultimately killed by 20° of frost.[4] At Abbotsbury this species is very thriving, surviving the severest winters and freely producing seed from which seedlings have been raised.

At Myddelton House,[4] Waltham Cross, Herts, it produces flowers, and, though killed every ten years or so, is so beautiful that it is well worth growing. There is a specimen in the Arboretum Herbarium, Kew, sent some years ago from Vicar's Hill, Lymington, by Mr. E. H. Pember ; but this does not appear to be now living, and was probably killed by frost.

On the west coast of Scotland this is one of the hardiest species. It has thriven at Kinloch Hourn,[5] where, planted in 1894, a tree which had twice lost its top in a storm, was 34 ft. by 1 ft. 10 in. in 1905. At Cromla,[6] in Arran, a tree planted in 1894, was 26 ft. by 13 in. in 1905. Here the flower buds form in August, and expand early in March of the following year, this tree being one of the first to flower in spring. At Inverewe on the west coast of Ross-shire this species is very thriving, and of all the species which flowered early in 1910, this was the only one which was in fruit in December of that year. The trees here were raised by Mr. Osgood H. Mackenzie from seed ripened at Abbotsbury ; and one of them in 1911 was 33 ft. high by 18 in. in girth, at ten years old.

In Ireland, Elwes saw a tree in flower at Castlewellan,[6] about 20 ft. high in 1908. At St. Anne's, near Dublin, a specimen,[7] which was planted out in 1904, has passed through one or two severe winters, and was 20 feet high in 1909. There are trees at Rostrevor of the same size, which flower freely. (A. H.)

---

[1] Raised from seed received from Hobart in 1888.
[2] *Gard. Chron.* xxiv. 191 (1898) and xlvi. 422 (1909).
[3] *Ibid.* xlvi. 403 (1909) and xlvii. 168, *Supply. Illust.* (1910).
[4] *Ibid.* xlvi. 422 (1909), and *Journ. R. Hort. Soc.* xxxi. p. xci. (1906).
[5] Landsborough, in *Trans. Bot. Soc. Edin.* xx. 520 (1896), and xxiii. 147 (1905). Mr. John Paterson, who sent a branch, informed me that this tree was 30 feet by 23 in. in June 1911.
[6] The late Earl Annesley sent a branch with flower buds to Dr. Masters in 1889. Cf. *Gard. Chron.* xxv. 58 (1889).
[7] *Gard. Chron.* xlvi. 403 (1909).

EUCALYPTUS PULVERULENTA, Australian Heart-leaved Gum

Eucalyptus pulverulenta, Sims, *Bot. Mag.* t. 2087 (1819); Bentham and Mueller, *Flora Austral.* iii. 224 (1866); Mueller, *Eucalyptographia*, Dec. viii. (1882); Howitt, in *Rep. Aust. Assoc. Advance. Sci.*, Sydney, 1898, p. 517, pl. xxvi.-xxix.; Deane and Maiden, in *Proc. Linn. Soc. N.S. Wales*, 1899, p. 465, and 1900, p. 110; Maiden, in *Proc. Linn. Soc. N.S. Wales*, 1901, p. 547; R. T. Baker, in *Rep. Aust. Assoc. Advance. Sci.*, Hobart, 1902, p. 345.

Eucalyptus cordata, Loddiges, *Bot. Cab.* t. 328 (1819) (not Labillardière).

Eucalyptus pulvigera, A. Cunningham, in Field, *Geog. Mem. N.S. Wales*, 350 (1825).

A tree, attaining in Australia 50 ft. in height and 10 ft. in girth; but flowering when in a shrubby state. Bark light brown, persistent, somewhat fibrous. Young branchlets, terete or slightly angled towards the tip, covered with a white glaucous bloom. Leaves (Plate 365, Fig. 3) on adult trees, opposite, sessile, in decussate pairs, ovate, about $1\frac{1}{2}$ in. long and broad; cordate and clasping at the base; rounded or acute at the apex, which is tipped with a short triangular point; similar in colour on both surfaces, which are more or less covered with a whitish bloom; lateral veins slender, spreading from the midrib at an angle of 60°; margin reddish, entire or slightly undulate; oil-glands numerous, mostly pellucid, not prominent or roughening the surface as in *E. cordata*.

Flowers glaucous, in axillary umbels of threes; peduncle stout, glaucous, $\frac{1}{5}$ in. long; calyx-tube sessile, turbinate, dotted with oil-glands, about $\frac{1}{5}$ in. long; operculum conic, tipped with an acuminate point; stamens all perfect, inflexed in the bud; anthers ovate, with distinct parallel cells. Fruit turbinate, about $\frac{1}{3}$ in. broad at the summit, glaucous, dotted with oil-glands; rim thick and convex; capsule slightly included, the three or four valves when open protruding beyond the orifice.

This species has thinner and smoother leaves than *E. cordata*, with their margin not crenate; the operculum of the flower bud is conical and long, while that of *E. cordata* is shorter and slightly convex; the fruits are smaller with more protruding valves than in the last-named species. The bark of the two trees is quite different, that of *E. pulverulenta* persistent and fibrous, while that of *E. cordata* is deciduous, smooth, and peeling off in ribbons.

1. Var. *lanceolata*, Howitt, in *Austr. Assoc. Advance. Sci.*, Sydney, 1898, p. 518.

Eucalyptus cinerea,[1] Mueller, in Bentham, *Fl. Austral.* iii. 239 (1866).

Eucalyptus nova-anglica, Deane and Maiden, in *Proc. Linn. Soc. N.S. Wales*, 1899, p. 616.

Eucalyptus Stuartiana, Mueller, var. *cordata*, Baker and Smith, *Researches on Eucalypts*, 105 (1902).

Leaves on old trees, lanceolate, 4 in. long, usually opposite, occasionally alternate.

E. pulverulenta is a native of Australia, where it was discovered by A. Cunningham on the Lachlan and Cox's rivers in New South Wales. The typical form of the species is widely diffused in the southern mountainous part of this colony from

---

[1] The type of *E. cinerea* comprises Cunningham's specimen from Lachlan river near Bathurst, and Mueller's specimen from Lake George, both localities in New South Wales. These specimens are, I think, intermediate between the typical form of *E. pulverulenta* and var. *lanceolata*.

Tumut to Berrima, and thence westerly to the Bathurst district. It also occurs near Lake George, and in Argyle and Camden counties; and is common in the Goulburn district. The lanceolate - leaved variety, which is known as Black Peppermint, is common over the greater portion of New England, and also occurs on the summit of Ben Lomond. Both the typical form and the variety occur in Gippsland in Victoria, at Buchan, near Walhalla, around Lake Omeo, and between the Avon and Mitchell's rivers.

Specimens of the lanceolate-leaved variety have been collected at Killarney and Stanthorpe in Queensland. This species appears to be always a small tree, of no economic value.

E. pulverulenta was introduced some years before 1819, when it was figured as *E. cordata* by Loddiges, who states that it only requires ordinary greenhouse protection in winter. It is, however, cultivated in the open air at Abbotsbury,[1] where it produces fruit freely, and is said to be hardy. It is also grown at Menabilly, where the best of five specimens is 32 ft. by 1 ft. 3 in.

E. pulverulenta has been much confused with *E. cordata*; and the trees recorded[2] at Braxted Park, Essex, which were said to have been 30 to 40 ft. high, and flowering freely in 1849-1851, were in all probability *E. cordata*.

E. pulverulenta did not succeed at Kinloch Hourn; and the tree[3] planted in 1856 at Pirnmill, Isle of Arran, must have died, as Dr. Landsborough does not mention it in his list of the Arran species in 1895.

Sir F. W. Moore tells us that a small plant at Kilmacurragh in Co. Wicklow appears to be perfectly hardy; and another at Mount Usher, about eight years old, was thriving and producing flowers in 1911. At Castlewellan,[1] Elwes collected a specimen from a tree 20 ft. high in 1908. (A. H.)

EUCALYPTUS GLOBULUS, BLUE GUM

Eucalyptus globulus, Labillardière, *Relation Voyage Recherche La Pérouse*, i. 153, *Atlas*, t. 13 (1799); J. D. Hooker, *Fl. Tasm.* i. 133 (1860); Bentham and Mueller, *Fl. Austral.* iii. 225 (1866); Mueller, *Eucalyptographia*, Dec. vi. with two plates (1880), and in *Gard. Chron.* xiv. 137, 213, 233 (1880); Bentley and Trimen, *Medic. Plants*, ii. t. 109 (1880); Hemsley, in *Gard. Chron.* ii. 784, *Supply. Illust.* (1887); Maiden, in *Austr. Assoc. Advance. Science, Hobart*, 1902, p. 372; Rodway, *Tasmanian Flora*, 58 (1903); Pinchot, *U.S. Dept. Agric. Forest Service, Circ.* No. 59 (1907).

Eucalyptus cordata, Miquel, in *Nederl. Kruidk. Arch.* iv. 140 (1859) (not Labillardière).

Eucalyptus diversifolia, Miquel, *loc. cit.*

Eucalyptus gigantea, Dehnhardt, *Cat. Pl. Hort. Camald.* 20 (1832) (not J. D. Hooker); Mueller, in *Nuovo Giorn. Bot. Ital.* xii. 47 (1880).

A tree, commonly attaining in Australia about 200 ft., rarely nearly 300 ft. in height. Bark smooth, greyish or bluish white, the outer layer peeling off in long ribbons. Young branchlets green, quadrangular, with four prominent ridges; older

---

[1] Both at Abbotsbury and Castlewellan, where both species occur, it was wrongly labelled *E. cordata*.
[2] *Gard. Chron.* xi. 469 (1892).   [3] Landsborough, in *Trans. Bot. Soc. Edin.* xx. 524 (1896).

branchlets terete, reddish brown.   Leaves (Plate 365, Fig. 10), on adult trees, alternate, lanceolate, about 6 to 9 in. long and 1 to $1\frac{1}{2}$ in. broad, but often greater or less than these dimensions ; falcate, thick and leathery, unequal and cuneate at the base, gradually tapering to an acuminate apex, which ends in a long curved slender point ; margin revolute, whitish, undulate ; both surfaces pale green and slightly shining ; lateral veins numerous, arising at an angle of about 45°, with the circumferential vein regularly distant about $\frac{1}{16}$ in. from the margin ; oil-dots scattered, unequal, mostly concealed ; petiole twisted, flattened above, $\frac{3}{4}$ to $1\frac{1}{2}$ in. long.   Leaves on young plants and suckers, opposite, decussate, with a short and not twisted petiole ; 2 to 5 in. long, 1 to $2\frac{1}{2}$ in. broad ; ovate, cordate at the base ; acute or rounded at the apex, which is tipped with a cuspidate point ; circumferential vein $\frac{1}{12}$ to $\frac{1}{6}$ in. distant from the white revolute crenate margin ; both surfaces more or less covered with a glaucous bloom, which is also present on the slender four-angled branchlets.

Flowers[1] axillary, solitary or rarely two or three in an umbel, borne on a short laterally compressed stalk ; flower-buds large and covered with a glaucous bloom ; calyx-tube pyramidal, nearly $\frac{1}{2}$ in. long, $\frac{3}{4}$ in. wide at the distal end, with four longitudinal ridges, the lateral pair of which are very prominent ; operculum hemispherical, exceeding the calyx-tube in breadth and length, very tuberculate on the surface, with a short triangular umbo ; stamens all fertile ; anthers ovate, with parallel cells ; filaments inflexed in the bud.   Fruit hemispheric or obpyramidate, glaucous, $\frac{3}{4}$ to 1 in. in diameter, with four longitudinal ridges, and an overhanging very broad rim, separated from the calyx by a furrow ; capsule not sunk, the valves being nearly level with the rim.

The seedling[2] has a stem, terete at the base, and quadrangular towards the tip, warty on the surface and covered with a glaucous whitish bloom.   Cotyledons deeply bifid, with obovate diverging lobes, about $\frac{1}{5}$ in. long and $\frac{2}{5}$ in. wide, three-nerved, the middle nerve ending in the sinus.   Primary leaves, opposite, decussate, sessile, the first and second pairs linear-lanceolate and acute, tapering at both ends ; third and fourth pairs lanceolate-oblong ; fifth to eighth pairs oblong.   Leaves on the subsequently developed branches, oblong, and subcordate at the base.

VARIETIES

According to Mueller,[3] most of the Victoria specimens have smaller flowers and fruits, the latter more convex than those in Tasmania.   According to Rodway,[4] the common form in eastern Victoria bears three-flowered umbels ; while in Tasmania the flowers are usually solitary.   In cultivated specimens, however, the number of flowers in the umbel appears to be an inconstant character, though solitary flowers are usually borne in England.   Rodway further states[4] that in Tasmania, where this

---

[1] This species frequently flowers when very young.   In *Gard. Chron.* xi. 180 (1892), a specimen three years old and 7 ft. high is said to have flowered in Thomson's nursery at Sparkhill, Birmingham.

[2] Lubbock, *Seedlings*, i. 530, fig. 339 (1892).   [3] *Fl. Austr.* iii. 225 (1866).

[4] In J. C. Penny, *Tasmanian Forestry*, 15 (905).

species and *E. viminalis* are mixed, a form is occasionally found, in which the flowers are in threes, with the operculum and fruit quite smooth, the latter only $\frac{1}{3}$ to $\frac{1}{2}$ in. in diameter, with the valves much protruding. This is possibly a hybrid.

DISTRIBUTION

The Blue Gum is a native of Tasmania, Victoria, and New South Wales. In Victoria, it occurs in valleys as well as on ridges and mountain slopes, chiefly in the humid southern and eastern districts from Cape Otway to Wilson's promontory, and extending northwards to the Hume and Tumut rivers in the southern part of New South Wales, where it is also met with between Braidwood and Araluen. It appears, however, to be very rare in New South Wales, as J. V. De Coque states[1] that he only knows of it in this colony in the Tumberumba district at 2500 to 3000 ft. elevation. There it is called *Eurabbie*, a purely local name, and is highly valued, and largely used for mining purposes, and for bridge-decking and girders. Except in the above district, it is little known in New South Wales, where it does not attain anything like the size and height that it does in Victoria or Tasmania; and recently fell into disrepute, owing to *E. amygdalina*,[2] which lacks strength and durability, having been mistaken for it. In Tasmania, it appears to be almost confined to the south-eastern part of the island near the coast, extending inland, according to Sir J. D. Hooker, from Hobart about forty miles. Rodway describes it as a tall erect tree, even in exposed situations, with few and acutely diverging branches. A tree of full growth will average 7 feet in diameter at the butt, 100 ft. to the lowest branch, and from 200 to 250 ft. in extreme height. In youth it grows rapidly; but when approaching maturity the growth is almost imperceptible; and Rodway supposes that the tree takes 300 to 400 years to attain its full dimensions. The Rev. J. E. Tenison-Woods says[3] that in Tasmania the forests of blue gum are limited in extent, and are confined to a few localities, from sea-level to about 1000 ft. altitude. It is only on steep slopes in the deep mountain valleys and gorges that the trees attain a great height. He states that experiments have shown that the trees make two rings of growth each year; and that Mr. Hill, who cut up thousands of trees at his saw-mills at Honey-wood, affirmed that he never found one over 75 years old; but this statement seems doubtful.

HISTORY

E. globulus was discovered in 1792 by Labillardière in Tasmania, and was described by him in 1799. It was introduced into continental conservatories early in the 19th century, where, on account of its different foliage in the young state it was known under several names.[4] It appears to have been first cultivated out of

---

[1] In *Journ. Proc. Roy. Soc. N.S. Wales*, xxviii. 212 (1894).

[2] *E. Maideni*, Mueller, another species also known as Blue Gum, has until recently been often confused with it.

[3] In *Journ. Roy. Soc. N.S. Wales*, xii. 17-28 (1879). Cf. also *Gard. Chron.* xiv. 179 and 187 (1880).

[4] *E. glauca*, De Candolle, *Prod.* iii. 221 (1828); *E. pulverulenta*, Link, *Enum. Pl. Berol.* ii. 31 (1822) (not Sims); *E. perfoliata*, Noisette, *ex* Steudel, *Nomencl.* 320 (1821); and possibly *E. perfoliata*, Desfontaines, *Cat. Hort. Paris*, 408 (1829).

doors in Europe at Camaldoli, near Naples, where the German gardener, Dehnhardt, employed at the Naples Botanic Garden, published it as *E. gigantea* in 1832. The tree, however, was only known as a curiosity, till Mueller sent large quantities of seed, gathered at the base of Mount Butler in 1853, to nearly all the botanic gardens of Europe. Probably from this seed, most of the earliest plants[1] were raised in Algeria, where the blue gum produced flowers for the first time in 1863. M. Prosper Ramel began in 1858 to extend the cultivation of this tree in the south of France, receiving large quantities of seed from Mueller.

The well-known plantations in the Campagna near Rome were commenced in 1879 by the Trappist monks of Tre Fontana, who received the first seed in that year from Mueller, through Dr. Goold, R.C. Archbishop of Melbourne. The blue gum was introduced in 1865 into Spain, where it became known as the fever tree, because it was believed in Spain, as in Italy, "to purify marshy regions that engender fever." The tree is also largely planted in the Nilgiri hills in India, where it is said to have been introduced as early as 1843. According to Mr. M'Clatchie, it was the first species of Eucalyptus planted in California, and the one that has been most successfully grown. The date of introduction there appears to have been 1856, when Mr. Walker planted fourteen species; but the cultivation of Eucalypti on a large scale in California is due to Mr. Nolan, a nurseryman of Oakland, who received a large supply of seed of several species from Australia in 1861. It is commonly planted in Chile, where in some parts of the country it has attained a great size, and is an important timber and shade tree.

With regard to the growth of the blue gum in foreign countries, Mueller states,[2] that a tree at Gaeta, planted in 1854, was 100 ft. high and 11 ft. in girth at the base in 1878. Dr. W. von Hamm of Vienna saw larger trees on Lake Maggiore, one[3] of which, supposed to be twenty-eight years old, was 120 ft. high in 1878. At Hyères, a tree raised from seed in 1857, had attained in 1875 a height of 67 feet with a girth of 7 ft. near the base. Farther south, the growth is still more rapid, a tree at Malaga having attained 65 feet in six years. The greatest rapidity of growth occurs at considerable altitudes on the mountains of the tropics, where the climate is temperate and equable, combined with continuous humidity of the atmosphere. At Arambi, near Ootacamund in the Nilgiris in India, the blue gum attains a height of 107 ft. in nineteen years, and yields 8696 cubic feet per acre, equal to 457 cubic feet annually per acre at 7426 feet above sea-level in lat. 11° N.[4] In California, according to M'Clatchie,[5] trees about thirty years old have attained 150 ft. in height and 3 to 6 ft. in diameter. He gives a plate of an avenue at Santa Barbara, where the trees, planted thirty years ago, range from 3 to 5 ft. in diameter. M'Clatchie also says there are many instances of blue gums attaining 50 to 75 ft. in from five to ten years; and adds that it is without doubt the fastest-growing tree in the world.[6]

---

[1] Planchon, according to Mueller, states that the first plants in Algeria were raised in 1854, from seed sent by the Jardin des Plantes, Paris, which may have come from Hobart direct.

[2] *Eucalyptographia*, dec. vi. (1880).   [3] We are unaware of the existence of any such trees now.

[4] Sir D. Brandis, quoted in *Kew Bull.* 1895, p. 3.   [5] *Eucalypts cultivated in the U.S.* p. 61, fig. 21 (1902).

[6] Ingham, in *Agric. Exp. Stn. Berkeley, Calif., Bull.* No. 196, p. 77 (1908), gives equally remarkable records; a plantation of 319 trees, set out 8 ft. apart on medium loam at Vacaville, averaged, at twenty-five years old, 125 ft. high and 14¼ in. in diameter at breast height.

The Eucalyptus was first planted in the Campagna and in other malarious districts on account of its supposed febrifuge action. At that time, it was not known that malaria is due to a parasite in the blood, which is carried to man by the bites of mosquitoes. The Eucalyptus plantations have indirectly, however, done good in two ways, by drying up the pools in which the mosquitoes breed, and by forming a screen, which hinders their flight.[1]

Hutchins[2] says that it is not worth planting in Cyprus, as it will not succeed in the drier parts, and is second rate for its timber in wetter soils. Its wood burns badly in an open fireplace. (A. H.)

REMARKABLE TREES

E. globulus is perhaps not absolutely hardy in any part of Britain or Ireland; but it may survive many years and attain large dimensions in favourable localities.

The largest tree of which we have any record, grew at Rozel Bay, Jersey, about 200 yards from the sea on rock covered with little soil, but in a warm and dry situation. It was planted in 1862, and was reported[3] by Mr. T. Sharman to have been 110 ft. high and 10 ft. 3 in. in girth in 1892. It produced flowers and fruit freely, from which seedlings were raised several times. It was killed[4] by the severe frost of 1894-1895.

The oldest tree was one at Tresco, Scilly Isles, which was planted about 1850. It never reached more than 40 ft. in height, as the top was blown off from time to time, but it was 9 ft. in girth in 1891, when it was blown down.[5] One of the best that I have seen in England is at Coombe Royal, Devonshire. Though only twelve years old in 1906, it was a tall straight tree, about 50 ft. by 2½ ft.[6]

At Menabilly, Mr. Bennett tells us of a tree, which was 75 ft. high by 3 ft. in girth in January 1911. There are two well-shaped trees in Sir Thomas Bazley's garden at Kilmorie, Torquay, which in October 1910 measured 58 ft. by 2 ft. 9 in. and 53 ft. by 4 ft. The latter is figured (Plate 359) from a photograph taken by Sir Thomas Bazley, who informed me that they were planted in 1897.

At Penmere, near Falmouth, the residence of Mr. Horton Bolitho, there is a remarkably fine tree[7] (Plate 360) which I measured in 1911, as 95 ft. high by 7 ft. 9 in. in girth, with the bole clear of branches for about 40 ft. Mrs. W. L. Fox informs me that this tree was planted between 1864 and 1867.

At Trevarno, Cornwall, a tree about eight years old was 25 ft. high in 1890, and bore flowers for the first time, in great profusion.[8] Trees flowered and produced

[1] Cf. Sir W. T. Thiselton-Dyer, in *Kew Bull.* 1903, pp. 1-10.
[2] *Report on Cyprus Forestry*, 65 (1909).   [3] *Gard. Chron.* xi. 468, and xii. 408 (1892).
[4] *Ibid.* xxxix. 281 (1906). In *Gard. Chron.* xxv. 145 (1899), it is stated that the genus Eucalyptus was practically exterminated from Jersey in the severe winter of 1894-1895.
[5] T. A. Dorrien-Smith, in *Gard. Chron.* x. 737, fig. 107 (1891).
[6] The statement in *Gard. Chron.* xxxiv. 292 (1903), that the blue gum flourishes at Powderham, and makes a growth of 60 ft. in ten years, is erroneous, as the only species cultivated there now is *E. coccifera* (cf. p. 1636). *E. globulus* was killed at Powderham in 1878-79, by 12° of frost, according to *Gard. Chron.* xii. 113 (1879).
[7] In *Gard. Chron.* xiii. 268 (1893), Mr. Howard Fox reported several trees at Penmere, 50 to 70 ft. high and 7 ft. in girth in 1893, which were flourishing in exposed positions. Mr. Fox says that at his own garden at Rosehill, which is sheltered, the trees grow too fast for their roots, and are blown down by a strong gale.
[8] *Gard. Chron.* viii. 138 (1890).

fruit in 1898 at Polgwin,[1] Cornwall, and at Huntly,[2] Bishops Teignton ; and in 1888 at Beaconhill House, Exmouth.[3]

In the Isle of Wight, according to the Rev. H. Ewbank,[4] the blue gum was very early planted, and had attained a large size at Ryde and elsewhere ; but all these trees were killed in the severe winter of 1881. The same applies, he says, to Bournemouth, where the trees thrive for a time, but are ultimately killed by a severe frost. At Ventnor, however, there were said[5] to be specimens 40 ft. high in 1890, which produced flowers and fruit regularly.

At Kew, seed was obtained[6] from Tasmania in 1888, which had been gathered from a few trees[7] growing in a sheltered gully near Tullochgorum, the only spot in Tasmania where this species occurs naturally in a cold climate. The seedlings which were raised did not prove as hardy as those from the ordinary form of the species ; and all died in 1889, though they were protected by canvas screens.

At Tan-y-bwlch, North Wales, several trees[8] had attained 50 feet in height at nine years from seed, and flowered freely, but were seriously injured in 1890-91. Seedlings were reported to have been raised from their seed, when they were only seven years old.[9] At Colwyn Bay, a tree was reported[10] to be 33 ft. high in 1894. Another, at Cefnamwich, Nevin, North Wales, sown in the spring of 1894 by Mr. Hugh G. Jones, was 59 ft. by 4 ft. 8 in. in December 1909.

At Belvoir Castle, this species[11] was killed when 12 ft. high on 8th February 1900, when the thermometer fell to 10° Fahr.

In Scotland, the tree only lives for a few years, even in the warmest places. The Rev. Dr. Landsborough, in his account[12] of the Eucalypti in Arran, says that the severe winter of 1880-1881 killed all the blue gums on the mainland of Scotland, including those at Stonefield, Loch Fyne, and all in Arran, except a tree at Lamlash, which, however, was blown down in 1892, when it was more than 40 ft. high. A tree,[13] however, at Tighnabruaich, Kyles of Bute, which was cut to near the ground in 1894-95, when the temperature fell to 19° Fahr., made fresh growth, and was 54 ft. by 3 ft. 9 in. in 1905. At Logan House, near Stranraer, a tree[14] planted against the south wall of the house, in 1884, had covered the whole wall in 1899 ; and produced flowers and fruit in 1894, from which numerous seedlings were raised. Two trees planted here in the woods in 1897 were reported to be thriving, although 17° of frost was registered in November 1898. At Castle Kennedy, I saw in 1906 a tree about

---

[1] *Gard. Chron.* xxiv. 322 (1898).

[2] *Ibid.* xxiii. 346 (1898). The large trees at Huntly, being too near the house, have been cut down. Mrs. Carpenter tells us that they flowered three times.   [3] *Ibid.* iv. 133 (1888).

[4] *Ibid.* xxv. 19 (1899). In *Gard. Chron.* xi. 212 (1892), *E. globulus* is said to have flowered abundantly in the shrubberies at Bournemouth when only 12 to 15 ft. high. These were all destroyed, when six to ten years old, by the severe winter of 1890-91.   [5] *Ibid.* viii. 694 (1890).

[6] Cf. *Kew Bull.* 1889, p. 61, and *Gard. Chron.* xii. 728 (1892).

[7] From a letter to Kew from the Botanic Garden, Hobart, which gives these particulars, it appears that the seed behaved differently in Tasmania, as seedlings planted at Tullochgorum grew into large trees, when native to the district ; whilst those raised from seed from the warmer part of the island all died.

[8] *Gard. Chron.* xi. 247 (1892).   [9] *Ibid.* vii. 170 (1890).   [10] *Ibid.* xvi. 74 (1894).

[11] W. H. Divers, *Spring Flowers of Belvoir Castle*, 83 (1909).

[12] *Trans. Bot. Soc. Edin.* xvii. 23 (1889), and xx. 523 (1896). An account is given in *ibid.* xvi. 162 (1886) of a tree at Colintraive on the Kyles of Bute, which was 47 ft. high in 1881, when it succumbed to the severe frost of that winter.

[13] *Ibid.* xxiii. 148 (1905).   [14] *Gard. Chron.* xxv. 138 (1899).

25 ft. by 2 ft., which had grown from the stump of one killed by frost in 1895-1896. At Kinloch Hourn, *E. globulus*, when planted out, speedily succumbs, as, according to Mr. Birkbeck, it is always killed by 15° of frost.

In Ireland, the tallest specimen is at Dinas Island, Muckross, Killarney, and measured in 1909 about 77 ft. by 6½ ft. The gardener informed me that it was severely injured by cold in 1879; but it was, when I saw it, a healthy tree in full flower in August. A finer tree, planted about 1870, grows on the lawn of the Marquess of Lansdowne's house at Derreen, Co. Kerry, and measured about 60 ft. by 7¼ ft. in August 1910. At Rossdohan, Mr. S. T. Heard reports a tree, 50 ft. by 5 ft. 8 in. in 1910.

At Garron Tower, on the coast of Antrim, near Larne, at 250 ft. elevation above the sea, a tree,[1] planted in 1857, was 60 ft. high and 12 ft. in girth at a foot from the ground in 1897; and 75 ft. high by 13 ft. at four feet from the ground in 1911. It branches at five feet up into two main stems; and produces flowers and fruit regularly, and from its seed numerous seedlings have been raised.

Sir Jocelyn Coghill, Bart., sent to Kew a branch from a tree grown at Glen Barrahane, Castle Townsend, Co. Cork, which, at fifteen years old, was about 40 ft. high in 1889. In Co. Wicklow Henry found a tree at Clonmannon about fifteen years old, which was 53 ft. by 4 ft. in 1905; and another at Dunran, 49 ft. by 4 ft. 4 in., and in flower in August 1904.

The largest tree that I have seen in Europe is in a sheltered ravine in the forest of Bussaco, Portugal, a little below the hotel. It measured in 1909 about 140 ft. by 12 ft.

Timber

In an account[2] of the forests of Tasmania, compiled by Mr. J. C. Penny, it is stated that the blue gum is the most valuable timber tree of Tasmania, having wood of great durability, hardness, and weight. It is said to be superior to anything produced in the Australian States for wharf and bridge construction. A specimen of this timber 146 ft. long, 18 in. wide, and 6 in. thick, sawn clear of heart and sap, was sent to the Exhibition of 1851 from Long Bay, Tasmania.

Mr. Harold J. Shepstone, in an article in the *Scientific American*,[3] says: The erection of the great national harbour at Dover has called attention to the wonderful properties of the Tasmanian blue gum. It is at once one of the strongest timbers in the world, as well as the densest and most durable. It is so heavy that it will sink like a piece of lead, whilst also practically immune from the attacks of the sea-worm (*Teredo navalis*). It has a specific gravity of 75 lbs. to the cubic foot, and being heavier than water, piles 100 ft. long and 18 to 20 in. square can be sunk in deep water, without weighting them at the bottom, which is necessary in the case of Oregon piles (Douglas fir) weighing only 48 lbs. to the cubic foot. Tests

[1] G. Porteous, in *Gardening*, 13th November 1897. In *Journ. Roy. Hort. Soc.* viii. 189 (1887), the tree was reported to have been about 50 ft. high and quite uninjured by the severe winter of 1879-1880.
[2] *Tasmanian Forestry*, 3 (Hobart, 1905).
[3] *Scientific American*, Jan. 21, 1905, quoted by Penny, *Tasmanian Forestry*, 36 (1905).

carefully made have shown that it will bear about double the weight of English oak, and will remain sound under water for a very long period. These particulars have been confirmed; but I think the statement as to the comparative strength of English oak requires some explanation.

An instance is quoted of an old ferry-boat built of blue gum in 1818, that has been lying a wreck on the banks of the Derwent in Tasmania for more than fifty years, between high- and low-water mark, the timber of which is still quite sound.

For mining purposes, Mr. Griffen, Inspector of Mines at Launceston, Tasmania, says that it should occupy the first place; whilst Mr. Dudley of Hobart, who has forty years' experience of its use for bending and general wheelwrights' purposes, says that it is unsurpassed for felloes, spokes, shafts, and body-work. He prefers the timber of young trees, but states that it requires careful seasoning in closed sheds for two years after sawing, as it is very liable to crack and split if exposed to sun and wind when freshly cut. A diagram, showing the results of experiments made in 1899 for Messrs. Pearson by D. Kirkaldy and Son of London, is given by Penny in *Tasmanian Forestry*.

Experiments carried out [1] at Berkeley University, California, on the wood of the Eucalypti grown in that state, show that the fastest-growing species produce the strongest timber. Thirty-year-old blue gum proved stronger than hickory. In California the wood of *E. globulus* is hard, strong, and tough. It has competed with Robinia for insulator pins, and is used locally for waggon axles, spokes, hubs, and felloes.

With regard to the timber of trees grown in Europe, it is difficult to give a fair opinion, as reports of users vary extremely. Judging from what I saw in Portugal, however, the timber of *E. globulus*, in common with that of several other species, is extremely subject to split, warp, and twist, and requires a great deal of experience to season and convert it without great waste. None of the species [2] seem at present likely to produce timber of any commercial value in Great Britain, and it is therefore unnecessary to go into details. (H. J. E.)

[1] Ingham, *Berkeley Cal. Agric. Exp. Stn. Bull.* No. 196, pp. 111-112 (1908).
[2] *E. Muelleri*, T. B. Moore, is possibly an exception.

Eucalyptus 1631

EUCALYPTUS PAUCIFLORA, Weeping Gum

Eucalyptus pauciflora,[1] Sieber, in Sprengel, *Syst. IV. Cur. Post.* 195 (1827); Mueller, *Eucalyptographia*, Dec. iii. (1889); Rodway, *Tasmanian Flora*, 55 (1903).

Eucalyptus coriacea, A. Cunningham, *ex* Schauer in Walpers, *Rep.* ii. 925 (1843); J. D. Hooker, *Fl. Tasm.* i. 136 (1860); Bentham and Mueller, *Fl. Austral.* iii. 201 (1866); Maiden, in *Rep. Austr. Assoc. Advance. Science, Hobart*, 1902, p. 353, and *Revision Genus Eucalyptus*, i. 133, plates 26-28 (1904).

Eucalyptus submultiplinervis, Miquel, in *Nederl. Kruidk. Arch.* iv. 138 (1859).

Eucalyptus phlebophylla, Mueller, *ex* Miquel, in *Nederl. Kruidk. Arch.* iv. 140 (1859).

A tree, attaining in Australia and Tasmania, 100 ft. in height and 12 ft. in girth. Bark peeling off, smooth and white. Young branchlets green or more or less covered with a whitish bloom. Leaves (Plate 365, Fig. 7) on adult trees, alternate, lanceolate, about 5 in. long and ¾ in. wide, usually falcate, unequal and cuneate at the base, gradually tapering to an acuminate apex, which is tipped with a long slender hook-like curved point; green and shining on both surfaces, thick and firm in texture; margin entire or undulate; main lateral nerves apparently longitudinal, arising near the base, and running for a considerable distance parallel to the midrib; petiole twisted, about ½ in. long.

Flowers in axillary umbels of five to fourteen; peduncles ¼ to ⅓ in. long, slender; flower-bud with pedicel about ⅔ in. long; calyx-tube obconic gradually passing into the terete or quadrangular pedicel; operculum hemispheric or shortly conic, and ending at the apex in a short point; stamens usually all perfect; anthers reniform, with short divergent cells confluent at the apex. Fruit, on short slender pedicels (less than $\frac{1}{12}$ in. long), pyriform, smooth, narrowed near the summit, ⅓ to $\frac{5}{12}$ in. long and wide, with a broad rim and a narrow orifice; capsule included, with the valves when open extending nearly to the orifice.[2]

This species occurs in Tasmania, Victoria, New South Wales, and South Australia. In Tasmania, where it is known as weeping gum, on account of its pendulous branches and branchlets, it is a small tree, usually much branched, and attaining in favourable situations only 60 or 70 ft. in length. Mueller says that it grows both on the ridges of the lowlands and on the highlands.

In Australia it is known either as white gum from the colour of its bark, or cabbage gum on account of the softness of its timber. In Victoria, where it occurs in the south, north-east, and east, it appears to be essentially an alpine species, ascending in the Gippsland Alps, where it forms forests, to 5000 ft.; yet it is able to maintain itself to some extent in localities but little elevated above sea-level. In South Australia it only occurs in patches close to the sea-coast in the south-eastern district.

In New South Wales, it grows usually in the undulating grassy country in the

[1] Sieber's name, being the earliest and accompanied by a clear description, must be adopted, though it is not very appropriate, this species producing copious flowers.

[2] The cotyledons are figured by Kerner, *Nat. Hist. Plant*, Eng. Trans. i. 621, fig. 148 (1898).

VI 2 Q

mountains and high table-lands, frequently forming timber-line, which is about 6500 ft. on Mount Kosciusko. In the Braidwood district, however, it occurs at all levels up to the highest point, 5000 feet, and attains 80 ft. in height and 5 ft. in diameter ; and Maiden states that all reports from this district agree in stating that the timber is soft, durable underground, but of no use above it. De Coque says[1] that the timber is inferior, being only used locally for fencing purposes ; and recommends that it should be avoided by architects for use in work of any description. Cattle browse on the foliage in seasons of drought.

Mueller says that "the chief interest of this species concentrates in its quality to cope with rather severe frosts ; indeed, together with *E. Gunnii*, it constitutes miniature forests up to 5500 feet in the mountains, growing close to glaciers, which on the shady sides of glens do not wholly melt in Victoria, wherever situated over 6000 ft. high,—though in the cooler latitudes of Tasmania, the limit of eternal snow descends 1000 ft. lower,—it being understood, only in the wide crevices of chasms of rock, or in other places where the sun cannot exercise any direct effect. Thus the bare crests may be free of snow in the height of summer even at nearly 7000 feet ; and we have therefore nowhere in summer an absolute permanent snow-line in the strict sense of the word."

I am unaware of the date of the first introduction of this species, which has been tried only in a few places. It has very pendulous branches, and is an elegant tree producing abundance of flowers. At Abbotsbury, it was killed in 1908 when 16 ft. high. At Colwyn Bay, N. Wales, Mr. A. O. Walker says[2] that a tree about fifteen years old, had its leaves slightly injured by 19° of frost in January 1894.

The Rev. Dr. Landsborough states[3] that in Arran the species has borne 21° of frost without injury, producing flowers every year, and equalling *E. coccifera* in hardiness. A tree at Craigard, Lamlash, sown in 1879, was 25 ft. high after being topped in 1895, when it produced seed, from which plants were raised in the Edinburgh Botanic Garden. It is one of the species that Mr. Osgood H. Mackenzie cultivates at Inverewe.

At Fota, in the south of Ireland, there is an old tree, about 50 ft. by 6 ft. 8 in., which was blown down in 1903, and then replaced in position. The stem is now partially decayed. At Rossdohan, Kerry, Mr. Heard has several trees, the largest about 31 ft. by 1 ft. 8 in. He says that all his species of Eucalyptus grow as well in peat as on gravel, being more easily blown down in the latter soil.

(A. H.)

[1] *Journ. Proc. Roy. Soc. N.S. Wales*, xxviii. 214 (1894).
[2] *Gard. Chron.* xvi. 74 (1894). According to *Journ. Roy. Hort. Soc.* viii. 120 (1887), and *Gard. Chron.* ii. 784 (1887), this tree was not injured in the severe winters of 1878-1879 and 1879-1880.
[3] *Trans. Bot. Soc. Edin.* xx. 519 (1896), and xxiii. 147 (1905).

EUCALYPTUS VIMINALIS, Manna Gum

Eucalyptus viminalis, Labillardière, *Pl. Nov. Holl.* ii. 12, t. 151 (1806); J. D. Hooker, *Fl. Tasm.*
i. 134 (1860); Bentham and Mueller, *Fl. Austral.* iii. 239 (1866); Mueller, *Eucalyptographia*,
Dec. x. (1884); Masters, in *Gard. Chron.* iv. 596, fig. 82 (1888); Deane and Maiden, in
Proc. Linn. Soc. N.S. Wales, xxvi. 137 (1901); Deane, in *Rep. Austral. Assoc. Advance.
Science, Hobart*, 1902, p. 378; Rodway, *Tasmanian Flora*, 57 (1903).

A tree, usually of moderate size, but occasionally attaining 200 ft. in height.
Bark variable, sometimes peeling off in ribbons, and smooth and white from the base
upwards, and sometimes persistent and scaly even to the upper branches. Young
branchlets very slender, with four projecting ridges, green and not glaucous; older
branchlets terete, reddish brown. Leaves (Plate 365, Fig. 8) on adult trees,
alternate, narrowly lanceolate, averaging 5 to 6 in. long, and $\frac{1}{2}$ to $\frac{3}{4}$ in. wide; but
often larger or smaller than these dimensions; falcate; unequal and cuneate at the
base; gradually tapering to a long acuminate apex, prolonged into a long slender
curved or twisted filament-like tip; margin whitish, undulate, revolute; equally
light green and not glaucous on both surfaces; oil-dots numerous, unequal, pellucid
on young leaves, not conspicuous on old leaves; lateral veins numerous, arising from
the midrib at an angle of 45°, the circumferential vein being regularly distant $\frac{1}{25}$ in.
from the margin; petiole slender, twisted, about $\frac{1}{2}$ in. long.

Flowers in axillary umbels, usually in threes, rarely six to eight; peduncle $\frac{1}{5}$ in.
long; pedicels about $\frac{1}{12}$ in. long; calyx-tube hemispherical, about $\frac{1}{6}$ in. long, crowned
by an operculum, equal in length, conical, and tipped with a short point; stamens all
fertile, inflexed in the bud; anthers ovate, with parallel distinct cells. Fruit turbinate,
about $\frac{1}{5}$ in. long, and $\frac{1}{5}$ in. in width at the distal end, on short pedicels; rim broad
and convex; capsule not sunk, with usually four valves, protruding when open
beyond the orifice.

This species is very variable; but typical Tasmanian specimens have three-
flowered umbels, and very small fruits. Forms occur in which the flowers are more
numerous, four to eight in the umbel, and with long pedicels; and such trees are said
to have rough scaly bark. Forms also occur with three-flowered umbels and large
fruits. The leaves are also sometimes very narrow, almost quite linear.

This species occurs in Tasmania, Victoria, South Australia, and New South
Wales. In Tasmania, it is very abundant throughout the island, where it is known
commonly as white gum, according to Rodway,[1] who states that it seldom exceeds
the dimensions of a small tree, with a much-branched and spreading habit. In
Australia, where it is widely distributed, it occasionally reaches a great height,
being perhaps in rare cases as tall as any other species. It is usually known as
manna gum, owing to its being the chief species which produces mellitose manna.
This exudes from the bark in minute drops, and is supposed to be due to the
punctures of species of cicada, though often no trace of insect attack can be found.

[1] In J. C. Penny, *Tasmanian Forestry*, 15, 18, 20 (1905).

This peculiar substance encrusts the bark like a coating of white sugar, and falls off in lumps.   It is not known to possess any medicinal value ; and is called "lerp" by the aborigines, who eat it as an article of food.

According to De Coque,[1] the timber possesses no durability, and is of no utility to architects ; but is used extensively throughout New South Wales for cheap rough fencing.   It is not mentioned amongst the species with useful timber enumerated by Mr. Penny in his account of the Tasmanian forests.

Manna gum appears to be rare in collections in the British Isles, the only large specimens being two trees at Rossdohan, Co. Kerry, the larger of which produced flowers and fruit in 1910.   Mr. Heard informs us that it is now, after losing 15 ft. of the top, 37 ft. high by 4 ft. 3 in. in girth.

It is reported[2] to have stood the severe winter of 1885-1886 at Bradfield, Collumpton, Devon ; but it was killed at Abbotsbury in 1905.   Mr. G. F. Heath reports that at Silverton, Devon, a young plant now 10 ft. high bore in 1909-1910 over 12° of frost.   At Menabilly, a tree of this species was 15 ft. high in 1910.

At Colwyn Bay,[3] North Wales, it was 33 ft. high in January 1894, when it was killed back to the trunk by a severe frost, the temperature registered being 13° Fahr. ; but it afterwards sent out shoots.

In Scotland this species did not succeed out of doors at Kinloch Hourn, as all the seedlings raised in 1894 and 1895 had perished in 1899.   In Arran, a tree[4] raised from seed in 1871, and planted at Cromla in 1874, flowered in 1886 and subsequent years, and had attained 40 ft. by 2 ft. 7 in. in 1895, when it was killed by severe frost, the temperature falling to 22° Fahr. on the 9th February.   A tree[5] at Roseneath, planted in 1876, suffered severely in 1880-1881, and was killed in 1894-1895.   In Ireland, this species has grown very fast at Mount Usher, where a tree raised from seed in 1904, was 28 ft. in height and bearing fruit in October 1911.

At Pallanza in northern Italy, it is reputed to be the hardiest species ; and I measured in 1909 a fine specimen in Rovelli's nursery as 70 ft. by 5 ft.   Elwes saw a very large tree of *E. viminalis* in April 1909, growing on the west lawn at Monserrat in Portugal, which measured 100 ft. by $16\frac{1}{2}$ ft.   In California,[6] this species nearly equals the blue gum in rapidity of growth, and has been extensively cut for fuel.

(A. H.)

[1] In *Journ. Proc. Roy. Soc. N.S. Wales*, xxviii. 214 (1894).

[2] *Gard. Chron.* xxvi. 754 (1886).   [3] *Ibid.* xvi. 74 (1894).

[4] Landsborough in *Trans. Bot. Soc. Edin.* xx. 521 (1896).   This is the tree called *E. amygdalina regnans* by the same author, *ibid.* xvii. 25 (1887).   It is referred to in *Gard. Chron.* iv. 596, fig. 82 (1888).

[5] Called *E. amygdalina regnans* in *Gard. Chron.* xxvi. 715 (1886), but Dr. Landsborough informed Mr. Birkbeck that it was *E. viminalis*.

[6] M'Clatchie, *U.S. Forestry Bull.* No. 35, p. 37 (1902).

EUCALYPTUS COCCIFERA, Mountain Peppermint

Eucalyptus coccifera,[1] J. D. Hooker, in *London Journ. Bot.* vi. 477 *bis* (1847), and *Fl. Tasm.* i. 133, t. 25 (1860); Lindley, in *Journ. Hort. Soc.* vi. 221 (1851); W. J. Hooker, *Bot. Mag.* t. 4637 (1852); Bentham and Mueller, *Fl. Austral.* iii. 204 (1866); Masters, in *Gard. Chron.* xiii. 395, fig. 69 (1880), ii. 784, fig. 151 (1887), and iii. 798, figs. 108, 109 (1888); Rodway, *Tasmanian Flora*, 56 (1903); Maiden, in *Rep. Aust. Assoc. Advance. Science, Hobart*, 1902 p. 365, and *Revis. Genus Eucalyptus*, i. 142, pl. 28 (1904).

A small tree, attaining in the high mountains of Tasmania 20 ft. in height. Bark smooth, white. Young branchlets terete, glabrous, glaucous, covered with a dense whitish bloom. Leaves (Plate 365, Fig. 9) on adult plants, alternate, lanceolate, about $2\frac{1}{2}$ to 3 in. long, and $\frac{2}{3}$ to $\frac{3}{4}$ in. wide, unequal and tapering at the base, acuminate at the apex, which is tipped with a long slender curved hook; margin entire, whitish, revolute; equally green or glaucous on both surfaces; oil-dots numerous, pellucid, very unequal in size; thick in texture; lateral veins, arising at an acute angle (30°) from the midrib, not conspicuous; petiole twisted, $\frac{3}{4}$ to $1\frac{1}{2}$ in. long. Leaves on young plants and on suckers, opposite, sessile, elliptical, about $1\frac{1}{2}$ in. long and 1 in. broad, rounded at the base and apex, the latter tipped with a short sharp point; entire in margin; green or glaucous. Branchlets reddish, with numerous elevated globose oil-glands.

Flowers, in axillary umbels of three to seven; peduncles glaucous, thicker towards the distal end, $\frac{1}{3}$ to $\frac{1}{2}$ in. long; flower-buds sessile, glaucous, wrinkled, $\frac{1}{2}$ in. long; calyx-tube turbinate, compressed on the back and front, with the two sides narrow and sharply winged; operculum short, nearly flat, depressed in the centre and with a warty margin; stamens all perfect, inflexed in the bud; anthers reniform with diverging cells. Fruit obconic, glaucous, about $\frac{1}{2}$ in. long, and $\frac{2}{5}$ in. broad at the slightly contracted distal end; nearly smooth externally, the angles of the calyx having disappeared; rim $\frac{1}{8}$ in. broad, flat or convex; valves slightly included, three to five.

The seedling[2] has a terete scabrous dark purple stem; cotyledons obcordate, cuneate at the base and retuse at the apex; leaves ovate-lanceolate, thin, entire.

Our cultivated trees of *E. coccifera* in England agree[3] well with the type

[1] This species is named, not from the actual presence of insects on the tree, but because the branchlets have a conspicuous whitish bloom, like the waxy secretion of some species of *Coccus*.

[2] Lubbock, *Seedlings*, i. 532 (1892).

[3] *E. coccifera* has been confused with the following forms in Tasmania, which have been studied by Maiden, who calls them "gum-top stringy barks," *i.e.* trees with smooth bark above and stringy bark near the base of the trunk; but very variable in the amount of each kind of bark on individual trees. I suspect that the following are natural hybrids; there is no reason to suppose that they are in cultivation in England.

    A. *Eucalyptus Risdoni*, Hooker, var. *elata*, Bentham, *Fl. Austral.* iii. 203 (1866); Maiden, in *Rep. Aust. Assoc. Adv. Sc. Hobart*, 1902, p. 361, and *Rev. Genus Eucalyptus*, i. 69 (1903) and i. 144 (1904); Rodway, *Tasmanian Flora*, 56 (1903).

This name was given by Bentham to two specimens, both numbered 1095, which were collected by Gunn in 1847 on the west side of Lake St. Clair, where tall trees formed pure forest close to the water's edge. One specimen has glaucous branchlets, with alternate and very falcate leaves, unequal at the base, and ending at the apex in a hooked point; flower-buds, six to seven in an umbel, like those of *E. obliqua*, but glaucous and with conspicuous oil-dots. The other specimen has branchlets and flowers, which are not glaucous. These specimens which have leaves and flowers like *E. obliqua*, and glaucous branchlets specimens of the species, which were collected in Tasmania on the summit of Mount Wellington by Gunn, and on the western mountains by Lawrence; but have larger leaves, and are usually more glaucous in all their parts.[1]

E. coccifera is confined to Tasmania, where it is common on the summits of the mountains at 3000 to 4000 ft. elevation. Rodway describes it as a small erect tree, 8 to 12 ft. high; but it attains much larger dimensions in this country, and doubtless owes its small size in Tasmania to the exposed situations in which it is found. It is known as mountain peppermint, and has no economic value in its own country.

E. coccifera was discovered in 1840 by Gunn, and was probably introduced in the same year, as in 1851 there was a plant, said to be eleven years old, growing in Veitch's nursery at Exeter, which was then 20 ft. high and producing flowers. In the Chiswick Garden, according to Lindley, it lived for many years against a south wall, without being injured by frost; but plants growing in open borders dwindled away and died.   (A. H.)

REMARKABLE TREES

The finest Eucalyptus in Britain is the noble tree[2] of this species, which grows in the American garden at Powderham Castle, Devonshire (Plate 361). When I last measured it in 1907, it was 75 to 80 ft. high and 13 ft. in girth at 5 feet; but the gardener, Mr. Bolton, informs us that in January 1911 it was exactly 16 ft. in girth at 4 ft. from the ground; the spread of the horizontal branches being about 90 ft. It is perfectly sound and healthy, and regularly bears ripe fruit, from which seedlings have been raised. One of these, when planted out under a warm wall at Colesborne was almost killed in the winter of 1908-1909; and died in 1912. This tree probably dates from the original introduction of the species in 1840, as it produced flowers and fruit in 1852, from which the plate in the *Botanical Magazine*, t. 4637, was drawn. It grows in sandy loam on the Red Sandstone formation on a fairly high river bank, about ten yards from the water, and doubtless owes its vigour to this situation. It was not injured in 1878 or 1879, when the temperature fell to 16° and 9° Fahr. respectively. Masters[3] states that this tree has changed its time of flowering, as it was reported to have produced flowers in December and January, 1880; while in 1883, it was in full flower in the month of June.   (H. J. E.)

---

and flower-buds like *E. Risdoni* and *E. coccifera* in all probability are hybrids. Rodway, *Notes on E. Risdoni* in *Proc. Roy. Soc. Tasm.* 367-369, plates 10-12 (1910), should be consulted on the varieties of this species.

B. *Eucalyptus radiata*, Sieber, var. 4, J. D. Hooker, *Fl. Tasman.* i. 137 (1860).

This name was given by Hooker to specimens 1100 and 1110, collected by Gunn in 1840 at Hobart and Grass Tree Hill. These two specimens are not identical, and are different from *E. Risdoni*, var. *elata*, from Lake St. Clair. They are close to *E. Risdoni*, and are probably hybrids of it with some other species.

Note.—*E. Risdoni*, Hooker, the "drooping gum," a small tree abundant in the dry hills of the southern parts of Tasmania, bears on the adult plant opposite sessile leaves; and seems to be entirely distinct from *E. Risdoni*, var. *elata*. *E. Risdoni* does not seem to have ever been tried in cultivation in England.

[1] Maiden, to whom I sent specimens of the Powderham and other trees, agrees with me that they are undoubtedly *E. coccifera*.

[2] Figured in *Gard. Chron.* xii. 113, fig. 18 (1879), and ii. 784, fig. 152 (1897). It was reported to be 58 ft. high and 7 ft. 4 in. in girth at 3½ ft. from the ground in 1879. Cf. also *Gard. Chron.* xxxiv. 291 (1903) and xxxix. 411 (1906).

[3] *Gard. Chron.* xix. 730 (1883), and ii. 784 (1887).

At Tresco, Scilly Isles, this is reported to be the hardiest species, but dwarf in habit, a specimen, 35 years old, being only 12 ft. high in 1894.

At Menabilly the best specimen,[1] planted in 1884, was, according to Mr. Bennett, 65 ft. by 6 ft. 5 in. at 4 feet from the ground in 1911. This was 56 ft. by 5 ft. 8 in., when measured by Mr. A. B. Jackson in 1908. An older tree planted in 1879, was only 40 ft. by 4 ft. 8 in. At Coombe Royal, a tree thirty-five years old, was 30 ft. by 2½ ft. in 1904. Trees raised at Vicar's Hill, Lymington, from seed sent to Mr. E. H. Pember in 1896 by Sir Charles Barrington from trees grown by him near Limerick, were 31 ft. by 13 in. in girth in 1910. At Killerton, a tree, planted in 1891, was 33 ft. by 3 ft. 1 in. in 1911. At Osborne, a bushy specimen, in an exposed position, survived the winter of 1884, when 20° of frost were registered; but was only 10 ft. high in 1911, having lost its leader. At Tregothnan, there are trees about 20 to 25 ft. high, which were planted ten years ago. At Cuffnells, Lyndhurst, there is a tree about 35 ft. high, which bears fruit. At Abbotsbury, there are specimens, 20 to 30 ft. high, raised from an older tree, which are very hardy and thriving. At Bradfield, Devon, it[2] bore well the severe winter of 1885.

There is a tree, planted in 1896, which attained 16 ft. high in 1906, in a sheltered spot at The Holt, Harrow Weald, Hampshire. This tree is much frequented by blue tits; but Mr. A. Kingsmill could not find any insect on it to attract them.

It is said not to be quite hardy at Kew; but Nicholson says[3] that young plants protected by a wall were not injured in the severe winter of 1879-1880. Seed was sent to Kew, which was gathered in 1888 from trees growing on the summit of Mount Wellington in Tasmania, where the branches were covered with icicles a foot long; but the seedlings raised succumbed in the winter of 1889, though they were protected by a canvas screen.[4] It stood out of doors three years at Coombe Wood,[5] but succumbed to the winter of 1909. *E. coccifera* is, in Mr. Birkbeck's opinion, one of the four hardy species; but we have not seen any trees of it in the eastern counties.

At Wimbledon, some plants[6] endured without injury several cold winters, and attained 12 to 15 ft. in height; but becoming broken by a fall of snow, they were killed by a subsequent severe frost in 1893.

Mr. W. H. Divers, reports[7] that at Belvoir Castle there are two trees, planted in 1899, which have proved quite hardy, and were 25 ft. high in 1909. This is the only one, of about eighteen species that were tried out-of-doors at Belvoir Castle, which has survived.

At Wansfell House, Windermere, a dozen trees[8] had withstood the severe weather of the preceding five winters, and were 15 to 20 ft. high in 1893. Of these, three survive, and measured in February 1911, 38 to 45 ft. in height, and 19 to 22 in. in girth.

---

[1] This tree appears to have been incorrectly known at Menabilly as *E. Gunnii*, till 1900, when specimens were sent to Kew; and is probably the tree figured under that name in *Gard. Chron.* xi. 787, fig. 113 (1892) and xxxiii. 234, 97 (1903).   [2] *Gard. Chron.* xxvi. 754 (1886).   [3] *Journ. Roy. Hort. Soc.* viii. 208 (1887).

[4] *Kew Bulletin,* 1889, p. 61, and 1892, p. 309; and *Gard. Chron.* xii. 728 (1892).

[5] It was stated in *Gard. Chron.* xxvi. 306 (1886) to be hardy at Coombe Wood. Mr. Harrow informs me that "none of the species prove in any degree hardy at Coombe Wood."   [6] *Gard. Chron.* xiii. 237 (1893).

[7] *Spring Flowers of Belvoir Castle,* 83 (1909).   [8] *Gard. Chron.* xiii. 237 (1893).

At Nant-y-Glyn Hall, Colwyn Bay, North Wales, there is a tree about 20 ft. high.   At Cefnamwich, Nevin, Mr. Hugh S. Jones informs us that he has a tree, 45 ft. high, which he raised from seed sixteen years ago.

On the west coast of Scotland this is one of the most successful species.   A tree[1] at Stonefield, Loch Fyne, the residence of C. G. P. Campbell, Esq., was reported by the forester, Mr. R. Stewart, to be 47 ft. by 2 ft. 8 in. at 4 feet from the ground in 1910, and had never been injured by frost.   At Castle Kennedy, there is a healthy young tree, which was raised from seed of a tree erroneously named *E. amygdalina*, which died some years ago.   At Monreith, a young tree received from Mr. Birkbeck in 1899, is about 17 ft. high, and produces flowers and fruit regularly. At Dalkeith, this species,[2] growing in a sheltered spot, bore without injury the severe frost of 7th January 1894, when the thermometer fell to 4° Fahr.

At Kinloch Hourn,[3] this species is very hardy, and there are several fine specimens, none of which were touched by the severe frost in 1893-1894.   It was killed, however, in 1895 at Kilmarnock and at Whittingehame, where the tree was ten years old.   At Roseneath,[3] a tree planted in 1886, flowered in July 1891, when only 6½ ft. high, and was 19½ ft. high in 1905.   In Arran,[3] where it was planted in 1886 in a very exposed site at 250 feet above sea-level, it was 14 ft. in 1895, but succumbed in the following winter.   At Gadgirth,[3] Ayrshire, it also died, when about 5 ft. high.   At Inverewe, this species is reported to be very successful.

In Ireland, this species has not been tried extensively; but it is perfectly hardy and thriving at Kilmacurragh and Mount Usher in Wicklow, and at Rossdohan in Kerry.   At Coolfin, near Portlaw, Co. Waterford, the residence of Rev. W. W. Flemyng,[4] there is a fine specimen, which was planted quite small in 1898, and measured no less than 35 ft. high in 1907.   (A. H.)

EUCALYPTUS GUNNII, CIDER GUM

Eucalyptus Gunnii, J. D. Hooker, in *Lond. Journ. Bot.* iii. 499 (1844), *Fl. Tasm.* i. 134, t. 27, (1860), and *Bot. Mag.* t. 7808 (1901); Bentham and Mueller, *Fl. Austral.* iii. 246 (1866); Mueller, *Eucalyptographia*, Dec. iv. (1879); Hemsley, in *Gard. Chron.* ii. 784, fig. 150 (1887); Deane and Maiden, in *Proc. Linn. Soc. N.S. Wales*, xxvi. 134 (1901); Maiden, in *Proc. Linn. Soc. N.S. Wales*, xxvi. 561 (1901), and in *Rep. Austr. Assoc. Advance. Sci., Hobart*, 1902, p. 377; Rodway, *Tasmanian Flora*, 57 (1903).

Eucalyptus Gunnii, var. *montana*,[5] J. D. Hooker, *Bot. Mag.* t. 7808 (1901).

Eucalyptus Gunnii, var. *glauca*,[6] Deane and Maiden, in *Proc. Linn. Soc. N.S. Wales*, xxiv. 464 (1899), and xxvi. 134 and 561 (1901).

---

[1] Dr. Landsborough, in *Trans. Bot. Soc. Edin.* xx. 518 (1896) and xxiii. 145 (1905), states that it was sown in 1881; flowered in 1895, when it was 21 ft. high; and was 27 ft. by 2 ft. 4 in. in 1905.

[2] *Journ. Roy. Hort. Soc.* xviii. 76 (1895).

[3] Cf. Landsborough, in *Trans. Bot. Soc. Edin.* xx. 518 (1896), and xxxiii. 145 (1905).

[4] Cf. *The Garden*, lxxi. 591, fig. (1907).

[5] As Maiden points out, in *Proc. Linn. Soc. N.S. Wales*, xxvi. 588 (1901), this differs in no respect from the typical form of the species.

[6] Maiden, in *Rep. Austr. Assoc. Advance. Sc., Hobart*, 1902, p. 377, suppresses var. *glauca*, which is only the typical form of the species.

A small tree, usually 20 to 30 ft. high in Tasmania, but occasionally attaining in sheltered places a height of 80 ft. Bark smooth and whitish, the outer layers peeling off in thin strips. Young branchlets slender, terete, more or less covered with a glaucous bloom. Leaves (Plate 365, Fig. 5) on adult trees, alternate, averaging 2 to $2\frac{1}{2}$ in. long, and $\frac{5}{8}$ to $\frac{7}{8}$ in. broad, ovate-lanceolate, thick and firm in texture, rounded or cuneate and equal-sided at the base, usually straight and not falcate, gradually tapering to an acute apex, tipped with a short point; margin entire, revolute, whitish; equally greyish or glaucous green on both surfaces; oil-dots numerous, unequal, often concealed; lateral veins few, inconspicuous, arising at an angle of 70°; circumferential vein undulate, distant about $\frac{1}{25}$ in. from the edge of the blade; petiole stout, twisted, about $\frac{1}{3}$ in. long.

Young plants[1] have glaucous branchlets and foliage; leaves sessile, opposite in decussate pairs, oval or ovate, cordate at the base; rounded or acute at the apex, which is tipped with a short point; crenate in margin. As the plants grow older, and on suckers, the leaves become thicker in texture, and more ovate in outline.

Flowers in axillary umbels of threes; peduncle slender, $\frac{1}{4}$ in. long; pedicels distinct, about $\frac{1}{20}$ in. long; flower-buds glaucous; calyx-tube campanulate, $\frac{1}{8}$ in. long; operculum nearly hemispheric, much shorter than the calyx-tube, tipped with a short umbonate point; stamens all perfect, inflexed in the bud; anthers ovate, with parallel distinct cells. Fruit glaucous, pear-shaped, distinctly pedicellate, $\frac{1}{3}$ in. long, $\frac{1}{5}$ in. wide at the distal end, where it is contracted; rim narrow; capsule sunk, with three to four valves, which when open scarcely reach the orifice.

This species has scarcely any of the pungent odour, so common in many other species; and Mueller says that it is liable on this account to have the foliage browsed by cattle and sheep.

The typical form of *E. Gunnii* was discovered by Sir J. D. Hooker in 1840, "forming a forest of small trees in a swampy soil[2] at elevations of 3000 to 4000 ft. in the centre of Tasmania, where it is known to stock-keepers as yielding abundantly an agreeable sap; and hence is called cider tree or swamp gum." Rodway, who calls it cider gum, states that it is found in the midlands and lake country of Tasmania as a small tree seldom exceeding 20 ft.; but rarely in sheltered places attaining 80 ft. high.

It was subsequently found by Mueller on the summit of Mount Baw Baw in Victoria; and the same tree, according to Maiden, also grows in the Tingeringi and Snowy mountains of New South Wales at about 5000 ft. elevation.

This tree is considered by the colonists in Tasmania to be different from *E. acervula*, which is known as red gum, and only occurs at low levels. Both Sir J. D. Hooker and Rodway consider these two species to be quite distinct; and I see no reason for uniting them together under the same name (*E. Gunnii*), as has been done by Maiden and some other botanists. The transitional forms noticed by Maiden, of which I have seen no specimens, are more likely to be hybrids than varieties.

[1] Figured in *Gard. Chron.* xix. 437, fig. 65 (1883).

[2] Rodway, in a letter to Elwes, says *E. Gunnii* is almost entirely confined to exposed moorland at an altitude above 2000 ft., and will not live at low elevations.

This species was introduced, shortly after its discovery, into Kew Gardens, where a tree[1] lived for many years, being the first Australian tree that was cultivated in the open air in England. It was cut almost to the ground by cold in several severe winters; but as often, sent up from the base one or more stems, which grew to be 10 to 20 ft. in height. It died only three or four years ago.   (A. H.)

REMARKABLE TREES

The most remarkable plantation of Eucalyptus that we know of in England, was made by the late John Bateman, Esq., of Brightlingsea Hall, Essex, who raised seeds of *E. Gunnii*, which[2] were sent to him from southern Argentina in 1887, by Mr. Shennan, who had naturalised there the Tasmanian *E. Gunnii*. Planted in an exposed situation in light sandy soil within a mile of the estuary of the Colne, some of the trees (Plate 362) have now attained a height of 40 to 50 ft. with a girth of 3 or 4 ft., and are quite uninjured by wind or frost, the lowest temperature recorded for the locality being 5° Fahr. When I saw them on 9th December 1906, some were in full flower, and many were covered with capsules of different ages. Mr. Bateman told me that they ripened seed every year, which he sowed in the open ground in April. The seedlings attain a height of about a foot in the first year, and 3 to 6 ft. in the second year. A self-sown seedling growing behind the coach-house was about 12 ft. high at three years old. Mr. Bateman found the tree somewhat difficult to transplant, and preferred to do this in September after a heavy rain. This species also seems to like a wet soil; and a great number of seeds had germinated in the gutter of the coach-house. Though the value of the timber[3] is as yet unproved in England, this tree might be used for planting marsh land on the coast, where, owing to its evergreen character, it would afford excellent shelter to other trees. Seedlings from these trees, though showing no variation in their botanical characters, have varied considerably in their hardiness at Colesborne. Some were killed to the ground by a frost in December 1904, and grew again from the root; whilst others were little affected.

Mr. Bateman informed me that this species grows equally well on sandy upland, on loam, or on heavy marshy clay soil, if trenched. The trees do badly, if they have not moisture at the roots, at least till they are well established; or if they are planted in sheltered corners under the shade of other trees; or if planted close to a wall; or if nibbled by rabbits, hares, or stock. They can be safely planted up to the 1st May. At Brightlingsea, bees swarm on the flowers which are produced from May to January; but those opening after 15th September set no fruit.

[1] Cf. Hemsley, in *Gard. Chron.* ii. 784, fig. 150 (1887). Smith, *Records of Kew Gardens*, 265 (1880), states that it was 20 ft. high in 1863. Smith calls it *E. polyanthemos*, the erroneous name by which this specimen was known for many years at Kew.

[2] Cf. Bateman in *Gard. Chron.* xxv. 202 (1899). In an article, which he wrote in *The Garden*, lxi. 110 (1902), he states however that he received the seed from the extreme south of Tasmania in 1887, and planted out sixty seedlings in the autumn of the same year. These, when five years old, were 15 ft. high and produced flowers and seed.

[3] The timber, according to Mr. Bateman, who used it for posts and rails, is very heavy. The wood of *E. Gunnii* is of no value in Tasmania, owing to its small size; and that of the allied species, *E. acervula*, seems to be little used except for firewood.

Mr. Llewellyn Lloyd tells us that numerous seedlings of the Brightlingsea trees have been planted in Essex,—as in his own garden at Kirby, and in that of Mr. J. B. Hawkins at Wyvenhoe. A tree at the Rectory, Frating, planted twenty-eight years ago in light sandy soil, is over 30 ft. high and about 2 ft. in girth. There are also some trees about eighteen years old and 30 ft. high at Birch Hall, Colchester, from which the Right Hon. James Round has lately raised seedlings. Messrs. Abbott, nurserymen at Ardleigh, near Colchester, are said to have a stock for sale.

At Menabilly, there was a tree of this species, from which the figure in *Bot. Mag.* t. 7808, was drawn [1] in 1901. It is doubtful if this is the same as a tree which was 38 ft. by 1 ft. 4 in. in January 1911.

At Borde Hill, Sussex, there is a tree in a sheltered corner close to the house, which flowered in July 1909, and bore fruit in the previous year. It measured 56 ft. by 3 ft. 9 in. ; and Mr. Stephenson Clarke says that it has endured 22° of frost without injury. At Hemsted Park, Kent, a healthy tree was 33 ft. high in 1911. In Miss Breton's garden at Sandhurst, a tree eighteen years planted, was 48 ft. high by 2 ft. 5 in. in girth in 1911 ; and a seedling from this is 20 ft. high. At Colesborne I planted a tree which was raised by the late Sir Charles Strickland in 1903, under a high wall facing north, and here it grew to 15 feet high in four years. In the severe winter of 1908-1909, when the thermometer went below zero, the top of this tree was killed as far down as the sun struck it, but the remainder continued healthy, and flowered in August 1910. After its top was frozen it threw out many shoots at ground level. The buds on this tree are formed about August and remain unopened for about twelve months.

At Putley Court, Ledbury, there is a handsome tree, planted about 1887, which in 1910 measured 48 ft. by 2½ ft., and produced flowers and fruit.

Mr. J. P. Rogers sent us in 1910 a branch from a tree at Penrose, near Helston in Cornwall, which is about 45 ft. high, and 2 ft. 11 in. in girth at three feet from the ground. This was raised from seed about sixteen years ago ; and about thirty seedlings from the same batch were planted out in the woods ; but Mr. Rogers says that only two survived, and only one of these grew properly. This species is not rare in Cornwall, as it is growing also at Enys and Penjerrick.

This species is apparently quite hardy in the eastern counties, even at a considerable distance inland, where severe frost is not unusual. A tree in the Cambridge Botanic Garden, obtained from Dicksons of Chester in 1898, is very thriving and about 25 ft. high. Another, growing in a sheltered position in an old quarry at Furze Hill, North Walsham, which was planted, according to Mrs. Petre, in 1899, is now about 40 ft. by 2 ft. 9 in., and is thriving, though occasionally a few twigs are cut by severe frosts.

Mr. A. R. Wallace has written [2] an article on this species, which he cultivated in his garden at Parkstone, Dorset. Planted in 1889, it was 30 ft. high by 3½ ft. at a foot from the ground in 1902 ; but was cut down soon after.

---

[1] According to notes on a specimen in the Kew Herbarium, the tree mentioned in the text of the *Bot. Mag.* t. 7808, never existed. The tree, figured as *E. Gunnii* at Menabilly, in *Gard. Chron.* xi. 787, fig. 113 (1897), and xxxiii. 234, fig. 97 (1903), was not this species, but apparently *E. coccifera.* See p. 1637, note 1.

[2] *The Garden,* lxi. 57, and lxii. 47 (1902).

At Wisley, where several species, including *E. urnigera, coccifera, cordata*, and *Beauchampiana*, have lately been tried, Mr. Chittenden informs us that the only one which has survived is *E. Gunnii*, a small specimen of which is four years old. At Myddelton House, Herts, where the same species and some others were tried, *E. Gunnii* and *E. Whittingehamensis* were the only ones that bore without injury the winter of 1908-1909. The former was about 15 feet high in 1911.

In Scotland, this species has been considerably planted on the west coast, and thrives as far north as Inverewe, where there are several trees, 30 to 40 ft. high. At Kinloch Hourn, trees raised from Tasmanian seed are perfectly hardy, and did not suffer in the least in the severe winter of 1894-1895, when the temperature fell to 0° Fahr., and some of the branches of the Whittingehame seedlings were browned. In Arran, according to the Rev. Dr. Landsborough,[1] the species did not succeed very well, as one planted at Lamlash in 1884 was blown down in 1894, and another in an exposed situation at Whiting Bay had the twigs and leaves injured by frost. At Whitefarland, a tree was 19 ft. high in 1905.

Mr. R. Lindsay says[2] that in 1899 he raised at Kaimes Lodge, Midlothian, two trees from seed which he received from Mr. Bateman. These are now over 20 ft. high and have produced seed, from which he raised seedlings. In 1909-1910, these seedlings were completely killed, and the old trees were severely injured, but the latter have now recovered. At Dalkeith, it is reported[3] to have borne without injury, growing in a sheltered place, the severe frost of 7th January 1894, when the temperature fell to 4° Fahr.

At Kilmacurragh, Co. Wicklow, this species is about 15 ft. high and perfectly hardy.   (H. J. E.)

EUCALYPTUS WHITTINGEHAMENSIS, WHITTINGEHAME GUM

Eucalyptus whittingehamensis, Nicholson, in *Kew Hand-List Trees*, 395 (1902).
Eucalyptus whittingehameii, Landsborough, in *Trans. Bot. Soc. Edin.* xx. 516 (1896).
Eucalyptus urnigera,[4] Masters, in *Gard. Chron.* iii. 460, figs. 64, 65, and 798, fig. 110 (1888) (not
   J. D. Hooker).

A tree, raised at Whittingehame, Scotland, from seed, which was probably obtained in Tasmania. This differs usually from typical *E. Gunnii* in the absence of glaucous bloom on the branchlets and leaves of the adult plant. The leaves (Plate 365, Fig. 4) are lanceolate, longer in proportion to their width than those of *E. Gunnii*, and never ovate as in that species, about 3 to 3¾ in. long and ⅝

---

[1] In *Trans. Bot. Soc. Edin.* xx. 517 (1896), and xxiii. 144 (1905). The tree, reported by Dr. Landsborough to be *E. Gunnii* at Stonefield, is *E. urnigera*.

[2] *The Garden*, lxxiv. 286 (1910).   [3] *Journ. Roy. Hort. Soc.* xviii. 76 (1895).

[4] The identification of the Whittingehame tree with *E. urnigera* is inexplicable, as the foliage and flowers of the latter are very different, and its fruits are much larger. Naudin, in *Desc. et Empl. Eucalpt.* 35 (1891), followed Masters; but he had not seen adequate material of the Whittingehame tree. According to Mr. Birkbeck, the cotyledons of the Whittingehame seedlings are quite different from those of *E. urnigera*; and he adds that no one who has seen both in the young state could possibly confuse them.

to $\frac{3}{4}$ in. wide, straight or occasionally slightly falcate, cuneate at the base, and gradually tapering to an acuminate apex, tipped with a short point; greyish green on both surfaces; thick and firm, with most of the oil-glands concealed; entire in margin; venation as in *E. Gunnii*; petiole twisted, $\frac{1}{4}$ to $\frac{1}{2}$ in. long.

Flowers in umbels of threes, slightly glaucous, and with a more conical calyx-tube than in *E. Gunnii*. Fruit slightly larger than that of *E. Gunnii*, urn-shaped, with the rim overhanging the contracted part of the fruit just below it; capsule sunk, with three valves not extending when open to the orifice.

This tree is considered by Maiden,[1] who has seen branches both of the parent tree and of its seedlings, to be *E. Gunnii*; but it does not match any of the specimens in the Kew Herbarium, either of that species or of *E. acervula*, which is united with *E. Gunnii* by Maiden. It resembles *E. acervula* in the absence of glaucous bloom on the branchlets and in the size and shape of the leaves; but has only three flowers in each umbel and bears different fruit. It is certainly not *E. urnigera*. Seedlings of the parent tree show considerable variation in the foliage, which in some specimens, together with the branchlets and flowers, is glaucous; and this points to a hybrid origin. Moreover, some of the seedlings at least are considerably hardier than the parent. I have not been able to make a study of the seedling trees in a fruiting stage; but I suspect that the Whittingehame tree is a hybrid, with *E. Gunnii* as one of the parents. The peculiar urn-like shape of the fruit suggests the probability of *E. urnigera* being the other parent, though the tree at Whittingehame does not resemble the latter in foliage.   (A. H.)

This remarkable tree is growing at Whittingehame,[2] the seat of the Right Hon. Arthur Balfour in East Lothian, where it was planted about sixty years ago, and is believed to have been raised from seeds brought by the late Marquess of Salisbury from Australia or Tasmania. Lady Gwendolen Cecil informs me that her father made his voyage round the world in 1851-1852[3] and visited both Tasmania and New Zealand, as well as Australia, in which country she believes that his stay was very brief; so that there is little doubt that the tree is of Tasmanian origin. In February 1904, when I measured it, it was 60 to 63 ft. high and 13 ft. 5 in. in girth at two feet from the ground, where it divides into three stems, which afterwards divide into six main limbs of which the largest is over 5 feet in girth (Plate 363). I was informed by the late Mr. John Garrett, the gardener, that after the severe frost of 1861 it was killed down to 9 feet from the ground; but in 1894 it endured a temperature[4] of −2° without serious injury. It ripens seed almost every year about September, and the seeds germinate and grow equally well if sown in autumn or spring. The bark of the old tree is more or less scaly and can be heard cracking in hot weather, and it remains green all the winter.

---

[1] In a note on a specimen at Kew, Maiden wrote in 1901: "*E. Gunnii*, varying a little from the type under cultivation." I sent in 1908 a number of specimens to Maiden, some of which he identified as *E. Gunnii*, whilst others he named *E. Whittingehamensis*.

[2] Whittingehame is three and a half miles from the sea, and 384 ft. altitude. The subsoil is gravel.

[3] The Rev. Dr. Landsborough states that the seed was sown in 1845, but this is erroneous; 1852 is the probable date.

[4] Dr Landsborough, in *Trans. Bot. Soc. Edin.* xxiii. 144 (1905) quotes a letter from Garrett, stating that "in the year 1894-1895, when on two nights the mercury sank at Whittingehame to zero, the young plants did not lose a leaf, while all those of the parent tree were destroyed."

Mr. Garrett informed me that of the numerous other species of Eucalyptus tried at Whittingehame, all perished in the severe winter of 1894-1895, with the exception of *E. vernicosa*, which was only damaged to a slight extent. None of the seedlings of *E. Whittingehamensis* were in the least injured. Of these he raised hundreds, which were distributed over the United Kingdom, the earliest dating from 1887, one of which at Whittingehame was 45 ft. high in 1903, while another raised from seed in 1886 was 29 ft. high and 14½ in. in girth in 1904. The true *E. Gunnii*, which he considered to be the species most like it, differed in producing seed freely, when only four years old, at Whittingehame. There is a thriving seedling raised in 1888 from seed of the Whittingehame tree, in the rose garden near the pagoda at Kew, which was planted in 1896, and is now about 40 ft. high. It has never suffered in the least from frost; and flowered for the first time in 1911. Two seedlings which were kindly sent me by Miss Balfour in 1905 were uninjured by the severe frost of that autumn, though one of them has died since, and the survivor does not seem to like the calcareous soil of Colesborne.

At Blackmoor, Liss, Hants, there are two trees, which were received from Whittingehame as one year seedlings in 1903. One is about 22 ft. high. The other, which was pollarded a year ago, is about 15 ft. high, and was bearing flower-buds in January 1911. At Leonardslee, Sussex, there is a tree which was 40 ft. by 3 ft. in January 1911, and was also bearing flower-buds. Sir E. G. Loder[1] received it from the Edinburgh Botanic Garden in 1889. At Terling Place, Essex, three seedlings were planted in 1895-1896, of which two were killed by a severe frost in 1909. The third was killed to within five feet of the ground, but is producing new shoots. At Hatfield House, Herts, four seedlings were planted in 1890, the best of which, growing in an open and exposed position, is now 40 ft. by 1 ft. 8 in., but has not yet flowered. A smaller one sheltered from the north was bearing flower-buds in January 1911. At Sandhurst, a Whittingehame seedling in Miss Breton's garden, planted thirteen years, was 35 ft. by 1 ft. 11 in. in 1911. At Abbotsbury, one four years old was 20 ft. high in 1911.

At Kinloch Hourn, Mr. Birkbeck planted the first seedlings of this in 1890 and 1891, and these had their leaves slightly browned in the severe winter of 1894-1895, when the temperature fell to 0° Fahr.; but have thriven since. In Kilmarnock[2] the Whittingehame seedlings were uninjured in the same winter, when *E. coccifera* was killed. A seedling at Monreith, planted when 2 ft. high in 1908, is now about 13 ft. high. Another at Inverewe, planted in 1896, was 45 ft. high by 2 ft. in girth in 1911.

(H. J. E.)

[1] Sir E. G. Loder says that his tree is much hardier and not the same as *E. urnigera*, several plants of which were killed at Leonardslee, although 20 ft. high, by a severe winter.
[2] Dr. Landsborough in a letter to Mr. Birkbeck.

EUCALYPTUS ACERVULA, Swamp Gum

Eucalyptus acervula, Miquel, in *Nederl. Kruidk. Arch.* iv. 137 (1859) (not Sieber[1]); J. D. Hooker,
    Fl. Tasm. i. 135 (1860); Rodway, *Tasm. Flora*, 57 (1903).

Eucalyptus Stuartiana, Mueller,[2] *ex* Miquel, in *Nederl. Kruidk. Arch.* iv. 131 (1859); Mueller, in
    Bentham, *Fl. Austral.* iii. 243 (1866) (in part).

Eucalyptus persicifolia,[3] Miquel, in *Nederl. Kruidk. Arch.* iv. 137 (1859).

Eucalyptus Gunnii, J. D. Hooker, var. *acervula*, Deane and Maiden, in *Proc. Linn. Soc. N.S. Wales*,
    xxvi. 136 (1901).

Eucalyptus Gunnii, var. *elata*, J. D. Hooker, *Bot. Mag.* t. 7808 (1901).

A tree,[4] attaining in Tasmania and Australia a height of 200 ft. in very favourable situations, but usually not more than 100 ft. Bark peeling off and smooth, except on old trunks, which are scaly near the base. This species differs from *E. Gunnii* in its larger size, and in the branchlets, leaves, flowers and fruit being never glaucous. Leaves larger than in that species, up to 4 in. long and $1\frac{1}{4}$ in. broad, lanceolate or ovate-lanceolate, dull green on both surfaces, often undulate in margin, usually straight, and not falcate, commonly equal but sometimes oblique at the cuneate base, tapering to the acuminate or cuspidate apex, which ends in a short point; thick and firm in texture; venation and oil-dots inconspicuous; petiole twisted, $\frac{1}{2}$ to $\frac{3}{4}$ in. long.

Flowers, axillary or lateral, 4 to 8 in the umbel; peduncle $\frac{1}{4}$ to $\frac{2}{5}$ in.; pedicels usually short and thick; calyx-tube conical, shining, $\frac{1}{8}$ in. in diameter at the widest part; operculum shorter than or as long as the calyx-tube, conical, ending in a long point; stamens all perfect, inflexed in the bud; anthers ovate with parallel distinct cells. Fruit obconic, about $\frac{1}{4}$ in. long and $\frac{1}{4}$ in. broad at the distal end, with a narrow rim and a wide orifice; capsule scarcely sunk, with four valves protruding when open.

This is one of the most widely diffused species,[4] occurring abundantly at low elevations in Tasmania, Victoria, and New South Wales; and by no means rare in South Australia and Queensland. It is known in Tasmania as red gum, in Victoria as swamp gum, and in New South Wales as white gum, swamp gum, or hickory. De Coque referring to it, under the name *E. Gunnii*, says[5] that the swamp gum in New South Wales produces worthless timber, which is soft and spongy, open in the grain, and retentive of moisture. Totally unfit for any work whatever, it should never be used in any circumstances.

[1] *E. acervula*, Sieber, in De Candolle, *Prod.* iii. 217 (1828), is identified with *E. piperita*, Smith, by Bentham, *Fl. Austr.* iii. 207 (1866).

[2] *E. Stuartiana*, Mueller, was founded on a Tasmanian specimen collected by Stuart, which is *E. acervula*, Miquel. The name *E. Stuartiana*, Mueller, is now applied by Maiden to another species, the "Apple Eucalyptus" of Victoria and New South Wales. Cf. p. 1615.

[3] *E. persicifolia*, Loddiges, *Bot. Cab.* vi. t. 501 (1821) is, I think, correctly referred to *E. viminalis*, Labill., by Bentham and Mueller, *Fl. Austral.* iii. 240 (1866); but Maiden, in *Proc. Linn. Soc. N.S. Wales*, xxvi. 556 (1901) considers it to be *E. acervula*, Miquel.

[4] Maiden mentions several varieties, as *ovata* and *maculosa*, occurring in Australia, which it is not necessary further to allude to, as they would not be hardy in Britain.

[5] *Journ. Proc. Roy. Soc. N.S. Wales*, xxviii. 213 (1894). Mueller, however, speaking of this species as a form of *E. Gunnii*, says that the wood is hard, good for many purposes, if straight stems are obtainable; as a rule not splitting well, but fair for fuel.

The only specimens which we know in England are three trees at Menabilly,[1] the largest of which was 29 ft. by 1 ft. 8 in. in January 1911; and another at Holkham, about nine years old, which was 30 ft. high in 1911. This species did not succeed at Kinloch Hourn, where all the seedlings were killed in 1894-1895.

(A. H.)

EUCALYPTUS VERNICOSA, DWARF GUM

Eucalyptus vernicosa,[2] J. D. Hooker, in *Lond. Journ. Bot.* vi. 478 *bis* (1847), and *Fl. Tasm.* i. 135 (1860); Bentham and Mueller, *Fl. Austral.* iii. 232 (1866); Rodway, in *Proc. Roy. Soc. Tasmania*, 1898-1899, p. 104, and *Tasmanian Flora*, 58 (1903); Maiden, in *Rep. Austrl. Assoc. Advance. Sc., Hobart*, 1902, p. 376.

An erect shrub, with smooth bark, usually 4 to 6 ft., rarely 12 to 20 ft. in height in Tasmania. Young branchlets green and angled towards the top, reddish brown and terete elsewhere. Leaves (Plate 365, Fig. 2) on adult shrubs, alternate, opposite, or sub-opposite, narrowly ovate or elliptical, $\frac{1}{2}$ to 2 in. long, and $\frac{1}{2}$ to 1 in. wide, rounded or slightly tapering at the equal-sided base, rounded or acute at the apex, which is tipped with a short triangular sharp point; margin entire or faintly undulate; very thick and coriaceous in texture; equally green and shining as if varnished on both surfaces; oil-dots few, scattered, unequal, concealed in the thicker older leaves; lateral veins arising at an angle of 45° from the midrib; petiole $\frac{1}{6}$ to $\frac{1}{2}$ in. long.

Flowers in axillary umbels of three, two, or one; peduncle very short, not exceeding $\frac{1}{12}$ in. long; flower-buds sessile, $\frac{1}{4}$ to $\frac{1}{3}$ in. long, shining dark brown; calyx-tube turbinate, with two lateral ridges; operculum conical, shorter than the calyx-tube; stamens all perfect; anthers ovate, with closely contiguous but parallel and distinct cells. Fruit hemispheric, $\frac{1}{4}$ in. long, sub-sessile; rim narrow, flat or convex; capsule slightly sunk, with three to four valves protruding when open.

This species is confined to Tasmania, where it grows on the summits of the higher mountains, as on Mount Fatigue, where it was discovered by Gunn in 1842 at 4000 ft. altitude, on Mount La Perouse, Mount Sorell near Macquarie Harbour, Mount Geikie, and Mount Direction (2409 ft. altitude). It is called dwarf gum by Rodway, who says that it seems to occur only on the sub-alpine plains of the west and south-west of the island. He adds that on Mount La Perouse, where it attains 20 ft. in height, the leaves are all opposite and the flowers solitary; whereas on the west coast, where it remains bushy, the leaves are opposite and the flowers in threes; while taller plants on Mount Geikie bear large alternate leaves with flowers in threes. All these forms occur on the specimens raised from the same seed at Kinloch Hourn.

---

[1] A specimen branch of apparently another tree was sent from Menabilly to Kew in 1903, when it was reported to have been 33 ft. high. This tree cannot now be found.

[2] Apparently this is the *E. verrucosa*, mentioned by Landsborough in *Trans. Bot. Soc. Edin.* xx. 518 (1896). Mr. Birkbeck says in his notebook, that "they have two sorts at Edinburgh, one labelled *E. vernicosa*, the other *E. verrucosa*; except that the former has larger leaves, I see no difference." *E. verrucosa* is evidently a misprint, as no species has been described under that name.

We are unaware, when the species was first introduced.[1] Mr. Birkbeck considers it to be far the hardiest of the Eucalypts, forming a beautiful shrub, and not developing into a tree.

At Kinloch Hourn, it was planted in 1891, and had attained 13 ft. high in 1905, and has never been injured by frost. At Whittingehame, when about 5 ft. high it bore 26° of frost without injury, in February 1894; but it was killed to within three feet from the ground in the following severe winter.[2] It produced several shoots in the following year, one of which was preserved; Elwes found this 20 ft. high and very thriving in February 1905. It ripens seed freely at both Kinloch Hourn and Whittingehame. It is one of the species which is cultivated successfully at Inverewe on the west coast of Ross-shire. We have seen no specimens in England or Ireland, but, Mr. Kempshall states that at Abbotsbury it has passed through the last three severe winters without injury.

There is no specimen at Kew, a seedling, which was planted out in 1907, having died in 1910 without any apparent cause. (A. H.)

EUCALYPTUS MUELLERI, MUELLER'S RED GUM

Eucalyptus Muelleri, T. B. Moore, in *Proc. Roy. Soc. Tasm.* 1886, p. 208 (not Miquel,[3] not Naudin[4]);
Mueller in *Proc. Roy. Soc. Tasm.* 1886, p. 209; Maiden, in *Rep. Aust. Assoc. Advance. Sci.*,
Hobart, 1902, p. 376; Rodway, in *Proc. Roy. Soc. Tasm.* 1898-1899, p. 105, and *Tasm. Flora*,
58 (1903).

A tall erect tree, attaining in Tasmania in favourable situations a height of 200 ft.; but in some localities much smaller. Bark smooth from the base, greenish, blotched with reddish brown. Young branchlets glabrous, reddish, with prominent oil glands. Leaves (Plate 365, Fig. 6) alternate, thick and firm in texture, lanceolate, averaging 3 to 4 in. long and $\frac{3}{4}$ to 1 in. wide, straight or slightly falcate, unequal at the cuneate base, gradually tapering to an acuminate apex, ending in a short blunt point; margin white with remote and very shallow crenations; equally shining green on both surfaces; oil-dots numerous, irregular in size, mostly not pellucid; lateral nerves inconspicuous, arising from the midrib at an angle of 30°; petiole twisted, $\frac{1}{2}$ to $\frac{3}{4}$ in. long.

Flowers in umbels of threes; peduncle $\frac{3}{8}$ in. long; flower-buds sessile; calyx-tube angled; operculum cap-shaped, tuberculate, with an umbonate point; anther-cells parallel. Fruit-peduncle short, stout, $\frac{1}{4}$ to $\frac{1}{6}$ in. long, thickened at the distal

[1] Mr. Birkbeck received his first plant from the Edinburgh Botanic Garden in 1891.

[2] Landsborough, in *Trans. Bot. Soc. Edin.* xxiii. 145 (1905), who states that a seedling was planted at Cromla, Arran, in 1906.

[3] *E. Muelleri*, Miquel, in *Ned. Kruidk. Arch.* iv. 130 (1859), is a synonym of *E. incrassata*, Labillardière, a species occurring in Victoria.

[4] *E. Muelleri*, Naudin, in *Rev. Hort.* lvii. 406 (1885), and in *Descrip. Eucalypt.* 45 (1821), described from a tree cultivated at Antibes, is unknown to me, and is a synonym of a previously described Australian species, possibly *E. salmonophloia*, Mueller. *E. Muelleriana*, A. W. Howitt, in *Trans. Roy. Soc. Victoria*, ii. pt. 1, p. 89 (1891) is a native of Gippsland in Victoria, and is possibly identical with *E. dextropinea*, R. T. Baker, in *Proc. Linn. Soc. N.S. Wales*, 1898, p. 417, t. 11.

end ; fruit turbinate, $\frac{1}{3}$ in. long, about $\frac{5}{12}$ in. in diameter, reticulate-tuberculate on the surface, wide at the orifice with a convex rim ; capsule slightly sunk, but with the three to four valves protruding when open.

This species[1] has been considered to be the lowland form of *E. vernicosa* ; but it differs from that species in the shape and larger size of the leaves, which are not so thick or quite so shining as in that species, which has smaller fruit.

E. Muelleri is confined to Tasmania, where it was discovered by Mr. T. B. Moore in 1886, on the dividing range between the Huon and Derwent watersheds. Here on high bleak land, at 2000 feet elevation, there is a forest extending in a narrow strip for three miles along the southern side of the range, with trees averaging 100 ft. in height.   Moore found trees 200 ft. in height in less exposed situations and at much lower elevations.   Rodway,[2] who calls this tree mountain red gum, says that it is common on the mountains of south-western Tasmania at 2000 ft. elevation. Mueller states that this species is of great importance, as it is the only one of large size with good timber, which bears considerable frost.   Both Mr. T. B. Moore and A. O. Green[3] agree that the timber is valuable, being heavy, hard and strong ; but the tree is not found in quantity near any shipping port.

This species is very rare in cultivation ; and when found is generally labelled with an incorrect name.   There are two trees at Abbotsbury, one about 40 ft. high, which were bearing fruit and flower buds in February 1911.   These stand in an exposed position, and are considered to be very hardy.   There is a young tree about 20 ft. high at Logan, Stranraer.   (A. H.)

There are two fine trees in a plantation near Derreen, Co. Kerry, one of which (Plate 364) the Marquess of Lansdowne showed me in August 1910.   It was growing on wet peaty[4] soil amongst poplars and *Thuya plicata*, which crowd it on both sides, and measured 63 ft. in height by 4 ft. 3 in. in girth at four feet above the ground. The other tree is about 100 yards distant, and measured, in 1911, 64 ft. by 5 ft. They are said to have been planted about 1880 ; and, as indicating their rapid growth, Lord Lansdowne says that close by there is a *Cupressus macrocarpa* of the same height, girthing 6 ft. 10 in., and a *Tsuga Albertiana*, also equal in height and 5 ft. 1 in. in girth.   These two conifers were planted in 1878.   A small specimen of *E. Muelleri* at Mount Usher is about 15 ft. high and very thriving.   Seedlings of this valuable species, from seed kindly sent by Mr. Rodway, are now being raised at Holkham, Culford, Avondale, and other places.   (H. J. E.)

[1] There is no specimen of *E. Muelleri* in the Kew Herbarium, and my description is drawn up from fruiting specimens of the tree at Derreen, which were identified with this species by Maiden.

[2] In J. C. Penny, *Tasmanian Forestry*, 17, 19, 21 (Hobart, 1905).

[3] In *Proc. Roy. Soc. Tasm.* 1902, p. 44.

[4] Lord Lansdowne says the soil might be described as almost pure peat.

EUCALYPTUS URNIGERA, Urn-bearing Gum

Eucalyptus urnigera, J. D. Hooker, in *Lond. Journ. Bot.* vi. 477 *bis* (1847), and *Fl. Tasm.* i. 134,
t. 26 (1860); Bentham and Mueller, *Fl. Austral.* iii. 227 (1866); Maiden, in *Rep. Aust. Assoc.
Advance. Sc., Hobart*, 1902, p. 375; Rodway, *Tasmanian Flora*, 58 (1903).
Eucalyptus cornigera, Earl Annesley, *List of Plants hardy at Castlewellan*, 88 (1903).

A tree, occasionally attaining 50 ft. in height in Tasmania. Bark peeling off, smooth, green blotched with reddish brown. Young branchlets terete, glabrous, green often tinged with red, dotted with oil-glands not raised above the surface. Leaves (Plate 365, Fig. 11) on adult trees, alternate, lanceolate or ovate-lanceolate, averaging 3 to $3\frac{1}{2}$ in. long and 1 to $1\frac{1}{2}$ in. broad, often falcate, equal or unequal at the short cuneate or rounded base, gradually tapering to the apex, which ends in a fine long straight point; margin whitish, revolute, with remote irregular shallow crenations; thick in texture, with numerous translucent unequal oil-dots; equally dull green on both surfaces; lateral veins numerous, arising from the midrib at an angle of 40°; petiole twisted, $\frac{1}{2}$ to $\frac{3}{4}$ in. long.

Leaves on young trees and on suckers, opposite, sessile, orbicular or broadly ovate, $1\frac{1}{2}$ to 2 in. in diameter, deeply cordate at the base, emarginate at the apex, from which arises a short point; conspicuously crenulate in margin; green on both surfaces; lateral nerves 7 to 8 pairs, arising from the midrib at an angle of 70°. Branchlets reddish, covered with bright red raised oil-glands.

Flowers in axillary umbels of three; peduncle $\frac{3}{4}$ in. long, slender, slightly thickened at the distal end; pedicels $\frac{2}{5}$ to $\frac{1}{2}$ in. long; flower-buds, $\frac{1}{2}$ in. long; calyx-tube narrowly urn-shaped, swollen at the base, contracted in the middle, and expanded above; operculum wider than the calyx-tube, cap-shaped, with a projecting umbo in the centre; stamens all perfect, inflexed in the bud; anthers with distinct parallel cells. Fruit, on pedicels which are nearly $\frac{2}{5}$ in. long, urn-shaped, swollen below, narrowed in the upper third, with a projecting narrow rim; surface with raised oil-glands; valves deeply sunk. The fruit varies in size, long narrow and short broad forms occurring, varying from $\frac{1}{2}$ to $\frac{3}{4}$ in. long, and $\frac{3}{10}$ to $\frac{4}{10}$ across the distal end.

1. Var. *elongata*, Rodway, *ex* Maiden, in *Rep. Aust. Assoc. Advance. Sci., Hobart*, 1902, p. 376.

A taller tree than the type, attaining 200 ft. in height, with smooth ashy-white bark, and very long linear-lanceolate leaves (up to 6 to 9 in. in length). Flowers; operculum conical, umbonate, half the length of the calyx-tube. Fruit much shorter than in the type, pyriform-globose, slightly constricted. This occurs at 1000 to 2000 feet elevation in Tasmania, and, though I have seen no specimens, seems to me to be a distinct species, or possibly a hybrid of *E. urnigera* with *E. viminalis*. We have found nothing like this form in cultivation in England.

E. urnigera is confined to Tasmania, where it is common in the mountains, especially in the south.   Rodway,[1] who calls it urn-bearing gum, informs us that it has not been found at lower altitudes than 1000 feet, and extends up to 3000 feet.   At 2000 feet it is very similar to *E. Muelleri* in general appearance, both of bark and habit; but below this altitude, it is said that the bark becomes ashy-white, and the leaves very long and narrow, constituting var. *elongata*.   So far as we can ascertain this species has no economic value in Tasmania, the timber being considered, according to Rodway, brittle and worthless.

It was discovered[2] by R. Brown at the beginning of the nineteenth century; but was described by Hooker from specimens gathered by Gunn in 1842 on the summit of Mount Wellington and near Lake Echo.   The exact date of its introduction is uncertain; but Hooker states that it was in cultivation in England in 1860.

This is one of the hardiest species, but is by no means a handsome tree, often being spare of branches.   It is also devoid entirely of the glaucous bloom on the branchlets and leaves, which give a pleasing effect to *E. coccifera* and *E. Gunnii*. The late Earl Annesley considered it to be as hardy as the laurel; but it has suffered at several places from frost, especially when in a young stage.

The finest trees in England are two growing at Coombe Royal, Kingsbridge, Devon, the larger[3] of which was 72 ft. by 9 ft., and the other 60 ft. by 6½ ft. in 1904.

At Menabilly, the finest Eucalyptus is a tree of this species, with a clean straight stem, which Mr. W. H. Bennett reported to be 80 ft. by 3 ft. 9 in. at three feet from the ground in 1909.   It[4] was raised from seed about twenty years ago, and is hardier at Menabilly than *E. coccifera* or *E. cordata*.   Younger trees here, planted in 1901, were about 30 ft. high in 1911.

At Sidbury Manor, Sidmouth, there is a tree, of which Sir Charles Cave sent us a branch in 1910.   Miss Woolward measured it in 1904, as 43 ft. high, with three stems springing from near the base, the main one being 3 ft. 8 in. in girth.   It was then fifteen years old, and growing in a sheltered situation at 250 ft. above sea-level.

At Abbotsbury, this is considered to be one of the hardiest species.   Young plants have passed through three severe winters without injury, and are now growing freely.   There are several specimens, the largest of which is 50 ft. by 5 ft.[5]   At Leonardslee, Sussex, specimens of *E. urnigera*, though 20 ft. high, were killed in a severe winter.   At Ponfield, Herts, a small tree, which was 15 ft. high in 1906, died in 1910.   We have seen no specimens near London or in the eastern counties.   At Cefnamwich, Nevin, N. Wales, a tree raised from seed in the spring of 1894, was 45 ft. high and 2 ft. 6 in. in girth in December 1910.

In Scotland, the finest specimens are at Stonefield, Loch Fyne, one of which measured in December 1910, 81 ft. high by 5 ft. in girth at four feet from the ground.

---

[1] In J. C. Penny, *Tasmanian Forestry*, 17, 19, 21 (1905).

[2] A specimen in the Kew Herbarium is labelled : "R. Brown, *Iter Australiense*, 1802-5, No. 4775."

[3] From this tree, which was called *E. montana*, a specimen was sent to Kew, in 1874, when it was reported to be 50 ft. high.   This tree is probably the oldest in England, and may date back earlier than 1860.

[4] Mr. Rashleigh in a letter to Kew, reported it to be 60 ft. by 2½ ft. in 1903.   It was 85 ft. by 4 ft. in January 1911.

[5] *E. urnigera* has been grown at Abbotsbury under several names, as *E. hæmastoma*, *E. rudis* × *E. rostrata*, and *E. Stuartiana*.

The forester, Mr. R. Stewart, says that it has never been injured by frost since it was planted out. It [1] had long been known erroneously both as *E. Gunnii* and *E. coccifera*.

This is one of the hardy species at Kinloch Hourn, where there are several good specimens. There are also many trees [2] on the hill-sides up to 600 ft. above sea-level, which were raised from Tasmanian seed in 1895, and were planted out in 1896 to 1898. They were about 24 to 28 ft. high in 1907. In Arran, a specimen [3] planted in 1885 was killed in the winter of 1894-1895; but it was growing in a wood amidst higher trees, and never prospered. None of the Eucalyptus do well when shaded by other trees; and Mr. Bateman lays great stress on this fact, as borne out by his experiments in planting *E. Gunnii* at Brightlingsea. We have received a specimen from Whitefarland, Pirnmill, Arran, where there are several trees, the oldest of which, planted in 1895, is now 23 ft. by 3 ft. 4 in. at a foot from the ground. This species also thrives well at Inverewe, where a tree, twelve years old, was 35 ft. by 1 ft. 8 in. in 1912. It also succeeds at Logan, near Stranraer. A tree [3] at Roseneath Manse, planted in 1883, was 12 ft. high in 1895, and had never been injured by frost; but it subsequently died in a mild winter without any apparent cause. At Dalkeith, it is reported [4] to have borne without injury the severe frost of 7th January 1894, when the temperature fell to 4° Fahr.

In Ireland, this species proved very hardy at Castlewellan, where a tree [5] attained 65 ft. in height; but it was blown down some years ago. There are two or three younger trees now living raised from its seed; but Eucalyptus trees are no longer planted at Castlewellan, as they are always blown down when they grow to a large size. We have not seen any specimens of *E. urnigera* elsewhere in Ireland except at Mount Usher, where there are small trees, raised from seed sown in 1904. These are very thriving, and bore fruit in 1911.   (A. H.)

[1] This tree is erroneously called *E. Gunnii* by Dr. Landsborough, in *Trans. Bot. Soc. Edin.* xx. 517 (1896), and xxiii. 144 (1905). It was planted in 1881, had attained 38 ft. high in 1896, and measured 71 ft. by 4 ft. 4 in. in 1905.

[2] Erroneously called *E. acervula* by Dr. Landsborough, in *Trans. Bot. Soc. Edin.* xiii. 145 (1905).

[3] Dr. Landsborough, in *Trans. Bot. Soc. Edin.* xx. 519 (1896), and xxiii. 146 (1905).

[4] *Journ. Roy. Hort. Soc.* xviii. 76 (1895).

[5] A tree formerly cultivated at Castlewellan as *E. cornigera* was undoubtedly *E. urnigera*; and specimens of it were so named by Naudin. It appears that the trees of this species at Castlewellan differed in habit. One with numerous spreading branches was called *E. cornigera*; others with few branches and ungainly in habit were named *E. urnigera*.

END OF VOL. VI

Printed by R. & R. CLARK, LIMITED, *Edinburgh.*

COMMON SPRUCE IN MASSACHUSETTS

PLATE 341.

COMMON SPRUCE AT STUDLEY

PLATE 342.

SPRUCE AVENUE AT OAKLEY PARK

COMMON SPRUCE AT GWYDYR CASTLE

PLATE 343.

SPRUCE AT KILWORTH, IRELAND

PLATE 344

PLATE 345.

WESTERN HIMALAYAN SPRUCE AT MELBURY

PLATE 346.

BLACK SPRUCE AT COLESBORNE

PLATE 347.

SIKKIM SPRUCE AT CASTLEWELLAN

COMMON JUNIPER AT COLESBORNE

PLATE 348.

JUNIPERUS RECURVA AT CASTLEWELLAN

PLATE 349.

CATALPA AT HAM MANOR

PLATE 350.

PLATE 351.

ROBINIA AT FROGMORE

PLATE 352.

ROBINIA AT THE MOTE, MAIDSTONE

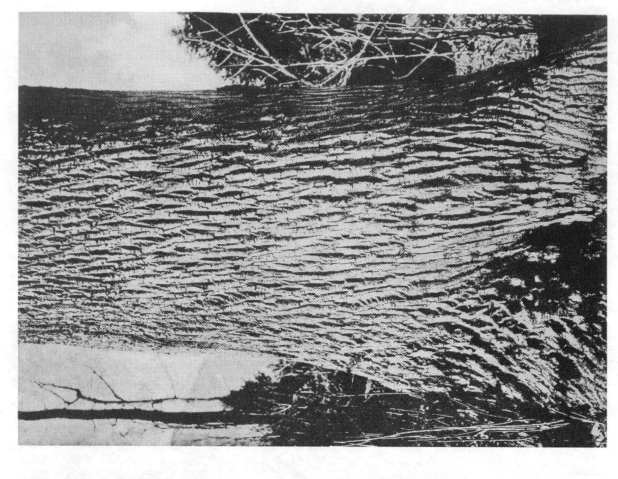

EVERGREEN MAGNOLIA IN NORTH CAROLINA

TRUNK OF ROBINIA IN AMERICA

PLATE 353.

PLATE 354.

WILD CHERRY IN SAVERNAKE FOREST

PLATE 355.

CHERRY AT GEORGE'S GREEN, SLOUGH

PLATE 356.

PEAR AT LASSINGTON, GLOUCESTER

PEAR AT STOCKTON, WORCESTER

PLATE 357.

PLATE 358.

MAGNOLIA AT WEST DEAN PARK

PLATE 359.

BLUE GUM AT TORQUAY

PLATE 360.

BLUE GUM AT PENMERE

PLATE 361.

EUCALYPTUS COCCIFERA AT POWDERHAM

PLATE 362.

EUCALYPTUS GUNNII AT BRIGHTLINGSEA

PLATE 363.

EUCALYPTUS WHITTINGEHAMENSIS AT WHITTINGEHAME

PLATE 364.

EUCALYPTUS MUELLERI AT DERREEN

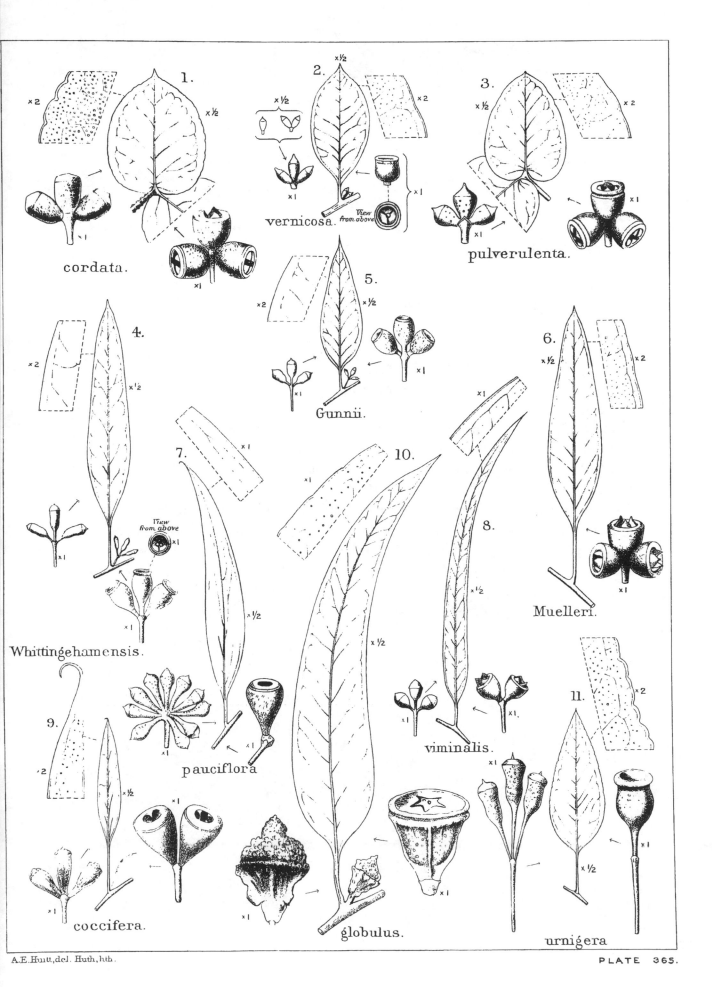

1. cordata.

2. vernicosa.

3. pulverulenta.

4. Whittingehamensis.

5. Gunnii.

6. Muelleri.

7. pauciflora

8. viminalis.

9. coccifera.

10. globulus.

11. urnigera

A.E.Hutt,del. Huth,lth.

PLATE 365.

EUCALYPTUS

WALNUT AT CAM-YR-ALYN PARK

THE magnificent Walnut tree here represented was unknown to us when Vol. II. was published in 1907. We are indebted to Mr. F. R. S. Balfour for its discovery, and to Mr. G. Cromar for the following particulars :—The tree stands in Cam-yr-Alyn Park, Denbighshire, the property of Wilson Sweetenham, Esq., and in 1910, when the photograph was taken by Mr. W. P. Wilkes, was about 70 ft. high, and $32\frac{1}{2}$ ft. in girth. It has five main branches, which measure as follows : 9 ft. 4 in., 10 ft. 3 in., $12\frac{1}{2}$ ft., 9 ft., and $11\frac{1}{2}$ ft. respectively, and cover an area of 88 ft. by 80 ft. The tree is healthy, and grows on a light loam and gravelly soil, at an elevation of about 100 feet above sea-level. The water in this district is full of lime, and there is a stream about 15 yards from the tree.

WALNUT AT CAM-YR-ALYN PARK

PLATE 366.

ORIENTAL PLANE AT WESTON PARK

THE Plane tree here figured was referred to on page 622 in the third volume of our work, as having measured, in 1875, 80 ft. by 18½ ft.   Until I saw it in 1909 I had not realised that though not quite so large as the tree at Ely, figured on Plate 174, it is in some respects a finer and better-shaped tree.   Though it was seriously damaged in March 1894 by a gale, it is now in perfect health, and measured, in August 1909, about 80 ft. by 20 ft. in girth, and 120 paces round the branches.   I am informed by the Earl of Bradford that he has no record as to the date when it was planted.   The leaves of this tree are unusually large, and though not so deeply cut as usual in the Oriental Plane, were quite free from the fungus *Glæosporium* (cf. p. 618).

ORIENTAL PLANE AT WESTON PARK.

PLATE 367.

SCOTS PINE IN GLEN MAILLIE

THE remarkable tree which is here figured, though the size of the plate is insufficient to do it justice, is a larger one than any of those mentioned in Vol. III. of this work. I am not sure whether it is the same tree of which Capt. Ellice had sent me a sketch, and which is mentioned on p. 588. Though much injured by the breaking of two of the large limbs, it was still healthy when I saw it in 1910. The trunk in the smallest place near the ground measured 18 ft. 1 in., and at five feet from the ground, below the fork, 22 ft. in girth. Its height is about 75 ft. An unusual feature in this tree is a young pine about 25 ft. high and 2 ft. 9 in. in girth, which has grown from a seed dropped in the fork of the old tree; and which has now become as completely united with the sound wood of the trunk as if it was a true branch. A good-sized birch and a small rowan are also growing as epiphytes on the trunk.

The primæval forest, in which this tree is probably the largest, is in my judgment the finest in Scotland, and extends from a little above sea-level up to 700 or 800 feet. The largest trees in it are probably over 300 years old, and grow on dry ridges among patches of peat covered with long heather and intersected by small watercourses. A few hollies, rowans, and birches are scattered among the pines; but few seedlings of the latter are visible owing to the presence of deer. There are many fine timber trees, as well as trees attractive to the naturalist. Among them is one 74 ft. high and 13 ft. in girth, which has three tall clean stems of equal size, dividing at about 10 feet from the ground and remaining close together for a considerable height.

In Lochiel's house at Achnacarry there is a beautiful water-colour, painted in 1847 by I. Giles, of a pine called "The Fir of Gusach," which formerly grew on the shore of Loch Arkaig, but has long ago disappeared. It was remarkably similar in form and size to the tree now figured.

PLATE 368.

SCOTS PINE IN GLEN MAILLIE

LARCH AT POLTALLOCH

THE plate shows a remarkable instance of witches' broom, growing on a larch, which I first heard of from Col. Malcolm of Poltalloch, Argyllshire, to whom I am much indebted for the photograph. This tree grows in a wood called Bar-na-sluid, about two miles from Poltalloch, at perhaps 200 ft. above sea-level, and is believed to be 60 or 70 years old. When I saw it in September 1911, it appeared to be quite healthy, and was about 48 ft. by 5 ft. The dense mass of twigs forming the witches' broom was about 15 ft. wide and 10 ft. deep. A stunted spruce grew close to the base of the tree, which was cut away in order that the photograph might be taken.

PLATE 369.

LARCH AT POLTALLOCH.

LABURNUM ALPINUM

THE finest specimen, that I have seen or heard of, grows at Countesswells near Aberdeen, the seat of Sydney J. Gammell, Esq., to whom I am indebted for the photograph taken by him on 30th June 1912, when the flowers were a little past their prime. The age of this tree is uncertain, but probably coeval with the oldest part of the house, built about 1700. Though several large limbs have fallen outwards, and the main trunk, which is split into three portions at the base, has been cut off at about ten feet from the ground ; yet the vitality of the tree is so great, and the old wood so resistant to decay, that the tree is covered with young branches ; and in some seasons the racemes of flower are so thickly set that the leaves can hardly be seen. The height of the tree is about 30 ft. ; the girth of the largest part of the trunk is 7 ft., and of the split trunk at the ground 10 ft. The spread of the branches covers an area 153 ft. round. The fertile granite soil of the district, where I saw many laburnums of large size, seems specially favourable to this species. A tree, self-sown, at the foot of a cherry at Balcraig farm, near Aboyne, was over 30 ft. high and almost without branches for 20 ft. Mr. H. B. Watt tells us of a fine tree in the grounds of Dr. R. Farquharson at Finzean, Aberdeenshire, which measured on 27th June 1912, when it was in full flower, about 25 ft. in height and 5 ft. 10 in. in girth at three feet from the ground. Here the altitude is 550 feet ; but Mr. Watt sent us specimens in flower of both species from Braemar at 1100 to 1200 feet. The alpine laburnum seems to grow and flower in shade better than the common laburnum.

ALPINE LABURNUM AT COUNTESSWELLS, ABERDEEN

PLATE 370.

PLATE 371.

SPRUCE PLANTATION AT RHINDBUCKIE HILL, DURRIS

Printed in the United States
By Bookmasters